P9-DTM-461

NOBEL LECTURES IN
PHYSIOLOGY OR MEDICINE
1922–1941

NOBEL LECTURES

Including Presentation Speeches
And Laureates' Biographies

PHYSICS

CHEMISTRY

PHYSIOLOGY OR MEDICINE

LITERATURE

PEACE

ECONOMIC SCIENCES

NOBEL LECTURES

Including Presentation Speeches
And Laureates' Biographies

PHYSIOLOGY
OR
MEDICINE

1922–1941

World Scientific
Singapore • New Jersey • London • Hong Kong

Published for the Nobel Foundation in 1999 by

World Scientific Publishing Co. Pte. Ltd.

P O Box 128, Farrer Road, Singapore 912805

USA office: Suite 1B, 1060 Main Street, River Edge, NJ 07661

UK office: 57 Shelton Street, Covent Garden, London WC2H 9HE

NOBEL LECTURES IN PHYSIOLOGY OR MEDICINE (1922–1941)

$\mathcal{TOC}$

ISBN 981-02-3410-4

Printed in Singapore.

Foreword

Since 1901 the Nobel Foundation has published annually *Les Prix Nobel* with reports from the Nobel award ceremonies in Stockholm and Oslo as well as the biographies and Nobel lectures of the Laureates. In order to make the lectures available to people with special interests in the different prize fields the Foundation gave Elsevier Publishing Company the right to publish in English the lectures for 1901–1970, which were published in 1964–1972 through the following volumes:

Physics 1901–1970	4 vols.
Chemistry 1901–1970	4 vols.
Physiology or Medicine 1901–1970	4 vols.
Literature 1901–1967	1 vol.
Peace 1901–1970	3 vols.

Since the Elsevier series has been out of print for many years the Nobel Foundation has given World Scientific Publishing Company the right to publish these Nobel lectures, biographies and presentation speeches. The Nobel Foundation is very pleased that the intellectual and spiritual message to the world laid down in the laureates' lectures will, thanks to the efforts of World Scientific, reach new readers all over the world.

Bengt Samuelsson   Michael Sohlman
Chairman of the Board   Executive Director

Stockholm, March 1998

Contents

Physiology or Medicine 1922

ARCHIBALD VIVIAN HILL

«for his discovery relating to the production of heat in the muscle»

OTTO FRITZ MEYERHOF

«for his discovery of the fixed relationship between the consumption of oxygen and the metabolism of lactic acid in the muscle»

Physiology or Medicine 1922

Presentation Speech by Professor J. E. Johansson, Chairman of the Nobel Committee for Physiology or Medicine of the Royal Caroline Institute

Your Majesty, Your Royal Highnesses, Ladies and Gentlemen.

The object of physiology is to endeavour to recognize in the vital processes well-known physical and chemical processes. Accordingly it has to give answers to such questions as these: what is it that takes place in a muscle that contracts, in a gland that emits a secretion, in a nerve when it transmits an impulse? In former times these processes were explained as being the work of what were called «life spirits» – beings who in their mode of existence possessed an unmistakable resemblance to the person who spoke of them. If the muscles of a recently killed animal were seen to twitch when cut or pierced, this was explained by saying that the life spirits had been irritated. From this way of looking at things there still remains the expression «irritation», which we use to denote the starting – or, as we also put it, the liberation – of an active process in an organ. It is a long time, however, since we learnt to regard living organs, muscles, nerves, etc., as mechanisms; and the expression «muscular machine» will probably not strike any educated person in our days as being strange or offensive.

In order to render clear the working of a mechanism it is customary to give a «simplified model» of it. A schematic drawing or an imaginary model may perform the same service, and is at any rate cheaper. The first model that was made of muscular mechanism had the steam-engine as its prototype. Very soon, however, it was perceived that the adoption of an engine of this type presupposes the existence of substances in the muscular fibres capable of sustaining temperatures far exceeding 100°C. The efficiency of muscular work can in fact amount to 20–30%; and such values cannot be obtained by a heat-engine unless the temperature in certain parts of the engine is raised to a considerable height. Hence the muscular machine cannot be referred to that group of motors that transform heat into mechanical work and that are based on the equalization of different temperatures. Theoretically, however, differences in osmotic pressure, surface tension, electrical potential, and so on, offer the same possibility of developing work; and consequently any chemical process whatever that takes place «spontaneously» and that gives rise to such differences in «potential», might be employed in a model of a muscular machine. Thus there is no lack of material for the construction of such a model. The difficulty is to select. In this case there was also a further difficulty, namely that of being able to emancipate oneself, in the design of such a model, from the old and discarded model of a heat-engine. One need not be a physiologist to recognize that muscular activity is essentially bound up with the *development of heat*, or even with *combustion*. Now as it is impossible to regard the muscle as a heat-engine, how is it possible to fit these phenomena into the course of action?

This problem has been successfully solved by the two investigators to each of whom the Professorial Staff of the Caroline Institute has this year resolved to award half of the Nobel Prize for 1922 in Physiology or Medicine, namely Professors Archibald Vivian Hill of London and Otto Meyerhof of Kiel. These two men have each worked independently and to a large extent with different methods. Hill has analysed, by means of an extremely elegant thermoelectrical method, the time relations of the heat production of the muscle; and Meyerhof has investigated by chemical methods the oxygen consumption by the muscle and the conversion of carbohydrates and lactic acid in the muscle. Both have made use of the same kind of experimental material, namely the surviving muscle excised from a frog – in fact, the classical frog muscle preparation.

Such a preparation remains alive for several hours, or even days. A suitable stimulus liberates a contraction or develops a state of tension, both of short duration. The twitch takes only one or two tenths of a second. If the stimulus be repeated, the muscle makes a new twitch, apparently resembling the preceding one; and if the muscle is attached to a suitable connecting lever, the several twitches give the same effect as the strokes of a piston in a steam-engine. What was more natural than to regard the muscular twitch as the expression of a circular process in the muscular elements? This process makes itself known in another way also, namely in the form of a development of heat in the muscle preparation. The amount of heat is very insignificant. It is measured in millionths of the usual unit of heat and is recorded in a thermoelectrical way in the form of readings on a galvanometer. Armed with technical resources for observing both the mechanical process and the development of heat in the twitch of an isolated muscle, investigators tried to penetrate more deeply into the muscular process proper. Our countryman Blix showed that everything that impedes the contraction of a muscle during the twitch – that is to say, impedes the diminution of surface of the muscular elements – increases the formation of heat, and from this concluded that the process sought is localized to the surface of certain structural elements which, owing to changed conditions in surface tension, acquire a tendency to pass from an ellipsoidical to a more spherical form. If the load of the muscle gives way to the tension thus created, external work is done. Hence the muscle is mainly to be regarded as a machine that converts chemical energy into tension energy.

In the first experiments that Hill carried out on this subject in 1910 he made use of a thermo-galvanometer designed by Blix. Here he noticed that the reading not only gives the total amount of heat developed, but also is to some extent affected by the period of time taken in the development of heat. He was able to distinguish between an «initial» and a «delayed» development of heat. A subsequent work contained the starting-point for a new method of investigation, which made it possible to trace the development of heat in muscular movements in their various stages. This technique may be described as having been completely developed by 1920; but some of the results that I shall mention had been obtained as early as 1913, that is to say before the outbreak of the World War.

The development of heat in the contraction of the muscle – which to preceding investigators appeared to be «one and indivisible», that is to say, was lumped together as a single phenomenon – can be divided by Hill's method into *several periods, the last* of which comes long *after the end of the mechanical process*, that of the twitch. To this must be added the fact that this delayed development of heat entirely *fails to appear if the supply of oxygen to the muscle be cut off*, while the development of heat *during the actual twitch* – tension and relaxation – is completely *independent of the presence of oxygen*. The process of combustion, which it had been customary to connect immediately with the contraction of the muscle, does not actually take place until afterwards. In the experimental arrangements with which we are now dealing (isometrical work) the development of heat during the actual twitch also includes the amount of energy which under other circumstances appears as external work.

Hill's discovery has had a veritably revolutionizing effect as regards the conception of the muscular process. The ordinary view of this process as divided into two phases, tension and relaxation, can, it is true, be retained with regard to the mechanical process, but with regard to the chemical process another division must be adopted – the *working phase proper*, independent of the supply of oxygen and corresponding to the whole of the mechanical process, and following it an oxidative *phase of recovery*. If previously in their speculations as to the muscular process physiologists had mainly shown an interest in the actual twitch, investigations now became directed towards the muscle in rest and especially the muscle after preceding exhaustion. Chemical considerations now attracted attention as well as the physical ones.

The earliest known chemical process in the muscle is the formation of lactic acid. This is mentioned as early as 1859 by Du Bois-Reymond. He had found that an excised muscle becomes acid on repeated stimulation even when the rigor mortis sets in. He supposed the cause of this to be the formation of lactic acid – owing, it is stated, to a communication from Berzelius, who had found great quantities of that acid in the flesh of a deer that had been killed in the chase. Since that time lactic acid has played a very important part in discussions as to rigor mortis and the fatigue of the muscle. Some years before Hill began his investigations, two of his countrymen, Fletcher and Hopkins, had shown that the excised muscle not only *forms* but also *converts* lactic acid, this depending on whether the muscle is shut off from oxygen or whether oxygen is supplied to it. Some observations also suggested that when the lactic acid disappears from the muscle, only part of it is burnt up, while the rest is re-transformed into the mother substance of lactic acid. In consequence of this there was reason to surmise that the part played by lactic acid in the muscles is not completely represented by such expressions as « by-product of the metabolism », « fatigue substance », « cause of rigor mortis », etc. In this connection Hill proposed that lactic acid should be included as a part of the actual muscle machine.

The formation of lactic acid in the muscle, according to Fletcher and Hopkins, and this development of heat in the muscle during its working phase, according to Hill, exhibit the striking accordance that they take place independent of the oxygen supply. According to Blix, the twitch came about due to the fact that along the surface of certain structural elements there suddenly appears some substance, the nature of which is not stated. If we suppose this substance to be lactic acid – formed either directly or with some intermediate stage from the muscles' well-known store of glycogen – we have a model which combines in itself the most valuable contributions of the investigations of the last few decades on this question. We make the stage of recovery, accompanied by the supply of oxygen, follow the working phase together with Hill's delayed development of heat and Fletcher's conversion of lactic acid. The fact is that lactic acid, when it has done its work, must be got rid of somehow in order that the machine may be kept going.

By a well-known calculation Hill tried to find support for the recently quoted supposition of Fletcher and Hopkins with regard to a reversion, in conjunction with the lactic acid combustion, of lactic acid to glycogen during the phase of recovery. It is easy to see that the correctness of this supposition forms a condition that the model cited should be acceptable from the point of view of energetics. But objections were made against the analyses and arguments of Fletcher und Hopkins. Moreover, there were adduced, from what were considered to be extremely competent quarters, direct observations which seemed to show that the lactic acid formed in the working phase was completely used in the process of recovery – a piece of wastefulness on the part of Nature which could only be explained by means of auxiliary hypotheses in the presence of which it would have been the simplest thing to let the whole of the attractive model take part in the combustion.

It is at this stage in the development of the question that Meyerhof's contribution comes in. In his investigations concerning the respiration of the tissues (1918) he came to devote his attention to the things that take place in the surviving muscle, and in this connection also to the objections that had been raised against the conclusions of Fletcher and Hopkins and their interpretation of the «lactic acid maximum» of the muscle. He showed that these objections do not really affect the result of the recently cited calculations of Hill. Most important of all, however, was his *parallel determination of the lactic acid metabolism and the oxygen consumption* during the recovery of the muscle, which yielded the result that the oxygen consumption does not correspond to more than $1/3-1/4$ of the simultaneous lactic acid metabolism. Evidently the greater part of the lactic acid disappears in some other way than through combustion. In another parallel determination – the development of heat and the oxygen consumption – the development of heat exhibited a deficit in comparison with what could be calculated from the simultaneously observed oxygen consumption. From this the conclusion may be drawn that the combustion of lactic acid in the muscle is combined with some other process, an endothermic one, in the course of which part of the heat developed in the combustion is used up. Meyerhof also made a parallel determination of the carbohydrates and lactic acid in the resting and in the working muscle, also in the recovery period after fatigue; he found: when lactic acid is stored in the muscle, an equivalent quantity of carbohydrates, chiefly glycogen, disappears, while when lactic acid disappears, the quantity of carbohydrates in the muscle is increased by an amount equivalent to the difference between the total amount of lactic acid that has disappeared and the quantity oxidized corresponding to the oxygen consumption.

Hence the processes which we have to take into account in the muscles are: (1) the formation of lactic acid from carbohydrates; (2) the combustion of lactic acid to carbonic acid and water; and (3) the reversion of lactic acid to carbohydrates. But these processes are not confined to the uninjured muscle. Meyerhof has also traced them in finely chopped muscle substance kept moist in a suitable liquid, and in that case found them take place 10–29 times more rapidly than in the well-known muscle preparation. In such a dilution it is also possible to study the effect of different factors such as the concentration of hydrogen ions, the presence of phosphates, etc.; and in particular it has been possible to make clear to what extent the various processes are connected with one another or can be varied in relation to one another. A matter of extremely great interest is the establishment of the fact that the combustion of lactic acid in the muscle cannot take place without a simultaneous formation of lactic acid from carbohydrates, and that the combustion of lactic acid is connected with the formation of carbohydrates in such a way that *out of four molecules of lactic acid one is oxidized, while the three others are reverted to carbohydrates*. It is not inconceivable that the reversion does not always extend so far as to produce carbohydrates; but the *ideal course* of the process may be regarded as precisely defined by Meyerhof, and it has been represented by him in the form of a scheme of chemical reaction. In this scheme, too, can well be fitted the *lactacidogen* discovered by Embden as a connecting link between glycogen and lactic acid.

The chemical processes just cited have to be fitted into the model of the muscle machine. Ignoring other considerations than those of *energy*, we can express the course of action in the following way: the change in the muscle which forms the basis of the mechanical process (the external work) presupposes a certain quantity of lactic acid, which comes from the muscle's store of glycogen. When this lactic acid has done its work, 1/4 is burnt into carbonic acid and water, while 3/4 return to the store of glycogen. The upper limit of the efficiency of the machine, calculated according to this scheme, will be 50%, which fully corresponds to the real state of things.

The combustion of lactic acid demands oxygen. The muscle preparation, however, can work even if the supply of oxygen is cut off. The lactic acid formed at every twitch spreads in the muscle out from the places where it is formed until the muscle substance finally becomes so impregnated with lactic acid that it is not relaxed between the twitches, and the impulses applied do not give rise to any further formation of lactic acid. The muscle is exhausted or, as one might also put it, poisoned with lactic acid. In the body the muscle is transfused with blood, which supplies oxygen in far greater abundance than that which the excised muscle preparation can obtain from its environment. Owing to its store of alkali, moreover, the blood itself provides room for a certain quantity of lactic acid from working muscles – a quantity of lactic acid that the blood can afterwards get rid of during a subsequent interval in the work. The possibility of thus distributing the combustion of lactic acid during a period that is longer than the work itself, provides us with an explanation of the immense amount of work achieved, especially in the sporting competitions of our day. Even with a volume per minute corresponding to the extreme working capacity of the heart there is not obtained in these cases a supply of oxygen corresponding to the formation of lactic acid in the muscles; and consequently the individual exposes himself to an accumulation of lactic acid in the blood and in all the tissues of the body – an accumulation that must be characterized as poisoning. When we are dealing with competitions for children and young people who are not yet grown up, there is good reason to think about this detail with regard to the muscle machine.

Professors Hill and Meyerhof. Your brilliant discoveries concerning the vital phenomena of muscles supplement each other in a most happy manner. It has given a special satisfaction to be able to reward these two series of discoveries at the same time, since it gives a clear expression of one of the ideas upon which the will of Alfred Nobel was founded, that is, the conception that the greatest cultural advances are independent of the splitting-up of mankind into contending nations. I also feel confident that you will be glad to know that the proposition which has led to this award of the Nobel Prize originated from a German scientist who, in spite of all difficulties and disasters, has clearly recognized the main object of Alfred Nobel.

In conferring upon both of you the sincere congratulations of the Caroline Institute, I have the honour of asking you to receive from His Majesty the King the Nobel Prize for 1922 in Physiology or Medicine.

ARCHIBALD V. HILL

The mechanism of muscular contraction

Nobel Lecture, December 12, 1923

In investigating the mechanism involved in the activity of striated muscle two points must be borne in mind, firstly, that the mechanism, whatever it be, exists separately inside each individual fibre, and secondly, that this fibre is in principle an isothermal machine, i.e. working practically at a constant temperature. There are several ways of studying this machine – the mechanical, the thermodynamic, the chemical, and the electrical, and of course any combination of these. In the case of nerve, much information has been acquired from a study of its electrical properties especially recently in the hands of Keith Lucas, Adrian, and Erlanger and Gasser. In the muscle fibre, however, as distinguished from muscular tissue in general, comparatively little has been discovered by studying the electric change. The apparent exception of the heart is not a real one, since here the electric change has been used to analyse, not so much the behaviour of the individual fibre – that is, of the ultimate mechanism itself – as the distribution and interconnection of the fibres to form the complete working organ. We will not therefore consider the electrical phenomena further in our discussion of the mechanism of the muscle fibre.

Much investigation has been devoted in the past to the mechanical response and characteristics of muscles, and it might have appeared that this subject is to some degree worked out and unlikely to yield further results of value. Actually, however, in combination with thermodynamical observations and considerations, much information has been derived recently from a study of the mechanical output and behaviour of muscles. Indeed, in the last two or three years the investigation of the connection between the rate of shortening, the work done and the heat production has pointed to new, hitherto unsuspected, mechanisms in muscle which are of considerable theoretical interest and practical importance.

The chief advances, however, during recent years have come from a study of the thermal and the chemical changes which occur in excised muscle. These two sides of the investigation are the ones which Professor Meyerhof and I will discuss today.

One of the fundamental characteristics of striated muscle, and the one involving the greatest difficulty in investigation, is the great rapidity with which changes take place in it. There is no doubt that ultimately the muscle is a chemical mechanism, in the same way for example as a Daniell's cell or an accumulator is a chemical mechanism. If we were aware of all the chemical events, we should know all that was necessary about the machine which we are studying. Unfortunately, the investigation of chemical events is a slow and laborious process. Undoubtedly lactic acid, as Fletcher and Hopkins showed some fifteen years ago, is an essential part of the machinery, but it is impossible to measure the production and removal of lactic acid instantly and contemporaneously during and after a single unit of muscular response. It is necessary to evoke a long series of responses and finally to study the gross changes of accumulation or removal of the acid. Attempts have been made to follow the chemical processes involved in muscular activity by studying the changes of hydrogen-ion concentration by physical instruments. This would appear to be more hopeful than the possibility of quickening up the study of the reactions by ordinary chemical means, but unfortunately – so far – it has been completely unsuccessful. The instantaneous and contemporary study of the events occurring inside the muscle fibre appears to be possible only in two ways, the mechanical and thermal. As regards the mechanical, the technique is obvious, namely, connecting the muscle to suitable recorders, ergometers and levers. The mechanical changes occurring in muscle are, however, only the end-products of activity, and if we wish to get inside the mechanism, it is necessary to study some intermediate process, something occurring between the stimulus and the response itself, something associated with the chemical events which evoke contraction. This is provided by the investigation of the heat production. It is obvious of course that the picture provided by thermodynamics is only a partial one; the certainty, however, with which the principle of the conservation of energy may be applied, gives us firm ground on which to start our investigations, and there would seem to be no doubt that the outline provided by thermodynamics, once established, must remain, and must finally have the complete chemical picture painted into it. The advantage of the study of the thermodynamics of muscle is that heat may be measured in absolute units, rapidly and at once, and the time-course of its evolution analysed by suitable means.

In the study of the thermal changes the most consistent and valuable results have been obtained by utilizing the isometric contraction of the sartorius muscle of the frog. The sartorius muscle is a very suitable medium for this investigation, insofar as it is practically of uniform cross-section and consists of straight fibres running along its length. The isometric contraction has the advantage, firstly, that energy is not liberated in it in any other form than heat, so that no complications arise by having to sum the thermal changes with the mechanical work, and secondly, that in it movements of the instruments are prohibited, which on the small scale of temperature with which we are dealing is of value in avoiding errors due to temperature differences along the muscle.

The fundamental difficulty in myothermic observations is the smallness of the changes involved and their rapidity. In the muscle twitch of a frog's sartorius at 20°C the rise of temperature is not more than 0.003°, and the time occupied in the earlier phases (as distinguished from the recovery process) is only a few hundredths of a second. The first requisite therefore is a very sensitive thermometric apparatus and great freedom from temperature changes, the second is extreme rapidity and lightness in the recording instruments. Neglecting for obvious reasons the use of ordinary mercury thermometers, there are two possible methods available, those of the resistance thermometer and the thermopile respectively. The resistance thermometer has not been employed successfully in myothermic observations. Calculation with the requisite physical constants shows the existence of a certain fundamental difficulty, which is confirmed by actual experiment. Were it possible to use the resistance thermometer, it would be exceedingly advantageous, as such an instrument may be made very small and light, so that it will respond with great quickness to the temperature changes in its neighbourhood. Moreover, unlimited sensitivity may be obtained by increasing the current in the resistance wire. The fundamental difficulty is that heat is thereby produced in that wire, which is conducted into the muscle and warms it up, causing serious disturbances of the zero, and enormous negative deflections when the smallest movement of the muscle occurs. I have attempted recently to use a resistance thermometer for myothermic observations and found it completely impossible, owing to the errors produced by the heat production in the resistance thermometer itself. There remains, therefore, only one method, that of the thermopile, which we will now discuss.

It is possible to make a small light thermopile, suitable for myothermic observations, containing one hundred couples, each providing 50 microvolts for a difference of temperature of 1°C. With a suitable sensitive galvanometer this gives us a scale of temperature in which 1°C is about one kilometre in length. Ample sensitivity, therefore, is available. Such thermopiles are laborious to make by the ordinary method of soldering the wires to-gether, though practically all the work hitherto recorded has been done with instruments thus constructed. Recently, however, an ingenious device has been published by Hamilton Wilson, by which a continuous piece of con-stantan wire is coated with silver in successive sections by electroplating, and each pair of junctions between a silver-plated and an unplated portion acts as a silver-constantan couple. By this device very small, light thermo-piles may be constructed with great ease, and recently Drs. Fenn and Azuma have used such thermopiles in their work with me.

The developments of the technique appear to lie in refining and lightening thermopiles, in improving their sensitivity, and in the use of more sensitive and rapidly moving galvanometers. It may conceivably be possible to am-plify the electromotive force produced by such thermopiles, and so to use a galvanometer of short period in order to quicken the recording. The pos-sibilities already available of myothermic investigation are mainly due to improvements in galvanometers and thermopiles.

It is not practicable to register the changes of temperature in a muscle as they occur, without lag or loss: the recording instruments are too slow, they possess too great a heat capacity, the flow of heat is not rapid enough and the galvanometer – which must be sensitive – has too long a period. It is nec-essary to carry out an analysis of the results obtained, preferably after photo-graphic recording of the galvanometer deflection. The possibility of ana-lysing the time-course of the evolution of heat in a muscle contraction de-pends upon certain fortunate physical properties of the system employed. The conduction of heat, the relation between e.m.f. and temperature, the movements of the galvanometer, its dynamics, the control on it for small movements, and its damping, indeed, all the factors connected with the deflection of the galvanometer resulting from the change of temperature of an object in contact with the thermopile, are governed by linear differential equations with constant coefficients. It is not necessary to know what these coefficients are, as I shall show shortly – they can be eliminated by a control experiment – it is essential only that the system employed should possess this general property. Its consequence may be expressed in a very simple form. If there be two productions of heat recorded separately, then the deflections produced, if summed together, are the same as the single deflection which would have been produced if the two heat productions had been summed together. If the damping had been proportional to the square of the angular velocity of the magnet system, or if the conduction of heat had been proportional to the square of the temperature difference, this fact would not have been true. Expressing the matter mathematically: if $y = f(t)$ be the equation relating the deflection y to the time t for a certain heat distribution* H, and $y' = f'(t)$ be the equation for another heat distribution H', then for a combined distribution of heat, $H + H'$, the deflection will be $Y = y + y' = f(t) + f'(t)$. This property of the system seems to be a very exact one, and it makes possible an analysis of the evolution of heat which would otherwise have been very difficult.

Owing to the heat capacity of the recording instruments and the lag in the conduction of heat and in the galvanometer deflection, the readings actually obtained with myothermic apparatus are considerably smaller than those calculated from the physical data of thermoelectromotive force, number of junctions, and galvanometer sensitivity. Indeed the calculation of heat production directly from such data gives results very far from the truth, and it is impossible without some direct method of calibration to compare work expressed in mechanical units with heat expressed in galvanometer deflections. Fortunately it is possible to make a direct calibration of the instruments by liberating in the same muscle, in the identical position on the thermopile at the close of an experiment, a known amount of heat. A muscle must first be killed by chloroform, and then an alternating current of known strength must be passed for a known time through the measured resistance of the muscle, so that a definite amount of heat may be liberated, and the resulting deflection of the galvanometer read. A comparison of the two gives h the value in absolute units (calories or ergs) of one scale division of the galvanometer. There are various technical difficulties in this calibration which need not be discussed now, but in one modification or another it serves as the basis of all experiments relating heat to work. To some of these I will refer later. I wish to speak at present rather of the analysis of the time-course of the evolution of heat in muscle when stimulated. An amount of heat liberated suddenly at a given time leads to a certain deflection of the galvanometer, $y = f(t)$, which rises to a maximum and falls slowly to zero again owing to conduction away of the heat, etc. A live muscle would give the same shape of deflection, if the heat produced in it were liberated at, or immediately after, the moment of stimulus. In point of fact the «live» deflection differs largely from the «control» one, a fact which can be explained only by supposing that the heat production in the live muscle is not instantaneous at the beginning, but distributed in time, there being production

* H also must be regarded as a function of the time.

or absorption of heat later on. I may say at once that neither I nor any of my colleagues have ever found an absorption of heat at any stage in the process, so that apparently nothing but production of heat need be taken account of. One of my earliest observations on the subject was that the galvanometer deflection persists much longer in a live muscle than in a control experiment: it remains away from its zero for periods up to ten minutes, whereas, if the production of heat were instantaneous at the beginning, it would be back to its zero in two or three minutes. This phenomenon can be due only to a delayed production of heat, and I found that this «recovery» heat, as we called it, is appreciable only in oxygen, being abolished by keeping the muscle in nitrogen, or by previous exercise violent enough to use up the oxygen dissolved in the muscle. My original analysis was rough and revealed little but the facts themselves and their order of quantities. It certainly, however, did establish the existence of a recovery heat production, which it was natural to associate – rightly, as it has proved since – with the oxidative removal of lactic acid discovered by Fletcher and Hopkins. A rough estimate of the magnitude of the recovery heat production made it approximately equal to the total initial heat. This estimate appeared to answer unequivocally a question long debated, on the fate of lactic acid in the recovery process. Fletcher and Hopkins had found that lactic acid is removed in the presence of oxygen, though the same muscle at the end of the recovery process can liberate during exercise or rigor the same amount of lactic acid as before. Was lactic acid removed by oxidation, or by restoration to the precursor from which it came? Previous experiments of my own had shown that the production of one gram of lactic acid in rigor leads to the liberation of about 500 calories. Experiments by Peters had proved that the production of 1 gram of lactic acid in exercise, or in exercise followed by rigor, leads to the liberation of about the same quantity of heat. Hence, if the recovery heat were equal to the initial heat, the oxidative removal of one gram of lactic acid would lead to the production of about 500 calories, which is less than 1/7th of the heat of oxidation of the acid. The conclusion, which was not universally accepted at first, seemed to me to be inevitable – that the lactic acid is not removed by oxidation. This conclusion has been amply confirmed by the later experiments of Meyerhof which he will himself describe.

In a large mass of muscle deprived of its circulation, the rate at which the recovery process can go on, after severe stimulation, depends on the rate at which oxygen can reach the fibres by diffusion. Such are the conditions in calorimetrical experiments. The myothermic method, however, employing small thin muscles subjected only to very light stimulation makes it possible to follow the recovery process under conditions where the rate is quite uninfluenced by any possible lack of oxygen. In such small contractions there is sufficient oxygen already dissolved inside the muscle to account for far more heat than the total amount liberated in recovery. It is possible therefore to follow the chemical dynamics of the recovery process uninterfered with by considerations of oxygen supply. The study of the actual time-course of the recovery heat production has required more elaborate and careful experiments than I made alone in 1913, and since the War, I have had the cooperation in this and other matters of my friend W. Hartree, whose skill in experimental work and calculation has made it possible to reach a degree of certainty in the analysis which I could never have attained alone. By photographic recording and accurate numerical analysis of the deflection for ten minutes after stimulation, it has been possible to describe the whole of the time-course of the heat production in recovery. Its absolute magnitude is 1.5 times the total initial heat; its rate is influenced by temperature in the way usual in biochemical reactions. It starts from a low rate, rises to a maximum, and declines to zero again. It seems to occur with a velocity approximately proportional to the square of the concentration of the bodies whose removal constitutes recovery. Taking Meyerhof's value for the heat per gram of lactic acid, we find that 1/5 to 1/6 only of the lactic acid is oxidized in recovery.

It is clear that the recovery process is a fundamental part of the whole mechanism of muscle. As we shall see later, oxygen is used only in the recovery process, a fact which has led to a new conception of the nature of the muscular machine.

The analysis of the recovery heat production is comparatively simple because its evolution is slow. It is desirable, however, also to analyse the production of heat in the earlier phases of contraction, a far more difficult matter. After a single shock the muscle shows two phases of contraction, namely, the development and the disappearance of the mechanical response – contraction and relaxation. Tetanus shows three phases, contraction, relaxation, and maintenance. It was desirable to analyse the chemical break-down processes associated with each of these phases. It will be simpler if I describe phenomena actually seen and recorded with the instruments.

A muscle is stimulated, the galvanometer deflects, a curve is recorded on a moving photographic paper. A control curve is made later, after the muscle is dead, for which a known small amount of heat is liberated rapidly, e.g. in 0.1 second, and another deflection is recorded. It is found even in the earliest phases, long before the development of the recovery process, that the two curves differ from one another. The curve for the live muscle rises less steeply and falls less steeply than the control curve, a fact which can be due only to a spreading out of the heat production. The curves do not differ largely from one another because the whole time occupied in the initial phase is so short compared with the time relations of the galvanometer and thermopile themselves. The difference, however, is sufficient to make the analysis practicable, and by keeping the muscle at a low temperature it is possible to slow down the events and to attain a greater degree of accuracy in the analysis. The result that appears is that during the development of the contraction there is a large outburst of heat, during the maintenance of the contraction there is a continued production of heat, reaching a constant rate as the contraction is prolonged, and a comparatively large and sudden evolution of heat during relaxation. An excited muscle behaves like an excited electromagnet: energy is required to magnetize the iron and to make it develop its pull; energy is required to cause it to maintain its magnetic condition and to continue its pull; heat is liberated when the magnetic condition, and the energy associated with it, disappear as the current is cut off. In the muscle, energy is required to set up the contraction, and, as we shall see later, more energy is required if work be done. Energy is needed to maintain a contraction, as, indeed, we know in our own bodies when we try to hold a weight for a long time. During relaxation the potential energy of strain possessed by a muscle during contraction has to disappear, and we find it as heat.

The most important point brought out by this analysis of the initial heat production is that relating to the influence, or rather to the absence of influence, of oxygen. The essential conclusion can be drawn without any analysis, merely by comparing the curves of deflection with and without oxygen. The proof that oxygen has no effect whatever on the time-course of the initial heat production complements an observation by my friend Weizsäcker, working with me in 1914, that the presence or absence of oxygen has equally no effect on the magnitude of the initial heat production. No difference whatever can be detected between the curves obtained: (*a*) from a muscle in pure oxygen, and (*b*) from one which has been deprived of oxygen in the most rigorous manner for several hours. The conclusion is important and supplements the observations previously described on the recovery heat production. Oxygen is not used in the primary break-down at all: it is used simply in the recovery process. A muscle is like an accumulator, which can be discharged without any kind of combustion or any kind of provision of energy from without: it requires external energy only when it is being re-charged. It has long been known, of course, that muscles can go on working for some time in the absence of oxygen, but it was open to anybody to suppose that the processes in the presence and in the absence of oxygen were different. So they are, to the extent that recovery takes place in the one case but not in the other, but the complete absence of any effect of oxygen on the initial processes of contraction shows that those initial processes are identical in both cases. The analogy of the accumulator is exact.

These conceptions arrived at by studying isolated muscle have an obvious application to man. One knows that after violent exercise one breathes heavily for some time: the more violent the exercise, the longer one's respiration is laboured. If a man runs a hundred metres as fast as he possibly can, he requires about four litres of oxygen extra, in the succeeding five minutes, to enable him to recover from his effort. In other words 11 seconds of exercise has caused him to use in recovery as much oxygen as he would require for about fifteen minutes at rest, as much oxygen as he could get during one minute of the most laboured respiration. The matter is a simple one to investigate by means of Douglas bags and gas analysis. One takes a reading of the resting oxygen consumption, and during the recovery process one measures the oxygen used in excess of the resting value. By employing a series of bags, and collecting the expired air for various short intervals during recovery, it is possible also to study the time-course of that process. Unlike the case of the isolated muscle on the thermopile the rate at which oxidative recovery occurs in man is determined largely by considerations of oxygen supply and not merely by the actual chemical reactions. At the end of exercise, when the oxygen consumption is at its maximum, it begins immediately to fall, and after moderate exertion reaches its resting value in about five minutes. After severe exercise, however, it remains high for a long time, only reaching its resting value after an hour or two of recovery. This condition of prolonged recovery from severe, or prolonged, exercise is associated with the presence of lactic acid in the blood, as can be shown by direct analysis. The lactic acid produced by the muscles in moderate exercise is oxidized there in recovery before it can diffuse into the blood in appreciable amount. Consequently, recovery is rapid. In severe exercise, however, the lactic acid accumulates in the muscles to a high value, escaping thence into the blood and other tissues of the body, from which it can only slowly be removed, after diffusion back into the muscles, during the process of recovery.

Similar methods may be employed for measuring the maximum intake of oxygen, and recent experiments have shown that a man may use as much as 4.2 litres of oxygen per minute, after his circulation and respiration have been worked up by running fast for three or four minutes. This, you will remember, is about the amount which the body requires in recovery from 11 seconds of severe exertion. Were it not for the fact that the body is able, so to speak, to take its exercise on « credit», instead of paying for it out of «income», it would be impossible for a man to take anything but quite moderate exercise. The body is capable of running up an oxygen «debt» which must be repaid during the recovery process. The maximum «debt» which we have found is 15 litres, which is nearly four minutes supply at the maximum rate. We have little doubt that a first-class athlete, in the height of training, could run up a debt considerably greater than this. There is every reason to suppose that this process of «running into debt» for oxygen is associated with the accumulation of lactic acid in the muscle. Lactic acid is found in blood in comparatively large quantities after severe exercise: this acid must have come from the muscles. The respiratory quotient reaches values which are quite impossible on any supposition other than that the CO_2 is at first driven out from bicarbonate by lactic acid, and later restored to bicarbonate as the lactic acid is removed in recovery. This removal of lactic acid requires oxygen and corresponds to the recovery process which we have found in frog's muscle. Experiments by my colleagues Long and Lupton have shown that the ratio of lactic acid removed to lactic acid oxidized is about the same as in isolated frog's muscle, namely about 6:1.

This application to human muscular exercise is perhaps a digression, but I feel rather an interesting one. It shows that the purely academic study of the isolated frog's muscle may be applied to the extremely important practical case of muscular exercise in man: it explains many of the well-known phenomena of athletics. We will return now to the frog and to the myothermic experiments made upon it.

If we take the case of the prolonged isometric contraction, heat is liberated in each of the three initial phases of contraction, maintenance, and relaxation. If the contraction be maintained for a long time, there is a steady heat production during the whole of that time, proportional to the tension developed. We may describe the phenomenon mathematically in the formula:

$$\frac{H}{Tl} = A + Bt$$

where H is total heat production, T is force developed, l is length of muscle, t is time during which the contraction is maintained, A and B are constants. It is found that A is independent of temperature, having, in the case of the sartorius muscle, always a value of about 1/6; B, which we may regard as the inverse of the «efficiency of maintaining a contraction», depends on the type of muscle, on temperature, on fatigue, and on many other factors. If we take the important practical case of human movements, so much energy is required to set them up, so much to maintain them for a given time. The amount required to set up a given contraction is constant, the amount required to maintain it is variable, depending on the nature of the muscle and its condition at the time. Anything which slows the single twitch of a muscle makes summation more easy and B smaller: the muscle becomes more efficient for maintaining a contraction. The quickest muscles of all are the least efficient when it comes to exerting a force for a long time. There is an important effect of temperature on the rate at which the total heat production increases as the muscle continues to be tetanized – the rate of heat production is greater at the higher temperature. It would appear as though there were some ready store of energy which is, so to speak, exploded by the first few shocks, and that later a steady stream of energy must be maintained to keep up the contraction. Possibly Embden's lactacidogen is the explosive substance used in the first few twitches of the summed contraction, the store which has been exploded being re-formed continually by chemical reactions, from its precursor glycogen. If so, we should expect the rate at which restoration can go on to have a temperature coefficient of a chemical kind.

We have dealt so far, for several reasons, with the simple case of the isometric contraction: firstly, because it involves us in one less variable, namely, the length of the muscle, and secondly, because myothermic experiments on the isometric contraction are so much simpler and freer from error. Recent developments, however, of the technique have made it possible to study the heat production in a muscle which is allowed to shorten as much as one desires, and provided that stringent precautions be taken, reliable results are obtained. The temperature equilibrium in the muscle chamber must be so good that shortening over the junctions of the thermopile does not bring cooler or warmer parts of the muscle in contact with them. It must be remembered that this constancy of temperature must be to the nearest 1/100,000 of a degree because our scale of temperature is so small. During the last year Dr. W. O. Fenn has studied the effects of shortening and doing work upon the total liberation of energy by a muscle. He has found that when work is done, there is an « excess » liberation of energy over and above that necessary in an isometric contraction of the same duration. The muscle is excited and allowed to shorten from one length to another, to lift various loads from one level to the other. The work done is proportional to the load, and the « excess » energy is also proportional to the load. If the load be held up at the end of the contraction so that the muscle relaxes unloaded, the excess energy is just about equal to the work done. If the muscle contracts freely and then lowers the load, excess energy is liberated of the same order of size as the work done by the load in falling: the muscle requires energy both to lower and to raise a weight. This is only one of the curious and unexpected results which Fenn has found. If a muscle be caused to lengthen during the development of its contraction, it gives out less energy than in an isometric twitch. If it be held fast and allowed to shorten only during relaxation, then again it will give out less heat. Shortening during contraction, lengthening during relaxation, appear to require excess liberation of energy. Lengthening during contraction, shortening during relaxation, appear to cause an excess « absorption » of energy, i.e. to lead to a total energy liberation less than that of the isometric twitch.

We are studying here the curious power which a muscle possesses of adapting its liberation of energy to the work it has to do. The subject is new and results are only now beginning to appear, but it is certain that we are dealing with a very fundamental property of the muscular machine. I should like shortly to illustrate the new principles which are emerging by some further examples.

Let us stimulate a muscle and allow its tension to develop isometrically until it has reached a maximum. Let us then, by a quick-release mechanism, allow the muscle to shorten suddenly and let us hinder its contraction by opposing it to the inertia of a mass; the mass employed may be either a flywheel, or an equilibrated beam, or a weight hanging on a long string and pulled horizontally. The greater the mass which opposes the contraction of the muscle, the more slowly will the muscle contract; the less the mass, the more rapidly will it contract. If a stimulated muscle were an elastic body, as Fick and others sometimes, as I always till recently, supposed, it would do the same work against a small mass as against a large one. It would turn the whole of the potential energy of strain which it possesses into kinetic energy in the mass. In muscle this is not so. The greater the speed of shortening, the less will be the work done. In the case of human-arm muscles the work does not attain anything near the maximum value unless the contraction has been opposed by a mass large enough to make it occupy at least two seconds. Even an unloaded contraction occupies about 0.25 seconds, in which case of course no work is done. Dependence of work on speed of movement I ascribed originally to the viscosity of the muscle substance: the greater the speed, the greater would be the fraction of the potential energy of the muscle which would be wasted in changing the muscle's own form. The original experiments were made on man, and it was possible that the form of the curve was to be ascribed to the central nervous system, adjusting the innervation of the muscle to the work which it had to perform. This possibility is eliminated by the fact that the same identical phenomenon is found in the frog, as Gasser recently has shown. A frog's muscle is stimulated and held fast, then suddenly released against the inertia of a mass. The kinetic energy produced is greater if the contraction be slow, and less if the contraction be fast, in a muscle subjected to a maximal tetanus. The phenomenon appears to be a genuine property of muscle. Gasser has found that the time relations of the frog's muscle in this respect are totally different from those of man. Relatively speaking, the frog's muscle is ten times quicker than that of man: it can do the same fraction of its maximum work when it contracts in about 1/10th of the time. This fact seems to indicate a curious and interesting application of the theory of dimensions, but I cannot discuss it further now. The difference in the time-scales of the two types of muscle makes one regard it as improbable that physical viscosity alone is the determining factor. One cannot see why viscosity should have ten times the effect in a human muscle than it has in a frog's, and probably one hundred or one thousand times as much as it has in a fly's. It would seem that we are dealing here with a fundamental mechanical property of a muscle which is actually shortening after stimulation.

Perhaps the facts we have just discussed are related to Fenn's work, who found that the greater the work, the greater the excess heat. They suggest an adaptation of the chemical break-down of the muscle to the work performed. There may be some kind of positive physiological machinery in the muscle to regulate the speed of contraction to the requirements of the animal. I have no time now to go into the matter further, but I believe that we are beginning here to see a further striking adaptation of the body to its needs, of the muscle to its load. For those who like mechanical and electrical analogies I will refer to the case of the electric motor. If the motor be unloaded, a certain amount of energy is used: increase the load and the motor automatically takes more current: try to drive the motor from without and a back e.m.f. is produced, causing less current to be taken. It is clear, in any case, that the « all-or-none » principle is not completely applicable to the muscle fibre, *if we state the principle in the form that the response depends only on the stimulus and on the initial circumstances.* The amount of energy which the muscle gives out obviously depends on factors which come in only after the muscle has begun to shorten, on the inertia opposing its contraction, on the load it has to lift. State it in the form that the response cannot be varied by varying the stimulus and it still holds. It is clear also that those of us who supposed that a stimulated muscle is simply a new elastic body were wrong. A muscle may possess some elastic properties, but it requires more energy to do more work, a fact which is fundamentally in opposition to the elastic body theory.

There are many sides of this entrancing subject, on which I have no time to touch: there are many experiments which must still be done. We are dealing here with a genuine physical problem, involving statics, dynamics, thermodynamics, the design and employment of instruments: and it would seem that in the study of the mechanism of muscle we have a better opportunity of exploiting the use of these exact tools than perhaps on any other side of the investigation of living, working material. Our study, however, needs light also from another aspect, it requires the skilled labours of the organic and biochemists. We who have worked on the physical problems provided by the muscles could never have progressed, had Fletcher and Hopkins not put us on the trail: we should have been lost and bewildered had not Meyerhof, in the brilliant researches which he will now describe, ably led us through a part of the forest where our own methods are helpless. To take another analogy, the completion of the drawing will rest with the chemists: we physicists can only provide a sketch; we can indicate the type of machine and its properties, the chemists must describe it in detail.

Biography

Archibald Vivian Hill was born in Bristol on September 26, 1886. His early education was at Blundell's School, Tiverton, whence he obtained scholarships to Trinity College, Cambridge. Here he studied mathematics and took the Mathematical Tripos, being Third Wrangler (1907). After graduating, he was urged to take up physiology by his teacher, Dr. (later Sir) Walter Morley Fletcher.

Hill started his research work in 1909. It was due to J. N. Langley, Head of the Department of Physiology at that time that Hill took up the study on the nature of muscular contraction. Langley drew his attention to the important (later to become classic) work carried out by Fletcher and Hopkins on the problem of lactic acid in muscle, particularly in relation to the effect of oxygen upon its removal in recovery. During his initial studies Hill made use of the Blix' apparatus, obtaining his first knowledge of the subject from papers of this Swedish physiologist. This led him to study the dependence of heat production on the length of muscle fibre (a relation later developed by Starling in his investigation of the mechanism of the heart beat).

After having obtained a Fellowship at Trinity in 1910, Hill spent the winter of 1910–1911 in Germany, working among others with Bürker (who taught him much about the technique of myothermic observations) and Paschen (who introduced the galvanometer to him, which he since used for his investigations). From 1911–1914, until the outbreak of World War I, he continued his work on the physiology of muscular contraction at Cambridge. During this for him important period, however, he also took up other studies: on the nervous impulse (with Keith Lucas), on haemoglobin (with Barcroft), and on calorimetry of animals (partly with T. B. Wood), having also as colleagues Gaskell, Anderson, W. B. Hardy, Mines, Adrian, Hartridge, and others.

In 1914 he tended to drift away from physiology and was actually appointed University Lecturer in Physical Chemistry at Cambridge. During the war he served for the entire period as captain and brevet-major, and as Director of the Anti-Aircraft Experimental Section, Munitions Inventions Department.

In 1919 he took up again his study of the physiology of muscle, and came into close contact with Meyerhof of Kiel who, approaching the problem from a different angle, has arrived at results closely analogous to his study. They have cooperated continuously ever since, by personal contact and through correspondence. In 1919 Hill's friend W. Hartree, mathematician and engineer, joined in the myothermic investigations – a cooperation which had rewarding results.

In 1920 Hill was appointed Brackenburg Professor of Physiology at Manchester University; there he continued the work on muscular activity and began to apply the results obtained on isolated muscles to the case of muscular exercise in man. From 1923 to 1925 he became Jodrell Professor of Physiology at University College, London, succeeding E. H. Starling. In 1926 he was appointed the Royal Society's Foulerton Research Professor and was in charge of the Biophysics Laboratory at University College until 1952. After retiring he returned to the Physiology Department, where he continues with his experiments to the present.

His work on muscle function, especially the observation and measurement of thermal changes associated with muscle function, was later extended to similar studies on the mechanism of the passage of nerve impulses. Very sensitive techniques had to be developed and he was eventually able to measure temperature changes of the order of $0.003\,°C$ over periods of only hundredths of a second. He was the discoverer of the phenomenon that heat was produced as a result of the passage of nerve impulses. His researches gave rise to an enthusiastic following in the field of biophysics, a subject whose growth owes much to him.

Dr. Hill is the author of many scientific papers, lectures, and books. Perhaps his best-known books are *Muscular Activity* (1926), *Muscular Movement in Man* (1927); also *Living Machinery* (1927), *The Ethical Dilemma of Science and Other Writings* (1960), and *Trails and Trials in Physiology* (1965).

He was elected a Fellow of the Royal Society in 1918, serving as Secretary for the period 1935–1945, and Foreign Secretary in 1946. He was awarded the Society's Copley Medal in 1948. He holds honorary degrees of many universities, British and foreign. He was decorated with the Order of the British Empire in 1918 and became a Companion of Honour in 1948. He also holds the Medal of Freedom with Silver Palm (U.S.A., 1947) and is a Chevalier of the Legion of Honour (1950). He has also been prominent in public life, being a Member of Parliament during the period 1940–1945, when he represented Cambridge University in the House of Commons as an Independent Conservative. He was a member of the University Grants Committee for 1937–1944 and served on the Science Committee of the British Council, 1946–1956. He was appointed a Trustee of the British Museum in 1947.

During World War II, he served on many commissions concerned with defence and scientific policy. He was a member of the War Cabinet Scientific Advisory Committee (1940–1946). He was Chairman of the Research Defence Society (1940–1951) and Chairman of the Executive Committee of the National Physical Laboratory (1940–1945).

He is also a member of the Society for the Protection of Science and Learning, and was President in 1952 of the British Society for the Advancement of Science.

Dr. Hill married Margaret Neville Keynes in 1913. They have two sons and two daughters.

OTTO F. MEYERHOF

Energy conversions in muscle

Nobel Lecture, December 12, 1923

This highest scientific honour, in the form of the Nobel Prize, which has been awarded to me for my investigations into the conversions of energy in muscle, gives me the pleasant duty of reporting to you on this problem and upon the results which my work has achieved. It is especially gratifying to me that this recognition is in part shared with me by my distinguished friend, the previous speaker, Professor A.V. Hill from London, with whom my work has had so many close points of contact and with whom, in spite of the present political unrest, I have worked in cooperation towards the mutual goal of explaining the process of muscle contraction.

The fact that chemical processes must be involved as a source of energy for muscle performance was already accepted as a necessary deduction from their own thesis by the discoverers of the law of the conservation of energy. In fact, the young Helmholtz had already made certain observations concerning the conversion of matter in muscles during activity which in themselves were correct but, on account of their incompleteness and the lack of knowledge concerning the chemical nature of the relevant substances, served no useful purpose. The previous speaker has already told you about the considerable progress achieved by the English scientists Fletcher and Hopkins by their recognition of the fact that lactic acid formation in the muscle is closely connected with the contraction process. These investigations were the first to throw light upon the highly paradoxical fact, already established by the physiologist Hermann, that the muscle can perform a considerable part of its external function in the complete absence of oxygen. As, on the other hand, it was indisputable that in the last resort the energy for muscle activity comes from the oxidation of nutriment, the connection between activity and combustion clearly had to be an indirect one. In fact, Fletcher and Hopkins observed that in the absence of oxygen in the muscle, lactic acid appears, slowly in the relaxed state and rapidly in the active state, and that this lactic acid disappears again in the presence of oxygen. Obviously, then, oxygen is involved not while the muscle is active, but only when it is in the relaxed state, and this assumption has been supported by further research on the part of Parnas and Verzár. In what relation the lactic acid stands to muscle performance, where it comes from and what becomes of it when it disappears in the presence of oxygen, was completely obscure. In fact, there were several different, irreconcilable interpretations current, all of which appeared nevertheless to be supported by experiment. It was at this point that I started to work on the problem. A bright light in the midst of this obscurity appeared when Professor Hill made the important discovery, about which he has just spoken to you, that the contraction heat of the muscle occurs in two distinct phases of approximately the same extent – one phase which is directly connected with the work and is the same in presence or absence of oxygen, which he called the «initial heat»; and a second phase, which basically only occurs in the presence of oxygen, and which he called «delayed heat» and quite rightly connected with the disappearance of the lactic acid. Apart from the pioneer work of Fletcher and Hopkins, it was this discovery above all which, shining out like a beacon light through a sea mist, made it possible for me to steer a safe course through the shallows.

If we now observe an excised frog muscle operating under maximum oxygen supply, chemical analysis will only prove that a certain quantity of glycogen in the muscle disappears, whereas an exactly sufficient quantity of oxygen necessary for its oxidation is assimilated, and the corresponding amount of carbon dioxide is given off. The connection between these processes can be more exactly analysed if the muscle is first allowed to work under anaerobic conditions, and subsequently brought out into oxygen. During the anaerobic phase, in fact, lactic acid accumulates in the muscle approximately in proportion to the amount of work performed. At the same time a corresponding quantity of glycogen disappears, while the quantity of lower carbohydrates, particularly free glucose and the hexose-phosphoric acid discovered in the muscle by Embden, is not noticeably altered. In the second, oxidative, phase the lactic acid which has formed disappears, while a specific quantity of extra oxygen is assimilated. In fact, the disappearance of lactic acid during this period is in exact proportion to the increased consumption of oxygen. However, the oxygen is only sufficient to oxidize a fraction of the disappearing lactic acid; the remainder, which in the case of complete fatigue is about three-quarters of the total lactic acid, is quantitatively reconverted into glycogen. I must state already here that this ratio of the lactic acid which disappears altogether to that burnt is not always constant under all conditions, and from the energetic point of view this is important, to which I must return later. To start with, however, we will concern ourselves with this figure obtained under suitable conditions of extreme anaerobic fatigue, and subsequent recuperation in oxygen. Of four molecules of lactic acid which disappear, three are then converted back into glycogen and one is oxidized. To be exact, we cannot even maintain with certainty that the lactic acid itself is burnt. We find only an oxidized carbohydrate-equivalent with the respiratory quotient 1. Whether this is sugar or lactic acid we cannot be certain. I have, therefore, chosen the formulation for the two phases which you can see on this board.

Anaerobic fatigue phase

$$5/n \, (C_6H_{10}O_5)_n + 5 \, H_2O + 8 \, H_3PO_4 \rightarrow$$
Glycogen

$$4 \, C_6H_{10}O_4(H_2PO_4)_2 + C_6H_{12}O_6 + 8 \, H_2O \rightarrow$$
Hexose-phosphoric acid   Glucose

$$8 \, C_3H_6O_3 + 8 \, H_3PO_4 + C_6H_{12}O_6$$
Lactic acid

Oxidative recovery phase

$$8 \, C_3H_6O_3 + 8 \, H_3PO_4 + C_6H_{12}O_6 + 6 \, O_2 \rightarrow$$
$$4/n(C_6H_{10}O_5)_n + 8 \, H_3PO_4 + 6 \, CO_2 + 10H_2O$$

In the anaerobic, active phase the glycogen is broken down into lactic acid via glucose and, I assume in agreement with Embden, by way of hexose-diphosphoric acid. On the board the decomposition of five sugar-equivalents of glycogen is assumed, of which four are esterified with phosphoric acid and form eight molecules of lactic acid.

In the second, aerobic phase these eight molecules of lactic acid disappear, while two of them, or alternatively, as we might equally well assume, one molecule of sugar, are burnt. The importance of this strangely coupled reaction can only be understood after a study of energetics. But before I turn to this, it is important to stress that this activity metabolism in the muscle is not a separate phenomenon, but is no more than an increase of the metabolism in the resting state. For even in the resting state the glycogen in an isolated muscle in oxygen disappears directly by way of oxidation into carbon dioxide and water. If, however, we keep a resting muscle in nitrogen for a considerable time, lactic acid is constantly accumulating in it during the anaerobiosis. If we now compare this lactic acid accumulation with the quantity of oxygen which the muscle would have assimilated in the same time under aerobic conditions, we find that approximately three times the amount of lactic acid has accumulated as could have been consumed by the oxygen in the same amount of time. Here also, then, the lactic acid is not just a simple intermediate product of the decomposition of the sugar. In fact, if we bring the muscle back into the air after extended anaerobiosis, it will assimilate a certain quantity of extra oxygen, approximately equivalent to the amount previously lost. At the same time, the lactic acid disappears once again in such a way that most of it is reconverted into glycogen, whereas only a fraction, or the corresponding quantity, of carbohydrate is consumed. The process is, therefore, exactly the same as when the muscle is active, only the accumulation of lactic acid progresses much more slowly. We can directly see from this the importance of muscle respiration in the resting state, in that it maintains a labile condition of lactic acid production and removal, which can be accelerated instantly on stimulation. Probably this explosive release of lactic acid during contraction occurs, because stimulation suddenly increases the permeability of membranes which have previously to a certain extent acted as a barrier between the participants in the reaction. Respiration in the muscle in the resting state can, therefore, be said to keep them in a state of readiness for activity.

We can establish that the lactic acid is directly associated with muscle contraction by an exact comparison of the work performed under anaerobic conditions with the formation of lactic acid. As the best expression of the activity potential of muscle we may choose here, following Fick and Professor Hill, the tension which the muscle develops on stimulation when prevented from shortening, i.e. the so-called isometric contraction. If we allow the muscle to go on working under anaerobic conditions until it is exhausted it produces a certain quantity of lactic acid and develops a degree of tension in proportion to this quantity. This total anaerobic work can be very considerable – for instance, a frog muscle of 1 g in weight in N_2 can produce 160 kg of tension in 1000 contractions.

I found that there is a very simple reason why there is any limit at all to this and why activity does not in fact continue until the available glycogen is used up. It was thought earlier, and in particular by Fletcher and Hopkins themselves, who were the first to become aware of the so-called fatigue-maximum, that this was conditioned by the exhaustion of an immediate preliminary stage of the lactic acid. This is, however, not the case – it stems rather from the accumulation of the acid in the muscle itself. If we remove a large part of the acid from the muscle by placing it in a Ringer's solution particularly rich in bicarbonate, it produces before total exhaustion not only very much more lactic acid, but also correspondingly more work. By the addition of various buffer mixtures to the muscle it was proved that the increase in performance due to this admixture corresponded almost exactly to the percentage of lactic acid which escaped from the muscle into the surrounding solution.

The significance of these chemical reactions only becomes clear when we consider the energetic conditions. In the anaerobic active phase lactic acid is formed from glycogen, at the rate of 1 g lactic acid from 0.9 g glycogen, since during the formation of every 180 g of lactic acid 18 g of water are absorbed

$$(C_6H_{10}O_5) \, (162) + H_2O \, (18) = C_6H_{12}O_6 \, (180) = 2 \, C_3H_6O_3 \, (2 \times 90)$$

The combustion heat of glycogen, according to Stohmann's readings, is 4191 cal/g – that is, 3772 per 0.9 g. As these readings were made about thirty years ago with still somewhat primitive instruments, a revised determination seemed desirable, especially as the American scientists Emery and Benedict had found a rather higher value of 4227 cal/g. This new determination was made at my suggestion in Germany by a pupil of Professor Roth in Brunswick, and at the same time in Manchester by Mr. Slater. In the first case Stohmann's readings were completely confirmed, resulting in 4188 cal/g or 3769 cal/0.9 g. Slater, however, using a differently produced glycogen, obtained a very much higher value, i.e. in relation to the above glycogen formula he obtained 3883 cal/0.9 g. I will come later to the reasons which for the time being have caused me to regard the values of Stohmann and Roth to be more accurate. The combustion heat of lactic acid I determined anew, since the values given in the literature appeared unreliable, and I obtained 3601 cal for dilute lactic acid, a value which was confirmed in Roth's Institute and which agreed also with the Americans Emery and Benedict. If the chemical process during contraction turned out to be as one was only recently tempted to imagine it – that is to say, if during activity lactic acid is formed from glycogen and this is evenly consumed during relaxation – then only the difference in combustion heat between 0.9 g of glycogen and 1 g of lactic acid – i.e. 170 cal – would be released by activity in the muscle. On the other hand, the combustion of the lactic acid at 3601 cal would take place in the oxidative recuperative period. Such a process, in which only 5% of the heat released would occur in the contractive phase, would appear to be extremely doubtful theoretically and would into the bargain contradict the fact established by Professor Hill that heat quantities in the active and recuperative phases are approximately the same. In fact, this consideration was really the beginning of my preoccupation with the problem of muscle. The process is actually quite different. It soon became evident to me from a great number of determinations that during the formation in the muscle of 1 g of lactic acid, not 170 cal but 380–390 cal were released, a figure which was not very far removed from older, slightly less accurate, results which Peters, a pupil of Professor Hill, had obtained. Before we discuss the reason for this very big divergence of the contraction heat from the difference between the combustion heats, we must first calculate the energy balance of the recovery phase. If, as is represented in the above equation and has on average been proved by my experiments, of a total of 4 molecules of lactic acid which disappear, one is burnt (or, which comes to the same thing, a carbohydrate equivalent of it), then altogether, for 1 g of sugar taking part in the reaction, or 0.9 g of glycogen, $3772/4 = 943$ cal must be released. As we have measured 385 cal in the active phase, the remainder – i.e. approx. 560 cal – must be expected during the recuperative phase. According to this, 40% of the heat must occur in the active phase, 60% in the recovery phase. In fact this was very well confirmed, at least with regard to the order of magnitude, by measurement of the total heat production in the recovery phase and comparison with the oxygen consumption. According to this: (1) the increase in heat produced in the oxidative recuperative phase was approximately as great as, or only slightly greater than, the anaerobic heat of the exhaustion state of the muscle; (2) this heat, reckoned according to oxygen assimilation, was smaller than the corresponding carhohydrate consumption which took place simultaneously. For every 1 c.c. of oxygen during carbohydrate oxidation 5 cal should have appeared; but there were only 3.5 cal, and altogether had vanished about the same amount of heat in the recovery phase as had appeared in the anaerobic phase.

We now find that this result agrees very well with Professor Hill's findings, about which he has just spoken to you: the heats of the active and the recovery phases are equal. But, as he went on to establish, this result can be confirmed even more exactly by the more accurate analysis of heat formation which is made possible by the myothermic method developed by Hill and Hartree, also by the study of the oxygen consumption in relation to the disappearance of lactic acid during recovery under various conditions. The result of these experiments shows that the quotient

$$\frac{\text{total disappearing lactic acid}}{\text{lactic acid burnt}}$$

is not constant – it is greater in completely fresh muscles, and can in fact amount to as much as $5:1-6:1$, and it is of approximately the same size in live humans as in live frogs. In the case of humans this was proved indirectly in Professor Hill's laboratory. I myself obtained a similar result on the whole frog by using the same direct methods as would have been used on the isolated muscle.

Such an increase in the quotient means, however, that a smaller part of the heat occurs in the recovery phase. For if out of six molecules, for instance, only one is burnt then there must arose from the conversion of 1 g of sugar $3772/6 = 630$ cal, of which 385 are in the fatigue phase, so that 245 must be in the recovery phase. In this case 60% must already be released in the active phase, and only 40% in the recovery phase. The quotient appears to lie between these two amounts according to the degree of exhaustion and the condition of the muscle – between $6:1$ and $4:1$; and in unfavourable conditions it is even smaller. The muscular mechanism operates so much the more economically the more lactic acid molecules can be transformed back into glycogen through the oxidation of one of them. Thus this figure represents the efficiency of the recovery process. It is an expression of how much of the oxidative energy is used in endothermal processes for the conversion of the material in the preliminary stages. In the case of the above equation the efficiency would be 40% – under the more favourable conditions of a completely fresh frog muscle or of a living animal it would be 50–60%. Curiously enough, the ratio is smaller in the case of respiration in rest, i.e. of 2–3 molecules lactic acid which disappear 1 is burnt. My more recent experiments have, in fact, shown that many poisons, and also traumatic damages, experienced by the animal before death, will cause the ratio to deteriorate still further. All these circumstances result in a squandering of energy.

There is a very important problem connected with the size of the anaerobic contraction heat itself, which, as we have seen, is in the region of 385 cal, whereas thermochemical data only give a difference in the combustion heat of 170 cal for the conversion of glycogen into dilute lactic acid. How does this difference arise? The following has been established: if the formation of heat and of lactic acid are compared, not in the working muscle, but in crushed muscular tissue suspended in phosphate solution, we then obtain about 200 cal/g instead of 385 cal, and at the same time the lactic acid passes into the phosphate solution. The heat of neutralization of lactic acid with biphosphate is, however, 19 cal/g. Added to this is the heat of the cleavage of glycogen into lactic acid, 170 cal, and these taken together amount to 190 cal, which agrees, allowing a margin of error, with our measured value of 200 cal.

Similarly, the heat even in an intact muscle can be reduced if a considerable amount of the lactic acid passes into the surrounding solution. This can be brought about if lactic acid is allowed to form in a resting muscle suspended in a carbonated Ringer's solution. Half of the lactic acid can escape into the surrounding solution, and the ratio will be 280 cal per g lactic acid instead of 385 cal, and finally only 230 cal.

This particular heat formation which lactic acid produces in the living muscle is, however, bound up with the hydrogen ion. In fact, other acids which we allow to penetrate from outside into the frog muscle cause considerable heat production which is independent of lactic acid formation within the muscle. In this way I observed that the penetration of valeric acid into the muscle caused the release of up to 11,000 cal/mol of acid assimilated by the muscle. Reckoned in terms of lactic acid this corresponds to a heat production of 120 cal/g. This heat, as close analysis shows, is dependent upon the reaction of the acid with the tissue protein. This tissue protein acts as a buffer substance and keeps the reaction within the muscle always more or less constant, even during heavy lactic acid production. Even in the case of maximum exhaustion, when approx. 0.4% of lactic acid is produced in the muscle, the index of the hydrogen-ion concentration (pH) is only displaced from 7.5 to 6.8. With this buffer reaction there is a characteristic heat formation – an inverse protein dissociation heat – which is bound up with the deionization of the protein.

These conditions are very clearly seen in relation to the amino acids, which behave in principle in the same way as protein, which is in fact composed of amino acids. If we start, for instance, with a solution of glycine with the addition of caustic soda solution there will be in the solution, apart from other substances, the salt sodium glycine, which we can regard as being completely dissociated, and which will be abbreviated here as NaG.

$$\text{Na}^{\cdot}\text{G}' + \text{H}^{\cdot}\text{L}' = \text{Na}^{\cdot}\text{L}' \,(\text{GH}) + 11{,}000 \text{ cal} \qquad (1)$$

(G = Glycine anion; L = Lactic acid anion)

If we now add hydrochloric acid or lactic acid or another not too weak acid, then a reaction will take place which is shown here in Formula (1). Out of the totally dissociated glycine-acid sodium there is formed the weak, non-dissociated glycine acid, and in this reaction a positive heat of approx. 11,000 cal can be measured. This is nothing else than the inverse heat from the electrolytic dissociation of the glycine. If a buffer solution is made of concentrated protein solution, free from basic salts, by the addition of caustic soda solution at about pH 8, and if lactic acid is now added in such a quantity that the H-ion concentration is barely altered, a corresponding reaction will obviously take place, which is shown here in the following equation:

$$Na^{\cdot}P' + H^{\cdot}L' = Na^{\cdot}L' + (HP) + 12{,}650 \ cal \qquad (2)$$
$$(P = Protein \ anion)$$

Here we find an even greater heat production – with muscle protein in the presence of ammonium salt it will be 12,650 cal. This dissociation heat of protein is the largest known dissociation heat of any acid. The reason for this may be connected with the fact that the deionization of amino acids and protein causes the production of internal ammonium salt, as shown in the following diagram:

$$
\begin{array}{c}
CH_2{}^{\cdot}NH_2 \\
| \\
COONa
\end{array}
+ H^{\cdot}L' =
\begin{array}{c}
CH_2{-}NH_3 \\
| \quad \diagup \\
COO
\end{array}
+ Na^{\cdot}L' \qquad (3)
$$

We can check this supposition by means of formaldehyde. Formaldehyde causes the formation of a methylene compound from the amino acids with stronger acid properties. At the same time as the addition of acids the very high dissociation heat disappears almost completely.

From the dissociation heat of 12,600 cal per equivalent can be calculated a heat production of 140 cal per g lactic acid. As, however, the lactic acid in the muscle to a certain reacts with phosphate and carbonate this figure should be reduced a little. I have already shown that of the 385 cal formed per 1 g of lactic acid in the anaerobic contraction phase 170 are due to the splitting of the glycogen into dilute lactic acid. There remain 215 cal. Of these, up to 140 can be explained by the dissociation heat of the protein. There remain, over and above the margin of possible error, 70–80 cal for which up till now we can only provide a hypothetical explanation. At first sight three such explanations appear possible: (*1*) the combustion heat of the glycogen may be higher than has been supposed; (*2*) secondary reactions about which we as yet know nothing may take place; (*3*) the dissociation heat of the protein may be greater in living muscle than in solution. The first possibility seemed to have found strong support from Slater's experiments, according to which glycogen is supposed to have a combustion heat 100 cal greater than the value I have given it. However, I was able to prove that this was unlikely, by bringing about a splitting of the glycogen into maltose and dextrose by means of a diastatic ferment which engendered combustion heats of 3752 and 3748 cal. To my surprise the heat of the cleavage of dissolved glycogen into dissolved maltose and dextrose was only about 10 cal/g of glucose. But these values in fact agree approximately with the glycogen combustion heat determined by Stohmann, as the solution heat of dextrose amounts to minus 12.5 cal/g. For the conversion of 0.9 g of glycogen into 1 g of solid dextrose the figure of 10 + 12.5 = 22.5 cal was thus determined by experiment. From this the combustion heat of the dissolved glycogen could be calculated as about 3770 cal per 0.9 g. That of anhydrous glycogen is incidentally much higher, on account of a very considerable hydration heat, but this need not concern us here, since the glycogen in the muscle is in hydrated form. At the same time, there is a very interesting conclusion to be drawn in connection with the result of these readings. We have good reason to believe that the actual process of muscular work does not begin with the splitting of the glycogen but with that of the dextrose or the phosphoric acid hexose. The energy released by the conversion of the glycogen into hexose is then lost to the activity process. Up till now it has been possible to assume that more than 30 cal/g of sugar – that is to say, no less than 8% of the total energy – was squandered in this way. It is now apparent, however, that as a result of the negative solution heat of glucose, the conversion of dissolved glycogen into dissolved dextrose only requires approx. 10 cal – that is to say, barely 3% of the contraction heat. It is possibly even lower in the case of the conversion into phosphoric acid hexose. This result is in harmony with the splendid economy which is shown in the reactions of the living organisms.

Of the possibilities just mentioned for the explanation of the difference of about 70 cal there now remain, therefore, only the two last. Up till now there has been no reason to suppose that, side by side with the carbohydrate metabolism, a fat or protein metabolism also plays a part in the contraction mechanism, nor that inorganic compounds such as phosphoric acid undergo permanent changes on account of anaerobic exhaustion. Therefore I incline to the third of the above hypotheses, namely that protein in living muscle has a higher dissociation heat than in solution: this supposition is supported by the consideration that the material of which the muscular machinery is composed is protein, whereas carbohydrates form the combustion material. Somehow the oxidation energy of the combustion material must play a part in the mechanism itself. This is in fact the case with the deionization of muscle protein. The oxidation energy first becomes active in the recovery phase; and then, as a result of the coupling of the oxidation with the re-synthesis of the lactic acid, not only is the endothermic process of rebuilding the glycogen brought about, but also the alkali is released from the vanishing alkali-lactate, and this leads to the endothermic, involuntary dissociation of the muscle protein. In this way the muscular apparatus is once again put into working order. We can compare this process to the charging of an accumulator, which Professor Hill has given as an image of the recovery reaction, or, if we prefer, to the winding-up of a watch, as I have more often described it. It would then appear theoretically quite logical to expect that a relatively large part of the total energy should be lost in this process. But we must leave the definitive explanation of this point to the future.

We can, however, say now that the deionization of the protein by the lactic acid produced by muscle activity plays, without any doubt, an important part in the contraction mechanism. It explains in the first place the flaccidity of the muscle which sets in after shortening under anaerobic conditions in spite of the presence of lactic acid. This flaccidity would be brought about by nothing more than the diminution of the lactic acid acidity, just as, conversely, we would hold the H-ions responsible for the release of the contraction. Certainly, the flaccidity heat discovered by Hill and Hartree, even though it must be considered to originate in the superimposition of various chemical and physical processes, is obviously the principal cause of the deionization heat of protein. On the other hand, the exhaustion maximum of lactic acid would be conditioned by the supply of alkali separable from the protein salt of the muscle.

I think, therefore, that in this way we have obtained a comparatively simple and satisfactory picture of energy conversions in muscle, the future shape of which will be of theoretical value and practical interest.

To remove the uncertainty concerning the combustion heat of glycogen mentioned in this lecture, I have since, with Dr. Meier, brought about the combustion of several glycogen preparations with various reservations. For anhydrous glycogen ($C_6H_{10}O_5$) from frog muscles we obtained 3806 cal, for glycogen hydrate ($C_6H_{10}O_5 \cdot H_2O$), 3786 cal, from which, taking into account the solution heat of dissolved glycogen, 3775 cal was found, so that the value used above is almost exactly correct.

It also became clear that during deionization of saturated solutions of amino acids in non-aqueous media (alcohol–water mixture) besides a deionization heat, a precipitation heat occurs, which amounts to just about 70 cal/mol – a pointer to the possible explanation for the cause of the unexpected remainder in the contraction heat.

Biography

Otto Fritz Meyerhof was born on April 12, 1884, in Hannover. He was the son of Felix Meyerhof, a merchant of that city and his wife Bettina May. Soon after his birth his family moved to Berlin, where he went to the Wilhelms Gymnasium (classical secondary school). Leaving school at the age of 14, he was attacked, at the age of 16, by kidney trouble and had to spend a long time in bed. During this period of enforced inactivity he was much influenced by his mother's constant companionship. He read much, wrote poetry, and went through a period of much artistic and mental development. After he had matriculated, he studied medicine at Freiburg, Berlin, Strasbourg, and Heidelberg.

In 1909 he graduated in medicine with a thesis on a psychiatric subject and devoted himself for a time to psychology and philosophy, publishing a book entitled *Beiträge zur psychologischen Theorie der Geistesstörungen* (Contributions to the psychological theory of mental disturbances) and an essay on *Goethes Methoden der Naturforschung* (Goethe's methods of scientific research). Under the influence of Otto Warburg, however, who was then at Heidelberg, he became more and more interested in cell physiology. After working for a short time on physical chemistry with Bredig at Heidelberg, Meyerhof spent some time in the laboratory of the Heidelberg Clinic and at the Zoological Station at Naples. In 1912 he went to Kiel, where he qualified in 1913, under Professor Bethe, as a university lecturer in physiology; and lectures which he delivered at Kiel, in England and the United States were published as *The Chemical Dynamics of Living Matter*. In 1915, when Professor Höber assumed the Directorship of the Institute of Physiology, Meyerhof was appointed Assistant. In 1918 he became Assistant Professor. In 1923 he was offered a Professorship of Biochemistry in the United States, but Germany was unwilling to lose him and in 1924 he was asked by the Kaiser Wilhelm Gesellschaft to join the group working at Berlin-Dahlem, which included C. Neuberg, F. Haber, M. Polyani, and H. Freundlich.

In 1929 he was asked to take charge of the newly founded Kaiser Wilhelm Institute for Medical Research at Heidelberg. In 1938 conditions became too difficult for him and he decided to leave Germany. From 1938 to 1940 he was Director of Research at the Institut de Biologie physico-chimique at Paris, where he was helped financially by the Josiah Macy, Jr. Foundation.

In June, 1940, however, when the Nazis invaded France, he had to flee from Paris. Driving with his family to Toulouse, he was befriended by the Medical Faculty there, but escape became essential and a tragic flight followed. Eventually, with the help of the Unitarian Service Committee, he reached Spain and ultimately, in October 1940, the United States, where the post of Research Professor of Physiological Chemistry had been created for him by the University of Pennsylvania and the Rockefeller Foundation.

Meyerhof's own account of his earlier work states that he was occupied chiefly with oxidation mechanisms in cells and with extending methods of gas analysis through the calorimetric measurement of heat production. In this manner he studied the metabolism of sea-urchin eggs, blood corpuscles, and various bacteria and especially the respiratory processes of nitrifying bacteria. He also studied the effects of narcotics and methylene blue on oxidation processes, and the respiration of killed cells. The physico-chemical analogy between oxygen respiration and alcoholic fermentation caused him to study both these processes in the same subject, namely, yeast extract. By this work he discovered a co-enzyme of respiration, which could be found in all the cells and tissues up till then investigated. At the same time he also found a co-enzyme of alcoholic fermentation. He also discovered the capacity of the SH-group to transfer oxygen; after Hopkins had isolated from cells the SH bodies concerned, Meyerhof showed that the unsaturated fatty acids in the cell are oxidized with the help of the sulphydryl group. After studying closer the respiration of muscle, Meyerhof investigated the energy changes in muscle.

Of Meyerhof's many achievements, perhaps the most important is his proof that, in isolated but otherwise intact frog muscle, the lactic acid formed is reconverted to carbohydrate in the presence of oxygen, and his preparation of a KCl extract of muscle which could carry out all the steps of glycolysis with added glycogen and hexose-diphosphate in the presence of hexokinase derived from yeast. In this system glucose was also glycolysed and this was the foundation of the Embden-Meyerhof theory of glycolysis. For his discovery of the fixed relationship between the consumption of oxygen and the metabolism of lactic acid in the muscle, Meyerhof was awarded, together with the English physiologist A. V. Hill, the Nobel Prize for Physiology or Medicine for 1922.

The discovery of Otto Meyerhof and his students that some phosphorylated compounds are rich in energy led to a revolution, not only of our concepts of muscular contraction, but of the entire significance of cellular metabolism. A continuously increasing number of enzymatic reactions are becoming known in which the energy of adenosine triphosphate, the compound isolated by his associate Lohmann, provides the energy for endergonic synthesis reactions. The importance of this discovery for the understanding of cellular mechanisms is generally recognized and can hardly be overestimated.

In 1925 Meyerhof succeeded in extracting the glycolytic enzyme system from muscle, retracing a pathway which Buchner and Harden and Young had explored in yeast. This proved to be a decisive step for the analysis of glycolysis. Meyerhof and his associates were able to reconstruct *in vitro* the main steps of the complicated chain of reactions leading from glycogen to lactic acid. They verified some, and extended other, parts of the scheme proposed by Gustav Embden in 1932, shortly before his death.

Among other honours and distinctions, Meyerhof was a Foreign Member of the Harvey Society and of the Royal Society of London, and a Member of the National Academy of Sciences of the U.S.A.

As a man Meyerhof was a fine experimenter and a master of physiological chemistry. By temperament he was most interested in theory and interpretation and he had a remarkable gift of integrating a variety of phenomena. He spent much time daily at his desk and in stimulating discussions with his pupils and collaborators. His chief scientific work was accomplished while he was at Heidelberg, but he also produced much while he was in America; and in America also he showed that he had never relinquished his active interest in philosophy by presenting to the Goethe Biennial Celebration of the Rudolf Virchow Society in New York a profound and critical evaluation of Goethe's scientific ideas. Throughout his life he retained a great love of art, literature, and poetry. His interest in painting was much stimulated by his wife Hedwig Schallenberg, herself a painter, whom he married in 1914. There were three children of this marriage.

In 1944 he suffered a heart attack; in 1951 another one which ended his life.

Physiology or Medicine 1923

FREDERICK GRANT BANTING

JOHN JAMES RICHARD MACLEOD

«for the discovery of insulin»

Physiology or Medicine 1923

Presentation Speech by Professor J. Sjöquist, member of the Nobel Committee for Physiology or Medicine of the Royal Caroline Institute

Your Majesty, Your Royal Highnesses, Ladies and Gentlemen.

The Professorial Staff of the Caroline Institute has resolved to award to Dr. Frederick Grant Banting and Professor John James Richard Macleod the Nobel Prize for 1923 in Physiology or Medicine for the discovery of insulin.

Although the disease which has received the name of « diabetes mellitus » has evidently been known from immemorial time – Celsus and Araeteus in their writings in the first century of our era described an illness which was characterized by an enormous secretion of urine, an unquenchable thirst and a considerable loss of flesh – it was not until the seventeenth century that the Englishman Thomas Willis made the important observation that the urine in this illness contains a sugar-like substance; and it was not until more than a hundred years later that his countryman Dobson was able to produce from such urine the kind of sugar in question. This discovery, it is true, led the study of the mysterious disease into the right paths; but nevertheless it was a long time before any real progress was made. At the time the sugar was regarded as being a substance foreign to the animal organism, which was formed only under diseased conditions. It is true that the observation by Tidemann and Gmelin in 1827, that starchy foods are under normal conditions transformed into sugar in the intestinal canal and that this is absorbed by the blood, marks an important advance; but really epoch-making was the discovery of the great French physiologist Claude Bernard in 1857 that the liver is an organ that contains a starch-like substance, glycogen, from which sugar is constantly being formed during life; in the words of Claude Bernard, the liver secretes sugar into the blood.

In connection with his investigations into the circumstances that affect the formation of sugar, Claude Bernard observed that in certain lesions of the nervous system the sugar content of the blood was increased and that the sugar passed into the urine of the animals in the experiments. For the first time, therefore, an appearance of sugar in the urine – a glycosuria, though of a transitory nature – was experimentally produced; and consequently this discovery by Claude Bernard may be characterized as the starting-point of a series of experimental researches into the causes and nature of diabetes.

Even before this, however, in the post-mortem examination of persons who had died of severe diabetes, pathologists had made the observation that the pancreas sometimes exhibits diseased changes. The attention of Claude Bernard was directed to this point, but he did not succeed in producing glycosuria by ligation of the duct which leads the secretion of the gland to the bowel or by injecting coagulating substances into it; the removal of the whole gland by operation he regarded as technically impracticable.

Hence it aroused an intense interest when in 1889 two German investigators, von Mering and Minkowski, succeeded in carrying out this operation on dogs. It was still more remarkable that the animals thus operated on, now not only excreted sugar in the urine but also became the victims of a lasting disease which in all essentials resembled the most acute form of diabetes in man, even to such an extent that the content of sugar in the blood rose above the normal and that the disease inevitably led to death with symptoms of poisoning. If a part of the gland was left behind or if a bit of it was sewn under the skin, diabetes failed to develop.

It thus became clear that the disturbance in the sugar economy of the body that appeared after the complete removal of the gland could not well be due to the failure of the pancreatic juice to pass into the bowel, but rather to the loss of some other function of the gland.

During the eighteen-eighties, above all through the investigations of the Frenchman Brown-Séquard, attention had been directed to the importance for the vital functions of certain ductless gland-like organs. Time permits me in this place only to point out that, according to the view now generally entertained, these glands exercise their effect through passing into the blood and tissue juices of certain chemically effective substances, which are called by the general name of hormones; the glands themselves, owing to the fact that they have no ducts, are called endocrine glands or glands with internal secretion. As regards the pancreas itself, it is true that it is a secreting gland, which by means of a duct pours secretion of the gland into the intestinal canal, where that secretion has certain important functions to perform in the process of digestion; but, as Langerhans showed as long ago as 1869, the pancreas also contains anatomical formations which have no direct connection with the duct, and which, after their discoverer are called « the cell islets of Langerhans » or « insulae ». In the beginning of the eighteen-nineties Laguesse expressed the surmise that it was just these cell islets that produce the inner secretion which is so important for the combustion of sugar.

Ever since the discovery by von Mering and Minkowski of the importance of the pancreas for the sugar economy of the organism and evidently also for the development of diabetes – that is to say, for more than a third of a century – a large number of investigations in different countries have devoted a great deal of work to discovering a remedy for diabetes from the pancreatic gland. It was natural to imagine, of course, that that disease was caused by the loss of power of the pancreatic gland to produce a hormone or to produce it in sufficient quantities, and that the introduction of this hormone in the diseased organism ought to be able to exercise a favourable influence on the disease, all the more as analogous conditions were well known with regard to other organs with internal secretion, especially the thyroid gland. Many of these investigations failed, while others succeeded in actually producing extracts or juices which, when injected into the blood of diabetic dogs and even human beings, showed themselves able to bring down the increased content of sugar in the blood, to diminish or even to stop altogether the excretion of sugar into the urine, and to bring about an increase in weight. Amongst these I should like especially to mention Zuelzer, who in 1908 produced an extract which was undoubtedly effective, but which also showed injurious by-effects – consequently it could not be used to any great extent therapeutically – and also Forschbach, Scott, Murlin, Kleiner, Paulesco, and many others.

The problem was in about this position when a young assistant in physiology at the Western University in London, Ontario, Frederick G. Banting, conceived an idea that was to prove of extraordinary importance for its further development. He thought to himself that the reason for the failure to produce effective pancreatic extract, was to be sought in an antagonistic or destructive effect on the hypothetical hormone of trypsin, the protein-splitting enzyme that is produced by the secreting cells of the gland, and that there would be a greater prospect of success if these cells were destroyed by ligation of the duct of the gland and the remaining part of the gland were then used as the original material. It had previously been observed by Schulze and by Ssobolev that the ligation of the duct involved the atrophy of the acini but not of insulae. He imparted his idea to Professor Macleod of Toronto, after which, together with several fellow-workers, among whom I should like especially to mention Best and Collip, he began to work under Macleod's guidance and in his laboratory in May 1921. The very first experiments in diabetic dogs were crowned with success. After the method of producing the effective extract, which at the suggestion of Sir Sharpley

Schafer had been called insulin, had been improved by Collip, and after its effect on the sugar content of the blood, on the respiratory quotient, and on the capacity of the liver for forming glycogen had been established, and also the dangers which might be produced by an overdose of the remedy through an excessive reduction of the sugar content of the blood had been determined by experiments on animals under Macleod's guidance, and after it had further been proved that the trypsin in an alkaline solution really destroys the hormone, the first injection of insulin was made in a youth of fourteen years, who suffered severely from diabetes, on 23 January and the following days in 1922. The result was that the sugar content of the blood of the patient fell to the normal, the passing of sugar into the urine was reduced to a minimum, and the general state of poisoning, acidosis, which is caused by certain injurious substances which are formed in this kind of diabetes mainly through disturbance in the fat metabolism, often in great quantities, was checked. Since then the new remedy, the production of which does not offer any great technical difficulties, has come into use in practically all countries and with favourable results.

We must not imagine that insulin is able to cure diabetes. How could that be possible if the cause of diabetes is to be found in the fact that the cells within our organism that produce the hormone necessary for the combustion of sugar are definitively destroyed? But insulin gives us the possibility of transforming the severe form to a milder one and thereby of restoring his capacity for work and a comparative state of health to the hopeless invalid who, despite the most trying and rigorous restrictions in diet, is constantly threatened by a fatal state of poisoning. Most striking is the effect of insulin in the cases in which the state of poisoning has already passed into that of diabetic coma, against which we have hitherto been helpless and which, before the days of insulin, inevitably led to death.

It could be prophesied with a very great degree of probability that such a substance as insulin some day would be produced from the pancreatic gland, and much of the work had been done beforehand by previous investigations, several of whom very nearly reached the goal. Consequently it also has been said that its discoverer was in a preeminent degree favoured by lucky circumstances. Even if this be so, yet there would seem to be cause to remember Pasteur's words: « La chance ne favorise que l'intelligence préparée. »

The Professorial Staff of the Caroline Institute has considered the work of Banting and Macleod to be of such importance, theoretically and practically, that it has resolved to award them the great distinction of the Nobel Prize.

Doctor Banting and Professor Macleod not having the opportunity of being present today, I have the honour of asking the British Minister to accept from His Majesty the King the prize, and to transfer it to the Laureates, together with the congratulations of the Professorial Staff of the Royal Caroline Institute.

FREDERICK G. BANTING

Diabetes and insulin

Nobel Lecture, September 15, 1925

Gentlemen. I very deeply appreciate the honour which you have conferred upon me in awarding the Nobel Prize for 1923 to me and Professor J.J. R. Macleod. I am fully aware of the responsibility which rests upon me to deliver an address in which certain aspects of the work on insulin may be placed before you. This I propose to do today and I regret that an earlier opportunity has not been afforded me of satisfying this obligation.

Diabetes and insulin

Since von Mering and Minkowski proved that removal of the pancreas produced severe and fatal diabetes in dogs, physiologists and clinicians have frequently endeavored to obtain from the pancreas an internal secretion which would be of value in the treatment of diabetes mellitus. Beginning with Minkowski himself, many observers tried various forms of extracts of the pancreas. Among the extractives used were water, saline, alcohol, and glycerin. The extracts thus obtained were administered by mouth, subcutaneously, intravenously, or by rectum, both to experimental animals and humans suffering from diabetes. Little or no improvement was obtained and any favorable results were overshadowed by their toxic effects. In 1908, Zuelzer tried alcoholic extracts on six cases of diabetes mellitus and obtained favorable results, one case of severe diabetes becoming sugar-free. His extracts were then tried by Forschbach in Minkowski's clinic with less favorable results, and the investigation was abandoned by this group of workers. Rennie found that the islet cells existed separate from the acinar cells in certain boney fishes and in conjunction with Fraser, extracts of the principal islet cells were tried both on animals and on the human. Their results, however, were not sufficiently convincing to warrant clinical application. The problem of the extraction of the antidiabetic principle from the pancreas was then taken up for the most part by physiologists among whom were Scott, Paulesco, Kleiner, and Murlin.

While these efforts were being made by the physiologists, valuable knowledge was being gained on carbohydrate metabolism. Lewis and Benedict, Folin and Wu, Schaffer and Hartman, and Ivar Bang had elaborated methods whereby the percentage of sugar in a small sample of blood might be accurately estimated. At the same time a vast amount of knowledge was accumulating on basal metabolism. Special attention was being given to the relative importance of the various foodstuffs, and emphasis was being put on dietetic treatment of diabetes. Guelpa, von Noorden, Allen, Joslin, and Woodyatt, had elaborated systems of diabetic diet.

On October 30th, 1920, I was attracted by an article by Moses Baron, in which he pointed out the similarity between the degenerative changes in the acinus cells of the pancreas following experimental ligation of the duct, and the changes following blockage of the duct with gall-stones. Having read this article, the idea presented itself that by ligating the duct and allowing time for the degeneration of the acinus cells, a means might be provided for obtaining an extract of the islet cells free from the destroying influence of trypsin and other pancreatic enzymes.

On April 14th, 1921, I began working on this idea in the Physiological Laboratory of the University of Toronto. Professor Macleod allotted me Dr. Charles Best as an associate. Our first step was to tie the pancreatic ducts in a number of dogs. At the end of seven weeks these dogs were chloroformed. The pancreas of each dog was removed and all were found to be shrivelled, fibrotic, and about one-third the original size. Histological examination showed that there were no healthy acinus cells. This material was cut into small pieces, ground with sand, and extracted with normal saline. This extract was tested on a dog rendered diabetic by the removal of the pancreas. Following the intravenous injection, the blood sugar of the depancreatized dogs was reduced to a normal or subnormal level, and the urine became sugar-free. There was a marked improvement in the general clinical condition as evidenced by the fact that the animals became stronger and more lively, the broken-down wounds healed more kindly, and the life of the animal was undoubtedly prolonged.

The beneficial results obtained from this first type of extract substantiated the view that trypsin destroyed the antidiabetic principle and suggested the idea that by getting rid of the trypsin, an active extract might be obtained. The second type of extract was made from the pancreas of dogs in which acinus cells had been exhausted of trypsin by the long-continued injection of secretin. Although many of the extracts made in this manner produced marked lowering of blood sugar and improvement in the general clinical condition it was not always possible to completely exhaust the gland; consequently toxic effects frequently resulted.

The third type of extract used in this series of experiments was made from the pancreas of foetal calves of less than four months development. Laguesse had found that the pancreas of new-born contained comparatively more islet cells than the pancreas of the adult. Since other glands of internal secretion are known to contain their active principle as soon as they are differentiated in their embryological development, it occurred to me that trypsin might not be present since it is not used till after the birth of the animal. Later I found that Ibrahim had shown that trypsin is not present till seven or eight months of intrauterine development. Foetal extracts could be prepared in a much more concentrated solution than the former two varieties of extract. It produced marked lowering of blood sugar, urine became sugar free and there was marked clinical improvement. Its greatest value however was that the abundance in which it could be obtained enabled us to investigate its chemical extraction.

Up to this time saline had been used as an extractive. We now found that alcohol slightly acidified extracted the active principle, and by applying this method of extraction to the whole adult beef pancreas, active extracts comparatively free from toxic properties were obtained.

Since all large-scale production methods for the preparation of insulin today have the acid-alcohol extraction as the first step in the process, it may be well to elaborate on the methods of preparation at this stage. Insulin was prepared by the extraction of fresh glands with faintly acid alcohol. The concentration of alcohol in the original experiments varied from 40 to 60 per cent. The alcoholic solution of pancreas was filtered and the filtrate concentrated by evaporation of the alcohol and water *in vacuo* or in a warm air current. Lipoid material was removed by extracting the residue with toluene or ether. The resulting product was the original whole gland extract. We were able to show that the active material contained in this extract was practically insoluble in 95% alcohol.

The extracts prepared in this way were tried on depancreatized dogs and in all cases the blood sugar was lowered. In one early case hypoglycaemic level was reached and the dog died from what we now know to be a hypoglycaemic reaction.

It had been known that depancreatized dogs were unable to store glycogen in the liver, and that glycogen disappears in three or four days after pancreatectomy. We found that by the administration of glucose and extract, the diabetic dog was enabled to store as much as 8% to 12% glycogen. Diabetic dogs seldom live more than 12 to 14 days. But with the daily administration of this whole gland extract we were able to keep a depancreatized dog alive and healthy for ten weeks. At the end of this time the dog was chloroformed and a careful autopsy failed to reveal any islet tissue.

The extract at this time was sufficiently purified to be tested on three cases of diabetes mellitus in the wards of the Toronto General Hospital. There was a marked reduction in blood sugar and the urine was rendered sugar-free. However the high protein content rendered the continuous use undesirable, due to formation of sterile abscesses.

At this stage in the investigation, February 1922, Professor Macleod abandoned his work on anoxaemia and turned his whole laboratory staff on the investigation of the physiological properties of what is now known as insulin.

Dr. Collip took up the biochemical purification of the active principle and ran the scale of fractional precipitation with 70–95% alcohol and succeeded in obtaining a more improved end product. But unfortunately his method was not applicable to large-scale production. Dr. Best then took up the large-scale production and contributed greatly to the establishment of the principles of production and purification. This work was carried out in the Connaught Laboratories under Prof. Fitzgerald who is kind enough to be here today.

It had been found that the final product obtained by the earlier methods was not sufficiently pure for prolonged clinical use, and efforts were made to secure a better product. The benzoic acid method of Maloney and Findlay which depends upon the fact that insulin is absorbed from watery solutions by benzoic acid was successfully used in Connaught Laboratories for several months.

Professor Shaffer of Washington University, St. Louis, and his collaborators, Somogyi and Doisy, introduced a method of purification which is known as the isoelectric process. This method depends upon the fact that if a watery solution of insulin is adjusted to approximately pH 5 a precipitate settles out which contains much of the potent material and relatively few impurities. Dudley has found that insulin was precipitated from water solutions by picric acid and he made use of this fact to devise a very ingenious method for the purification of the active material.

Best and Scott who are responsible for the preparation of insulin in the

Insulin Division of the Connaught Laboratories have tested all the available methods and have appropriated certain details from many of these; several new procedures which have been found advantageous have been introduced by them. The yield of insulin obtained by Best and Scott at the Connaught Laboratories, by a preliminary extraction with dilute sulphuric acid followed by alcohol, is 1,800 to 2,200 units per kg of pancreas.

The present method of preparation is as follows. The beef or pork pancreas is finely minced in a large grinder and the minced material is then treated with 5 cc of concentrated sulphuric acid, appropriately diluted, per pound of glands. The mixture is stirred for a period of three or four hours and 95 per cent alcohol is added until the concentration of alcohol is 60 to 70 per cent. Two extractions of the glands are made. The solid material is then partially removed by centrifuging the mixture and the solution is further clarified by filtering through paper. The filtrate is practically neutralized with NaOH. The clear filtrate is concentrated *in vacuo* to about 1/15 of its original volume. The concentrate is then heated to 50° C which results in the separation of lipoid and other materials, which are removed by filtration. Ammonium sulphate (37 g per 100 cc) is then added to the concentrate and a protein material containing all the insulin floats to the top of the liquid. The precipitate is skimmed off and dissolved in hot acid alcohol. When the precipitate has completely dissolved, 10 volumes of warm alcohol are added. The solution is then neutralized with NaOH and cooled to room temperature, and kept in a refrigerator at 5° C for two days. At the end of this time the dark-coloured supernatant alcohol is decanted off. The alcohol contains practically no potency. The precipitate is dried *in vacuo* to remove all trace of the alcohol. It is then dissolved in acid water, in which it is readily soluble. The solution is made alkaline with NaOH to pH 7.3 to 7.5. At this alkalinity a dark-coloured precipitate settles out, and is immediately centrifuged off. This precipitate is washed once or twice with alkaline water of pH 9.0 and the washings are added to the main liquid. It is important that this process be carried out fairly quickly as insulin is destroyed in alkaline solution. The acidity is adjusted to pH 5.0 and a white precipitate readily settles out. Tricresol is added to a concentration of 0.3% in order to assist in the iso-electric precipitation and to act as a preservative. After standing one week in the ice chest, the supernatant liquid is decanted off and the resultant liquid is removed by centrifuging. The precipitate is then dissolved in a small quantity of acid water. A second isoelectric precipitation is carried out by adjusting the acidity to a pH of approximately 5.0. After standing overnight the resultant precipitate is removed by centrifuging. The precipitate, which contains the active principle in a comparatively pure form, is dissolved in acid water and the hydrogen-ion concentration adjusted to pH 2.5. The material is carefully tested to determine the potency and is then diluted to the desired strength of 10, 20, 40, or 80 units per cc. Tricresol is added to secure a concentration of 0.1 per cent. Sufficient sodium chloride is added to make the solution isotonic. The insulin solution is passed through a Mandler filter. After passing through the filter the insulin is retested carefully to determine its potency. There is practically no loss in berkefelding. The tested insulin is poured into sterile glass vials with aseptic precautions and the sterility of the final product thoroughly tested by approved methods.

The method of estimating the potency of insulin solutions is based on the effect that insulin produces upon the blood sugar of normal animals. Rabbits serve as the test animal. They are starved for twenty-four hours before the administration of insulin. Their weight should be approximately 2 kg. Insulin is distributed in strengths of 10, 20, 40, and 80 units per cc. The unit is one third of the amount of material required to lower the blood sugar of a 2-kg rabbit which has fasted twenty-four hours from the normal level (0.118 per cent) to 0.045 per cent over a period of five hours. In a moderately severe case of diabetes, one unit causes about 2.5 grams of carbohydrate to be utilized. In earlier and milder cases, as a rule, one unit has a greater effect, accounting for three to five grams of carbohydrate.

With the improvement in the quality of insulin, the increased knowledge of its physiological action and the increased quantities at our disposal, we were now prepared for more extensive clinical investigation. In May 1922, a clinic was established in association with Dr. Gilchrist, at Christie Street Hospital for Returned Soldiers. Following this, a clinic was established in the Toronto General Hospital in association with Drs. Campbell and Fletcher, and at Toronto Hospital for Sick Children in association with Dr. Gladys Boyd. In general the routine followed in all these clinics was as follows.

After a careful history had been taken, the patient was given a complete physical examination. Special attention was directed to the finding of foci of possible infection. The teeth, tonsils, accessory sinuses, chest and digestive system were examined clinically, as well as by X-ray. Special consideration was given to biliary tract infection, constipation, and chronic appendicitis. If any source of septic absorption was located it was appropriately treated, since such conditions may lower carbohydrate tolerance. If indicated the eye grounds were examined for a possible diabetic retinitis or neuro-retinitis.

The daily routine urinalysis included the volume of the twenty-four hour specimen, the specific gravity, the reaction, and tests for albumen by heat or nitric acid. The acetone bodies were estimated by means of the Rothera and ferric chloride tests. Sugar determinations were done by means of the Benedict qualitative and quantitative solutions. In addition to the above, the blood sugars were estimated by means of the Schaffer-Hartman method and the respiratory quotients with the Douglas bag and Haldane gas-analysis apparatus.

At first the patient continued on the same diet as that previous to his admission to hospital in order to obtain some idea of the severity of his case, and to avoid complications from sudden change of diet. Coma will be discussed separately. On the second or third day he was placed upon a diet, the caloric value of which was calculated on his basal requirement. This was determined from Dubois' chart and Aub-Dubois' table. It has been estimated by Marsh, Newburgh, and Holly that the body requires two-thirds of a gram of protein per kilogram of body weight per day (1 kilo = 2.2 pounds) in order to maintain nitrogenous equilibrium. The remaining calories must be supplied by carbohydrate and fats in a ratio that will prevent the production of ketone bodies.

The patient remained on this basal requirement diet at least a week. During this time, blood sugar was estimated before, and three hours after, breakfast, in order to determine the fasting level and the effect of food. The quantity of sugar excreted was estimated daily, and this amount subtracted from the available carbohydrate ingested gives approximately the utilization. The available carbohydrate includes 58 per cent of the protein, 10 per cent of the fat, and the total carbohydrate in the diet. It may be noted that when a patient was placed upon a diet in which the protein, fat and carbohydrates were balanced, that the amount of sugar excreted soon approached a fairly constant amount, whereas if the diet was not well adjusted to the patient's requirements, there was wide variation in the amounts of sugar excreted.

If a patient became sugar-free and blood sugar normal on a basal requirement diet, the caloric intake was gradually increased until sugar appeared in the urine. The tolerance was thus ascertained. If a patient remained sugar-free and had a normal blood sugar when on a diet containing five hundred calories above his basal requirement he was not considered sufficiently severe for insulin treatment, since five hundred calories over and above the basal requirement are sufficient for daily activities. If, however, he was unable to metabolize this amount, insulin treatment was commenced.

Diabetes mellitus is due to a deficiency of the internal secretion of the pancreas. The main principle of treatment is, therefore, to correct this deficiency. If it is found that the patient is unable to keep sugar-free on a diet that is compatible with an active, useful life, sufficient insulin is administered to meet this requirement.

In severe cases insulin was administered subcutaneously three times a day, from one-half to three-quarters of an hour before meals. This was done so that the curve of hypoglycaemia produced by the insulin was superimposed on the curve of hyperglycaemia produced by the meal. In rare cases a small fourth dose was given at bed time to control nocturnal glycosuria. The less severe cases could be satisfactorily treated on a morning and evening dose or a single dose before breakfast.

When the insulin treatment was established, if sugar was present in the twenty-four hour specimen of urine, the dosage was gradually raised till the patient became sugar-free. If he was not receiving sufficient food for maintenance, diet and dosage of insulin were gradually raised. If small quantities of urinary sugar persist, it was desirable to find out at what period of the day this was excreted. In order to do this, each specimen in the twenty-four hours was analysed separately. An increase in the dose previous to the appearance of glycosuria will prevent its occurrence.

In severe cases it was found preferable to give the largest dose of insulin in the morning, and reduced doses throughout the day. For example, a patient may receive fifteen units in the morning, ten units at noon, and ten units at night. If three equal doses are given there may be morning glycosuria and evening hypoglycaemia, whereas the extremes of blood sugar causing these conditions may be prevented by the above distribution.

The effect of the same dosage of extract on different individuals was found to vary considerably. Five patients, whose weights varied from forty-six to sixty-seven kilograms, each received two cubic centimetres of the same lot of insulin, and in four hours the blood sugars had decreased 0.012%, 0.044%, 0.128%, 0.146%, and 0.0180% respectively. It was found, however, that one patient would persistently give marked decreases in blood sugar after insulin, while in another the fall in blood sugar was persistently less. In our experience, the more marked decreases in blood sugar occurred in the milder cases.

The blood sugars of some of the patients were followed throughout the twenty-four hours and it was found that it was possible to gauge the dosage of insulin so as to keep the blood sugar within normal limits and still avoid the dangers of hypoglycaemia.

Coincident with the maintenance of the blood sugar at normal level the cardinal symptoms of the disease disappear. The patient loses the irritating thirst and dryness of the mouth and throat, and does not desire the large amounts of fluid with which he had previously tried to combat these symptoms. The lowered fluid intake diminishes the polyuria and from a twenty-four hour excretion of three to five litres the output falls to normal. The appetite which has been voracious is now satisfied with a normal meal, the carbohydrate of which is utilized, and the patient loses the persistent craving for food.

We found that when a patient was given too large a dose of insulin there was a marked reaction, and the hypoglycaemia which developed gave rise to symptoms which were very similar to those observed in animals. The reaction began in from one and a half to six hours after the patient received the overdose. The average time was three to four hours. The interval varied with the individual, the dosage, and the food ingested. The first warning of hypoglycaemia was an unaccountable anxiety and a feeling of impending trouble associated with restlessness. This was frequently followed by profuse perspiration. The development of this symptom was not affected by atmospheric conditions. It appeared while the patient was in a frosty outside atmosphere, or in a heated room, and was independent of physical or mental activity. At this time there was usually a very great desire for food. No particular foodstuff was desired, but bulk of any kind seemed to give satisfaction. At times the appetite is almost unappeasable.

At this stage of the reaction the patient noticed a certain sensation as of clonic tremor in the muscles of the extremities. This could be controlled at first. Coordination, however, was impaired for the more delicate movements. Coincident with this there was a marked pallor of the skin with a rise in pulse rate to one hundred or one hundred and twenty beats per minute, and a dilatation of the pupils. The blood pressure during this period fell about fifteen to twenty-five millimetres of mercury, and the patient felt faint. The ability to do physical or mental work was greatly impaired. In a severe reaction there was often a considerable degree of aphasia, the patient having to grope for words. The memory for names and figures became quite faulty.

The onset of hypoglycaemic symptoms depends not only on the extent, but also on the rapidity of fall in blood sugar. The level at which symptoms occur is slightly higher in the diabetic with marked hyperglycaemia than in a patient whose blood sugar is normal. When the blood sugar is suddenly reduced from a high level premonitory symptoms may occur with a blood sugar between the normal levels of 0.100% and 0.080%, while the more marked symptoms of prostration, perspiration, and incoordination develop between 0.080% and 0.042%. As a patient becomes accustomed to a normal blood sugar the threshold of these reactions becomes lower. One patient who formerly had premonitory symptoms of hypoglycaemia at 0.096% now has no reaction at 0.076%, but symptoms commence between this level and 0.062%.

The ingestion of carbohydrate, in the form of orange juice (four to eight ounces), or of glucose, relieves these symptoms in from one-quarter to one-half hour. If the reaction is severe, or if coma or convulsions occur, epinephrin or intravenous glucose should be given. The former acts in from three to ten minutes, but in order that the symptoms should not recur, glucose must be given by mouth as soon as the patient has sufficiently recovered. The patients were warned that when these reactions occurred they were to obtain carbohydrate immediately.

«Fats only burn in the fire of carbohydrate.» The ability of the severe diabetic to burn glucose is markedly impaired, therefore the excess of fat is incompletely oxidized, giving rise to ketone bodies. These appear in the blood and urine as acetone, diacetic and betaoxybutyric acids. Insulin causes increased carbohydrate metabolism, and consequently fats are completely burned. This is substantiated by the fact that acetone and sugar disappear from the urine almost simultaneously following adequate amounts of insulin. When insulin is discontinued in these cases, acetone bodies and sugar reappear in the urine.

Since the Rothera test is exceedingly delicate (sensitive to 1 part of acetoacetic acid in 30,000), patients on a high fat diet may be sugar-free and still show traces of acetone bodies. A comparison with the ferric-chloride test (which is sensitive to only 1 part in 7,000) is, therefore, desirable. The persistence of ketone bodies in amounts which can be determined by the ferric-chloride test necessitates either an increase in the carbohydrate or a decrease in fat of the diet.

When the production of acetone bodies is more rapid than the excretion they accumulate in the blood, giving rise to air hunger, drowsiness, and coma. The need of insulin is then imperative. After its administration, the utilization of carbohydrate by the body gives complete combustion of the fats. When a patient was admitted to hospital in coma the blood-sugar tests and a urinalysis were done as soon as possible. (The urine was obtained by catheterization if necessary.) While these tests were being carried out, the large bowel was evacuated with copious enemata. If the blood sugar was high and acetone present in large amounts in the urine, from thirty to fifty units of insulin were given subcutaneously. Blood and urinary sugar were frequently estimated because of the danger of hypoglycaemia. To prevent this, from thirty to fifty grams of glucose in ten per cent solution were given intravenously. If the patient was profoundly comatose, the insulin was administered intravenously with the glucose.

The patient usually regained consciousness in from three to six hours. From this time on, fluids and glucose were administered by mouth if retained. The patient was urged to take at least two hundred cubic centimetres of fluid per hour. In from eight to ten hours, the ketone bodies were markedly reduced. On the following day protein was given every four hours as the white of one egg in two hundred cubic centimetres of orange juice. In two to three days, when ketone bodies had disappeared from the urine, fat was cautiously added, and the patient was slowly raised to a basal requirement diet. He was then treated as an ordinary diabetic. During the period of coma the patient was kept warm and toxic materials eliminated from the bowel by purgation and repeated enemata. A large amount of fluid was given to dilute the toxic bodies and promote their elimination. This was administered intravenously, subcutaneously, or per rectum. If signs of circulatory failure developed these were treated by appropriate stimulation.

Striking results were obtained with the above procedure. However, it was found that the longer the period of untreated coma the more grave was the prognosis and the slower the recovery if it occurred. Cases complicated by severe infection, gangrene, pneumonia, or intestinal intoxication may recover from acidosis and coma, but succumb to the complication.

Marked lipaemia was present in three cases. This disappeared in the course of a week to ten days after the patient was placed on insulin and on a diet in which the fat was restricted. The urine of one patient became acetone-free while lipaemia persisted.

The severe diabetic, whose ability to burn carbohydrate is markedly impaired, has a persistently low respiratory quotient, from 0.7 to 0.8, which is but little raised by the ingestion of glucose: when glucose and insulin are given together, the respiratory quotient is markedly increased, showing that carbohydrate is being metabolized. The highest values have been obtained when pure glucose was used with insulin. Less extensive rises have been secured when the patient, while on a mixed diet, received insulin.

All the patients gained in weight on the additional calories. There was an increase in sexual vigour and there was a greater ability to do mental and physical work. Nearly all of the patients have returned to their former employment, and while still under supervision, they administer their own insulin and arrange their own diets with satisfactory results.

All diabetics who have not an adequate knowledge of the dietetic treatment of their disease should be admitted to hospital in order that they may receive instruction in the preparation of their calculated and weighed diet – that they may learn the qualitative tests for sugar and acetone in the urine – that their carbohydrate tolerance may be accurately determined; and that the use of insulin, if required, may be safely instituted. Mild cases, especially if over fifty years of age, can be controlled by diet. Cases that cannot be adequately controlled by dietetic treatment alone should be given sufficient insulin to enable them to attain to a diet on which they may « carry on ».

One of the commonest complications of diabetes, especially in untreated patients over fifty, is gangrene. It is often associated with varying degrees of sclerosis of the leg arteries, which makes it extremely difficult to obtain healing. This may be accomplished by the use of insulin, but when permanent impairment has occurred it is advisable to amputate. Amputation is also advisable when an infection is so severe that the life of the patient is in jeopardy. Treatment of these cases is difficult because, due to the infection, there is a marked variation in the daily production of insulin by their own pancreas. But with careful treatment they can be rendered free from acetone and sugar, and their general condition improved. Operation is then performed preferably under nitrous oxide and oxygen anaesthetic. If the blood sugar is maintained normal, and acidosis is prevented, the wound heals kindly, provided that the amputation has been high enough to assure a good blood supply. For varying periods after the operation, the patient remains on insulin treatment. In nearly all cases at the end of three or four weeks, mild hypoglycaemic reactions indicate an overdose of insulin. It is then necessary to increase the diet or decrease the insulin. In some cases the tolerance improves sufficiently to warrant the discontinuance of insulin.

Diabetic patients requiring major operations, such as appendectomy, cholecystectomy, and tonsillectomy, or removal of teeth, are first rendered sugar- and acetone-free, unless the severity of symptoms demand immediate attention. Patients formerly considered bad surgical risks, if given proper dietetic treatment with insulin may be protected from the acidosis, hyperglycaemia, and glycosuria which otherwise usually result from the anaesthetic. In the diabetic, infections such as boils and carbuncles, and also intercurrent infections such as bronchitis, influenza, and fevers are favorably influenced by the normal blood sugar and increased metabolism which the administration of insulin permits. In the diabetic with tuberculosis, insulin allows the administration of proper nourishment to combat the tubercle infection.

During the past year and a half I have not been in active practice but have remained associated with the clinics. I have also kept in personal touch with the first fifteen patients who received insulin treatment. These patients were all extremely severe diabetics for whom diet had done its best. Of these fifteen patients, seven were children under fifteen years. It has been possible through the intelligent co-operation of the parents to continue a proper balance between diet and insulin dosage, and to maintain six of the seven children sugar-free. None of these have had to return to hospital, and all have gained in tolerance, and require from one-half to one-third less insulin than when they first began treatment. They have all gained in height and weight, and for the most part have developed into healthy normal children. The one child whose diet and insulin has not been properly controlled has been back in hospital repeatedly and is steadily losing in tolerance. Of the remaining eight cases there were four women and three men whose ages ranged from twenty-five to thirty-five years. The weight of the women varied from seventy-four to seventy-nine pounds. Two of the women, although they have gained to normal or overweight and now have no symptoms of disease, have not shown any increase in tolerance, due, perhaps, to the fact that they have not kept sugar-free. All the others, both men and women, have been able to reduce their dose of insulin from two-thirds to one-fifth of the original requirement. The one remaining case was admitted for amputation. She had had diabetes for six years, and at the time of admission, her blood sugar was 0.350%, and large amounts of acetone and sugar were being excreted in the urine. She was rendered sugar- and acetone-free by means of insulin before the operation was performed. Amputation was done at the middle third of the thigh. The stump was entirely healed in three weeks. Within six weeks of her operation, insulin was discontinued and her diet was increased without the return of diabetic symptoms. It is now three years since her operation and she is sugar-free on a liberal diet without insulin.

It may be of interest to mention a few cases in greater detail to further illustrate the improvement in carbohydrate tolerance following insulin treatment.

Case 1: male, aged 29 years, had suffered from chronic appendicitis. The urine of the patient in December, 1916, was sugar-free. About the middle of March, 1917, he suddenly developed polyuria, polyphagia, and polydipsia, and lost fourteen pounds in weight in a fortnight. There was marked weakness. Urinary sugar was discovered to be as high as eight per cent at this time. On April 4th, the patient was placed on Allen treatment, and slowly regained a tolerance of about two hundred grams available carbohydrate. He returned to his army duties in September 1917, and was able to carry on uninterruptedly until March, 1919. His tolerance had decreased during this time to about one hundred and fifty grams. Following discharge from the army in March, 1919, the course of the patient was slowly downhill until October, 1921, when a particularly severe form of influenza shattered his tolerance. Up to this time the patient was maintained practically sugar-free, but following the attack of influenza, his tolerance fell to about sixty-six grams of available carbohydrate. He began to lose weight rapidly. Thirst, hunger, and polyuria returned. His strength diminished and, owing to mental and physical lassitude, he found it impossible to continue his work. Glycosuria became persistent and acetone bodies made their appearance, and steadily increased. A distinct odour of acetone was at times distinguishable in the patient's breath.

On February 11th, 1922, this patient was taken to the Physiology Department of the University of Toronto, and the respiratory quotient was found to be 0.74, and unchanged by the ingestion of thirty grams of pure glucose. Then 5 cc of insulin were given subcutaneously, and within two hours the patient's respiratory quotient had risen to 0.90. The urine was sugar-free and he had shaken off his mental and physical torpor. Following this experiment, the patient did not again receive insulin until May 15th, as the product was being further improved. Since the latter date, the patient has been constantly on insulin.

During the first six months of insulin treatment it was impossible to maintain him sugar-free, although he received about 120 units per day. However, he gained in weight and his clinical condition improved. About January, 1923, with the improvement in the quality of insulin, the patient became sugar-free and has remained sugar-free with the exception of one or two occasions. During the first nine months he required no reduction in the dose of insulin, but since that time, on the average of every two months, he has had a series of hypoglycaemic reactions which necessitated the reduction of the dose. One exception to this occurred in June, 1924, at which time appendectomy was performed following a mild attack of appendicitis. An increased dose was required to maintain him sugar-free during this period. At the present time he requires but 20 units of insulin, or one-sixth of his original requirement. His diet has been practically constant during the whole period of observation. All symptoms attributable to diabetes have long since disappeared. He has gained twenty-five pounds in weight and apart from the necessity of taking insulin and controlling his diet he leads an active normal life.

This case is a striking example of the fact that it is only in cases who are maintained sugar-free over long periods of time that an improvement in tolerance is obtained with a consequent reduction in the dose of insulin.

Case 2: female, age 15 years. In the autumn of 1918, the patient had poly-dipsia and polyuria, and complained of weakness. During the winter she suffered from pains in the legs and back, and from insomnia. In March, 1919, these symptoms became more severe. The appetite became excessive and there was some pruritus. The weight by this time had fallen from seventy-five pounds to sixty-two pounds. Glycosuria was discovered and she was placed under the care of Dr. F. M. Allen, to whom we are very much indebted for complete record of the case from April, 1919, till August, 1922. During this period the diet was controlled so as to maintain the urine free from sugar. Despite this careful dietetic regime the patient's condition became progressively worse.

When she came under my care on August 16th, 1922, the examination showed: patient emaciated; skin dry; slight edema of ankles; hair brittle and thin; abdomen prominent; marked weakness. The patient was brought on a stretcher and weighed forty-five pounds. Nothing of note in the respiratory, cardiovascular, digestive, or nervous system.

At this time she was receiving a diet of protein 50 g, fat 71 g, carbohydrate 20 g (919 calories). Insulin treatment was started immediately. At this early stage, the unit of insulin had not been worked out, and it is therefore difficult to accurately estimate the dosage she received. The diet was increased daily so that, at the end of two weeks, she was receiving protein 63 g, fat 208 g, carbohydrate 97 g (2512 calories). This diet was continued up to January 1st, 1923. Insulin was given 15 to 30 minutes before the morning and evening meals. A sufficient amount was given to maintain the urine free of sugar. Each specimen of urine was examined and the dose was increased slightly if traces of sugar appeared. When hypoglycaemia occurred, orange juice or glucose candy was given. Between August 16th and January 1st, the urine was sugar-free, except on ten occasions when traces of sugar appeared, and on two other occasions when less than 2 g was excreted. Acetone was absent from the urine.

On this treatment the patient gained rapidly in strength, and was soon able to take vigorous exercise. Her weight increased from 45 to 105 pounds in the first six months. The diet included such foodstuffs as cereals, bread, potato, rice, corn, tapioca, corn starch, and even honey.

At present (June 1925) she is in the best of health, and to use her own words «never felt better in all my life». She has grown four inches and weighs 134 pounds. Her present diet which is only approximate because she has dispensed with the weighing of food, is 125 g carbohydrate, 50 g protein, 50 g fat. This diet is practically the same as that of December, 1922. The insulin required to maintain her sugar-free has been reduced about one-third.

Dr. Gladys Boyd, who is now in charge of the diabetics at the Hospital for Sick Children, Toronto, has been able to follow a number of cases of children under insulin treatment. She has estimated the insulin requirement per 10 g of carbohydrate in a number of cases, and in general her results show a decided increase in tolerance in all cases in which glycosuria and hypergly-caemia are adequately controlled. To illustrate – Case 1, which required 6.9 units per 10 g carbohydrate in March, 1923, only required 2.6 units in January, 1924. Case 2, which required 7.8 units per 10 g in January, 1925, in June 1925 required only 2.8 units. Case 3, which required 6.5 units per 10 g in April, 1922, required only 3.7 units in January, 1925.

From a review of the work, Dr. Boyd has found that all the patients had had hyperglycaemia or even glycosuria at times, but if such occurrences were only transitory and infrequent, improvement in tolerance occurred. Even short periods of rest to the pancreas by means of balanced diet and insulin resulted in improvement in tolerance. Two of our earliest cases, Fanny Z. and Elsie N. are the only exceptions to this rule. Fanny is to all appearances in the best of health with a blood sugar of 0.3% to 0.4%. She has been admitted in coma four times. During her stay in hospital she improves but does as she chooses on discharge. Her tolerance is becoming less all the time. Elsie keeps in touch with us but is looked after by another physician. He purposely allows her to have glycosuria at night. She is fine physically, but requires much more insulin than formerly.

Dr. Boyd has also found that in those cases who can handle sufficient food without insulin, although the disease has been kept under control there has not been such striking increase in tolerance.

The best evidence that there is regeneration of the pancreas with insulin treatment is provided by Drs. Boyd and Robinson. The following is the case reported by them.

Clinical history: B.N., white, male, aged 9 years. *Family history*: Father and one maternal uncle have diabetes. Diabetes diagnosed in this child when he was two years old. He was placed on a suitable Allen diet, which was strictly adhered to, and for a time did well except for recurrent attacks of dysentery, which lowered his tolerance. Failure to gain in stature or weight in any way commensurate with his age was noted and the general condition became worse each year until he was more or less a chronic invalid with increasingly frequent attacks of acidosis during the last year before starting insulin.

He was admitted to the Hospital for Sick Children, Toronto, the end of December, 1922. At this time he was an emaciated dwarf, more or less drowsy and unhappy. His weight was thirty pounds, and his height thirty-nine inches. His tolerance to carbohydrate had decreased until he was unable to utilize 15 g of such food. Insulin treatment was started at once and his diet increased to a diet suitable for a boy of his age. Sufficient insulin was given to keep him sugar-free and his blood sugar normal. He was discharged on an adequate diet plus insulin. Progress, both in general condition and in improvement of pancreatic function, was steady. His tolerance to carbohydrate trebled in the year, as shown either by the fact that 30 units of insulin controlled the disease as adequately as 90 units a year before, or, stated in another way, without insulin he could now handle 54 g carbohydrate instead of 15. From a chronic invalid in 1922 he became « the leader of the gang », in 1923. He was killed by fracturing his skull when sleigh riding. He lived for about three hours after receiving the injury and an immediate post-mortem examination was made. The pancreas was removed within thirty minutes of death.

From this clinical history one might expect the pancreas to show marked degeneration. However, on section there was little sign of degeneration, but on the other hand there was strong evidence to support the view of active regeneration both of acinar and islet tissue. These regenerative changes were more marked in the periphery and smaller lobules of the pancreas than in the central area.

The acinar cells were found to be actively proliferating in cords and clusters forming small lobules in some areas, and were in close association with newly formed functioning ducts.

The islets were greatly increased in number, particularly in the periphery, there being about four times as many per field as in the central area. These cells were large but might be overlooked with an ordinary stain. However, they could be identified as islet cells by Bowie's special granule-stain. This stain also demonstrated that these cells were almost entirely beta cells and were probably concerned in the increased carbohydrate tolerance. On the other hand, those islets in the central areas showed an increased number of cells all in an active state of nutrition, but closely packed together. The special stain showed a normal ratio of alpha and beta cells.

These sections were studied by Bensley, Opie, Allen, and others, who concurred in the opinion of Drs. Boyd and Robinson.

Dr. F. M. Allen, Morristown N. J., after using insulin for three years states as his belief, « That there has been improvement of tolerance in some cases beyond what was possible without insulin ». « This observation is trustworthy only in cases where prolonged strict control of symptoms by diet was previously employed. On the other hand, the marked increase of tolerance is limited to a minority of cases and has not proved to be continuous in any of them. In other words the improvement always stops short of a cure. There is certainly no decline of tolerance with the passage of time, provided the case is kept under proper control. »

This summary is the belief of the most conservative of the outstanding clinicians in the United States engaged in diabetic work on a large scale.

Dr. E. P. Joslin, Boston, Mass., who has one of the largest diabetic clinics in the world, has also found that, « The diabetic who is able to reduce his insulin is the diabetic who is absolutely faithful to diet and restricts gain in weight to a moderate degree. »

Joslin and his associates have carefully analysed the gain in weight and height of their thirty-two diabetic children under fifteen years of age. Their conclusions are:

(1) The gain in weight of the diabetic child treated with insulin resembles that of the normal child, but the diabetic child is still under weight for his age, though often not for his height.

(2) The increase in height of the diabetic child treated with insulin, though occasionally normal, is usually below that of the normal child. So far he has not grown tall like the normal child, either at the expense of growing thin or while being well nourished.

Of the 130 children treated with insulin, 120 are still living, while of the 164 who did not receive insulin, there are 152 dead. Of the 120 still living, 40% have either not increased or have actually decreased their insulin. Dr.

Joslin believes that if the 60% who have had to increase their insulin had received similar treatment, they too would have been able to reduce their insulin.

Sixteen children under ten years of age who have taken insulin under Dr. Joslin's care for an average of two years, are all alive, and now their duration of life is more than three times the duration of life of diabetic children of similar age treated by Dr. Joslin prior to 1915.

Regardless of the severity of the disease, it has been found that by carefully adjusting the *diet and the dose of insulin, all patients may be maintained sugar-free.* Since this is possible, it is to be strongly advocated, because we have abundant evidence for the belief that there is *regeneration of the islet cells of the pancreas* when the strain thrown upon them by a high blood sugar is relieved. The *increase in tolerance* is evidenced by the *decreasing dosage of artificially administered insulin. In fact, in some moderately severe cases*, the tolerance has increased sufficiently that they no longer require insulin.

Diabetes mellitus may be considered fundamentally as a *disordered metabolism, primarily of carbohydrates*, and *secondarily of protein and fat*. It is indisputably proven that for normal metabolism of carbohydrate in the body, adequate amounts of insulin are essential. It follows, therefore, that the treatment consists in giving just sufficient insulin to make up for the deficiency in the patient's pancreas.

Insulin enables the severe diabetic to burn carbohydrate, as shown by the rise in the respiratory quotient following the administration of glucose and insulin. It permits glucose to be stored as glycogen in the liver for future use. The burning of carbohydrate enables the complete oxidation of fats, and acidosis disappears. The normality of blood sugar relieves the depressing thirst, and consequently there is a diminished intake and output of fluid. Since the tissue cells are properly nourished by the increased diet, there is no longer the constant calling for food, hence *hunger pain* of the severe diabetic is replaced by *normal appetite*. On the increased caloric intake, the patients *gain rapidly in strength and weight*. With the relief of the symptoms of his disease, and with the increased strength and vigor resulting from the increased diet, *the pessimistic, melancholy diabetic becomes optimistic and cheerful*.

Insulin is not a cure for diabetes; it is a treatment. It enables the diabetic to burn sufficient carbohydrates, so that proteins and fats may be added to the diet in sufficient quantities to provide energy for the economic burdens of life.

Biography

Frederick Grant Banting was born on November 14, 1891, at Alliston, Ont., Canada. He was the youngest of five children of William Thompson Banting and Sarah Squire Grant. Educated at the Public and High Schools at Alliston, he later went to the University of Toronto to study divinity, but soon transferred to the study of medicine. In 1916 he took his M.B. degree and at once joined the Canadian Army Medical Corps, and served, during the First World War, in France. In 1918 he was wounded at the battle of Cambrai and in 1919 he was awarded the Military Cross for heroism under fire.

When the war ended in 1919, Banting returned to Canada and was for a short time a medical practitioner at London, Ontario. He studied orthopaedic medicine and was, during the year 1919–1920, Resident Surgeon at the Hospital for Sick Children, Toronto. From 1920 until 1921 he did part-time teaching in orthopaedics at the University of Western Ontario at London, Canada, besides his general practice, and from 1921 until 1922 he was Lecturer in Pharmacology at the University of Toronto. In 1922 he was awarded his M.D. degree, together with a gold medal.

Earlier, however, Banting had become deeply interested in diabetes. The work of Naunyn, Minkowski, Opie, Schafer, and others had indicated that diabetes was caused by lack of a protein hormone secreted by the islands of Langerhans in the pancreas. To this hormone Schafer had given the name insulin, and it was supposed that insulin controls the metabolism of sugar, so that lack of it results in the accumulation of sugar in the blood and the excretion of the excess of sugar in the urine. Attempts to supply the missing insulin by feeding patients with fresh pancreas, or extracts of it, had failed, presumably because the protein insulin in these had been destroyed by the proteolytic enzyme of the pancreas. The problem, therefore, was how to extract insulin form the pancreas before it had been thus destroyed.

While he was considering this problem, Banting read in a medical journal an article by Moses Baron, which pointed out that, when the pancreatic duct was experimentally closed by ligatures, the cells of the pancreas which secrete trypsin degenerate, but that the islands of Langerhans remain intact. This suggested to Banting the idea that ligation of the pancreatic duct would, by destroying the cells which secrete trypsin, avoid the destruction of the insulin, so that, after sufficient time had been allowed for the degeneration of the trypsin-secreting cells, insulin might be extracted from the intact islands of Langerhans.

Determined to investigate this possibility, Banting discussed it with various people, among whom was J. J. R. Macleod, Professor of Physiology at the University of Toronto, and Macleod gave him facilities for experimental work upon it. Dr. Charles Best, then a medical student, was appointed as Banting's assistant, and together, Banting and Best started the work which was to lead to the discovery of insulin.

In 1922 Banting had been appointed Senior Demonstrator in Medicine at the University of Toronto, and in 1923 he was elected to the Banting and Best Chair of Medical Research, which had been endowed by the Legislature of the Province of Ontario. He was also appointed Honorary Consulting Physician to the Toronto General Hospital, the Hospital for Sick Children, and the Toronto Western Hospital. In the Banting and Best Institute, Banting dealt with the problems of silicosis, cancer, the mechanism of drowning and how to counteract it. During the Second World War he became greatly interested in problems connected with flying (such as blackout).

In addition to his medical degree, Banting also obtained, in 1923, the LL.D. degree (Queens) and the D.Sc. degree (Toronto). Prior to the award of the Nobel Prize in Physiology or Medicine for 1923, which he shared with Macleod, he received the Reeve Prize of the University of Toronto (1922). In 1923, the Canadian Parliament granted him a Life Annuity of $7,500. In 1928 Banting gave the Cameron Lecture in Edinburgh. He was appointed member of numerous medical academies and societies in his country and abroad, including the British and American Physiological Societies, and the American Pharmacological Society. He was knighted (baronet) in 1934.

As a keen painter, Banting once took part of a painting expedition above the Arctic Circle, sponsored by the Government.

Banting married Marion Robertson in 1924; they had one child, William (b. 1928). This marriage ended in a divorce in 1932, and in 1937 Banting married Henrietta Ball.

When the Second World War broke out, he served as a liaison officer between the British and North American medical services and, while thus engaged, he was, in February 1941, killed in an air disaster in Newfoundland.

JOHN J. R. MACLEOD

The physiology of insulin and its source in the animal body

Nobel Lecture, May 26, 1925

The knowledge that the isles of Langerhans of the pancreas have the function of secreting into the blood a hormone which plays an essential role in the regulation of the metabolism of the carbohydrates, is the outcome of numerous investigations extending over many years, and to the development of this knowledge workers in various fields of medical science have contributed.

In 1889, when Minkowski and von Mering discovered that complete extirpation of the pancreas leads to fatal diabetes, practically nothing was known concerning the significance of the ductless glands, and few conceived that it would be possible to extract, from various of them, substances capable of replacing the lost function, when administered to animals from which some particular gland had been removed. Although, at this time, it was known that the thyroid gland is atrophied in myxoedema and in cretinism, it was not until 1892 that Murray discovered that administration of the gland removes the symptoms, and it was only later that the doctrine of internal secretions, first enunciated by Claude Bernard in 1856 in connection with the production of sugar by the liver, came to take its place in physiological teaching. Minkowski in the complete account of his researches, published in 1893, considered the antidiabetic function of the pancreas to be dependent upon its acting as a ductless gland, and no doubt he had in mind that it performed this through an internal secretion, although the positive statement that such exists was first made by Lépine, who thought that it took the form of a glycolytic enzyme. But, so far, there was no hint as to the actual structure within the pancreas upon which the antidiabetic influence of the gland depends and it is primarily to the anatomists, Laguesse and Diamare, that we owe the hypothesis that this must be the collection of cells, named after their discoverer, the isles of Langerhans. By careful studies of the cytological characteristics of the cells of these islets, as distinguished from those of the much more numerous secreting acini among which they lie, and by painstaking examination of the anatomical relationships of the two kinds of cells in different classes of vertebrates, Laguesse and Diamare concluded that the islets must be responsible for the antidiabetic influence.

As this anatomical work was in progress, the potent action of extracts of the suprarenal gland on the blood pressure and other physiological functions was discovered, in 1894, by Oliver and Schafer, thus adding strong support to the hypothesis that the ductless glands function by producing internal secretions. The hypothesis, that the islets of Langerhans of the pancreas must act in a similar manner, gained a firm hold among physiologists and clinical workers, with the result that many attempted to alleviate the symptoms of diabetes by administration of pancreas, or of extracts of the gland, to patients suffering from the disease. No success attended these attempts partly, we believe, because the antidiabetic principle was destroyed, either during the preparation of the extracts or by the action of the digestive juices, and partly, because of imperfect knowledge of the clinical course of the disease, particularly with regard to the relationship of diet to it. Notwithstanding the failure of these attempts, the hypothesis that the isles of Langerhans are the structures to which the pancreas owes its antidiabetic function was still maintained, and indeed strengthened, by the supporting evidence furnished by the graft experiments of Minkowski and Hédon. These workers showed that no diabetic symptoms supervene in dogs when a portion of the pancreas is transplanted into the wall of the abdomen prior to, or at the same time as, removal of the remainder of the gland, but immediately do so in full intensity when this graft is subsequently excised. Moreover, it was known that ligation of the ducts of the pancreas, or their injection by oil or paraffin, is not followed by diabetes. Since, in neither of these types of experiment, can any of the digestive secretion gain the intestine it was clear that the antidiabetic function of the pancreas must be independent of its digestive function. It may be well to point out also that the graft experiments once and for all disproved the view held by some (by Pflüger, for example), that damage to the nerve structures adjacent to the pancreas, or in the duodenal wall, is responsible for the diabetic symptoms.

A distinct step forward was taken in 1900 when Schulze and Ssobolev discovered that the degenerative changes which follow ligation of the ducts affect the cells of the acini much more markedly than those of the islets, and although among those who repeated these researches, there were some who failed to corroborate the findings, the conclusions of Schulze and Ssobolev were generally accepted. It was not long after this that the first, though unsuccessful, attempt was made to see whether an extract of the degenerated residue of duct-ligated pancreas might not relieve the symptoms of diabetes. About this time also (1906) – as was revealed in 1922 by the opening of a sealed package deposited with the Société de Biologie – Gley had found similar extracts to diminish the symptoms in diabetic dogs, and in the same year, Miss Dewitt had tried their effects on glycolysis.

The insular hypothesis of diabetes was meanwhile strongly supported by the careful histological studies of the pancreas of patients who had died from the disease (Opie) for, although it had been known, even prior to the experiments of Minkowski and von Mering, that the gland is often the seat of morbid change, it was not realized that the islets are the structures which are chiefly affected.

In 1903–1904, Rennie, by anatomical studies in certain Teleostei, gave strong support to the view of Diamare, that the islet cells in these fishes exist as separate glands of relatively large size and more or less independent of the pancreatic acini. Both workers attempted to demonstrate an effect of extracts of these glands on sugar or starch solutions, but without success. They administered them by mouth to diabetic patients with no favourable results, although in one case, in which an extract was given subcutaneously, there was decided alleviation of the diabetic symptoms (Rennie and Fraser).

About this time the significance of internal secretions in the control of animal functions was clearly demonstrated by the discovery of secretin by Bayliss and Starling (1902), and the term «hormone» came into use to designate their active principles. Many believed that the antidiabetic function of the pancreas must depend on a hormone secreted by the isles of Langerhans, but neither the graft experiments already referred to, nor the transfusion experiments of Hédon – in which it was found that when the blood of a normal dog was transfused in a diabetic one the symptoms were alleviated – could prove the hypothesis. To do this it was necessary to show that extracts of the islets, or at least of the pancreas, are capable of removing the symptoms of diabetes. In 1907 Zuelzer published results which must be considered, in the light of what we now know, as really demonstrating the presence of the antidiabetic hormone in alcoholic extracts of pancreas. But unfortunately, even although several diabetic patients were benefitted by administration of the extracts, the investigations were not sufficiently completed to convince others, and, apparently, Zuelzer himself was discouraged in continuing them because of toxic reactions in the treated patients.

To describe, even in mere outline, the further attempts to prepare active antidiabetic extracts of the pancreas would far exceed the limits of this essay.

To Knowlton and Starling, Meltzer and Kleiner, E. L. Scott, Murlin and Cramer, and to Clark, we owe much, for although none of these investigators succeeded in demonstrating beyond doubt that an extract having antidiabetic properties could be prepared from the pancreas, they all obtained results which were sufficiently positive to keep alive the hope that some day this would be possible. Special reference must also be made to the more recent work of Paulesco who prepared extracts having very decided effects on the sugar and the urea of the blood of diabetic animals.

Believing that the want of success to prepare extracts of uniform potency was due to the destruction of the antidiabetic hormone by the digestive enzymes also present in the gland, F. G. Banting suggested preparing them from duct-ligated pancreas, and with the aid of C. H. Best, and under my direction, he succeeded in 1922 in showing that such extracts reduced the hyperglycaemia and glycosuria in depancreatized dogs. The general symptoms of diabetes were also found to be alleviated and the duration of life of the depancreatized animal prolonged, by the repeated injection of alcoholic extracts of foetal, as well as of adult-ox pancreas. Later it was shown, in collaboration with Collip, that other symptoms of diabetes, namely the ketonuria and the absence of glycogen from the liver, were favourably influenced by the extracts and, with Hepburn, that the respiratory quotient became raised. These results on depancreatized dogs showed beyond doubt that the antidiabetic hormone was present in potent form in the extracts, and the time seemed ripe to investigate their action on the clinical forms of diabetes. This was done by Banting in a severe case under the care of W. R. Campbell, with the result that the hyperglycaemia and glycosuria were diminished. At the same time, however, it was found that it would be necessary to rid the extracts of irritating substances before the value of their repeated injection in the treatment of diabetes in man could be adequately put to the test. This was accomplished by Collip, and the name insulin was decided upon for the purified extract. This name had previously been suggested by Sir E. Sharpey Schafer (1916'), who had been one of the first to support the hypothesis of the insular derivation of the antidiabetic hormone. I need not here detail the rapid progress which it was now possible to make in studying the therapeutic value of insulin in the treatment of diabetes in man; for it is with experimental aspects of the subject that this essay is concerned.

The invariable lowering of the blood sugar which was observed to result from the administration of insulin in animals rendered diabetic by pancreatectomy, raised the question as to whether such would also occur in those forms of hyperglycaemia which can be induced by other experimental procedures, such as the injection of epinephrin, piqûre, or asphyxia. As the first step in the investigation of this question, Collip injected insulin into normal rabbits and found the blood sugar to become lowered, thus furnishing a valuable method for testing the potency of various preparations and, therefore, for affording a basis for their physiological assay. At the same time it was found that neither piqûre, nor epinephrin, nor asphyxia caused any hyperglycaemia in rabbits in which, as a result of injection with insulin, the blood sugar was at a low level to start with.

Peculiar symptoms (convulsions and coma) were observed in many of the injected animals, and it was soon possible to show that these were related to the lowering of the blood sugar and that they usually supervened when this was about 0.045 per cent. Sometimes the animals recovered spontaneously from these symptoms, but more frequently the coma became so profound, with marked fall of body temperature, that death occurred. That the lowering of blood sugar is closely related to the occurrence of the symptoms, was proved by finding that the subcutaneous injection of a solution of glucose was followed, almost immediately, by complete recovery, even in cases in which death was imminent from deep coma. It has been found, in collaboration with Noble, that glucose is remarkably specific in this regard, the only other sugar which approaches it being mannose and, in certain animals, such as the mouse, maltose. Laevulose and galactose are decidedly inferior in their antidoting action, the pentoses are entirely inactive and none of the disaccharides, other than maltose, has any effect. It is evident that this specificity in the action of glucose, in combating the hypoglycaemic symptoms, offers an opportunity to determine, not only what related substances are readily converted into glucose in the animal body, but also what groupings in the glucose molecule itself are significant for the effects. By substituting various side chains in the molecule, as for example, by methyl groups, it has been found, in collaboration with Herring and Irvine, that none of these substitution products is effective, even such compounds as the mono-methyl glucosides being entirely inactive.

The fall in blood sugar is dependent upon increased diffusion of sugar into the tissues and not to its more rapid destruction in the blood itself. Thus, Eadie and I could detect no change in the rate of glycolysis by adding insulin to blood incubated under sterile conditions outside the body, or in blood withdrawn from animals injected some minutes before death with insulin.

Hepburn and Latchford, on the other hand, demonstrated that the addition of insulin to the fluid perfused through the excised mammalian heart markedly increased the rate at which the percentage of sugar became diminished in it.

The striking relationship between the concentration of glucose in the blood and the normal functioning of the nervous system, which is revealed by these observations, had already been noted by Mann and Magath in their experiments on hepatectomized dogs. They observed that when the blood sugar fell to about 0.045 per cent, characteristic symptoms supervened which could be antidoted by glucose, and to a less extent, by laevulose and mannose. We must conclude that when the tension of glucose in the tissue cells falls below a certain level (glucatonia), a condition of irritability becomes developed; but little is known as to what the underlying cause for this may be. Olmsted and Logan have advanced some evidence that it may depend on interference with the process of oxidation in the nerve cells, or that these are irritated by substances produced elsewhere in the body by faulty oxidation. More recent experiments by Argyll Campbell on the tension of oxygen in the tissues lend support to this view.

These observations emphasize the great importance of a certain tension of glucose within the tissue cells. They help us to understand why it is that the concentration of this sugar in the circulating fluids of animals of every order and species in which it has been determined, varies only within narrow limits, even after prolonged periods of starvation, or following muscular exercise.

We must imagine that it is by lowering the tension of glucose within the tissue cells that insulin primarily acts, so that the glucose of the blood plasma, with which the tissue glucose is in equilibrium, diffuses into the cells to maintain the tension. With Eadie we have found that the free glucose extractable from the muscles by warm alcohol is reduced following the injecting of insulin, but we know nothing of the fate of the glucose which disappears. It is not converted into glycogen (McCormick, Noble and Macleod, Dudley and Marrian, Cori, etc.) nor is it immediately oxidized, since the respiratory metabolism (intake O_2 and respiratory quotient) does not become increased at the time when the blood sugar is falling (Eadie, Dickson, Macleod, and Pember; Trevan and Boock; Krogh; Boothby and Wilder, etc.)*.

* Practically all observers have confirmed the observation first made by Dickson and Pember that the R. Q. (respiratory quotient) rises somewhat in normal animals injected with insulin but the extent of this rise is not sufficient to indicate that increased combustion of glucose can be the significant cause for the rapid reduction in blood sugar.

Although the intake of oxygen may become greater in certain animals such as dogs, cats, and man when hypoglycaemic symptoms make their appearance, this does not occur when sugar is also administered. In the light of these results we have concluded that the glucose which disappears must become converted into some hitherto unidentified substance, but we have been unable to obtain any clue as to what this substance may be. Large amounts of it must be formed to account for the enormous quantities of glucose which may vanish from the blood, as when glucose is injected along with insulin. We have, for example, injected into rabbits, in the course of eight hours, as much as 10 grams of glucose per kilo body weight along with insulin, without finding, at the end, any increase in blood sugar, or in the free or the combined sugar of the muscles or liver. Burn and Dale have also shown that very large quantities of glucose can be injected along with insulin into eviscerated animals without increasing the percentage of the blood sugar. It is conceivable that between glucose and the material which is finally oxidized in the tissues there exists, not one, but a group of substances constantly changing from one into another in an equilibrated system, and that no one of them ever accumulates in sufficient quantity to make its identification possible by available chemical methods.

Be this as it may, it is significant that the percentage of inorganic phosphoric acid in the blood declines at the same rate as the sugar, although, in the recovery process, the phosphoric acid begins to rise decidedly before the sugar in animals injected with insulin. Accompanying this fall in the phosphates of the blood, those of the urine entirely disappear for several hours and then return to considerably above the normal level so that, in urine collected throughout the 24 hours, an excess is excreted, as compared with the amount on days during which no insulin is given. (Winter and Smith, Allan and Sokhey, etc.). These facts would seem to indicate that in the process responsible for the disappearance of glucose in the tissues there is a stage when compounds of phosphoric acid with sugar or its immediate breakdown products are formed. One immediately thinks of the possibility that an increase in the amount of the substance, described by Embden and his school, in muscle, and named lactacidogen, might be responsible, but we have been unable to demonstrate that this is the case (Eadie, Macleod, and Noble). At the present time we are entirely at a loss to account for the disappearing glucose. When this problem is solved it may be anticipated that a great advance will become possible in our knowledge of the intermediary metabolism of the carbohydrates.

Having outlined the known facts with regard to its physiological action, we may now turn to the interesting question of the source of insulin. The observations of Banting and Best, that simple extracts of the residue of pancreas remaining several weeks after the ducts are tied possess antidiabetic properties, does not necessarily prove that insulin is derived from the islets. As Bensley and others have shown there may still remain, at this period after duct-ligation, a considerable amount of more or less normal acinar tissue. Even were the gland allowed to degenerate for a sufficient time so that all acinar tissue had disappeared – which is considerably over a year in the rabbit – it would be difficult, in the event that extracts of the residue still contained insulin, to be certain that this insulin is not of the type which it is possible to extract from various materials, including even the tissues of depancreatized animals, as Best and Scott have shown.

Further investigation of the problem was therefore undertaken by continuing the work of Diamare and Rennie on certain of the Teleostei, such as Lophius and Myoxocephalus, in which the islet cells exist apart from the acinar tissue, as the so-called «principal islets». Extracts were made by alcohol from these structures, as well as from the acinar tissue, and it was found that, whereas very large yields of insulin are readily obtainable from the islets, little or none at all can usually be prepared from the pancreas itself. Indeed, extracts of the latter sometimes cause the blood sugar of rabbits to become raised, instead of lowered. The lowering of the blood sugar by acinar extracts when it occurred, may have been due to the presence of a few scattered microscopic islets, such as have been observed by Slater Jackson to exist in the pancreatic bands of Myoxocephalus. That the principal islet may have some acinar tissue associated with it does not detract from the value of the foregoing observations as evidence supporting the insular hypothesis, since it has been shown that extracts of the acinar tissue are comparatively impotent. In this connection it is of interest to note that insulin can be readily prepared from the pancreas of the Elasmobranchi (Raja and Squalus), which occurs as a compact gland with the islet tissue included in it, much in the same manner as in the mammalian pancreas.

But in view of the fact that insulin, or at least extracts capable of lowering the blood sugar in normal rabbits, can be prepared from other tissues than the pancreas or the principal islets, there still remains the possibility that it might be secreted internally from some of these. It is indeed possible that although this did not occur in the normal animal, in which a sufficient amount was coming from the pancreas, it might occur when the normal secretion was cut off, as in diabetes. By such a vicarious functioning of extra-pancreatic potential sources of insulin, the tolerance of the diabetic organism for carbohydrate might become raised. It seemed important, therefore, to see whether diabetes would result from excision of the principal islets alone, an operation which is possible in Myoxocephalus, the two principal islets being readily removable without exposure of the fish to air for more than 15 minutes.

In a large number of fish in which this operation was performed, it was found, in collaboration with McCormick, that the blood sugar became raised, often to ten times the normal value. At this high level it remained so long as the fish were kept alive – 11 days in one case, 5–10 days in others. The hyperglycaemia in itself was not sufficient to prove that the fish had become diabetic as a result of the isletectomy, for it was found, in other fish that were exposed to air for a period equal to that required for the operation (about 15 minutes), that the blood sugar rose, sometimes almost as much as in the operated ones. This asphyxial hyperglycaemia, however, was found to disappear within four days*, nor while it lasted was it so pronounced as in the isletectomized fish. There is no doubt that removal of the islets in Myoxocephalus causes pronounced diabetes, as judged by the behaviour of the blood sugar, and, it is of interest to add that there was, on an average, considerably more fat and less glycogen present in the liver of the operated fish than in those of the controls. It remains to determine whether, by giving insulin to the isletectomized fish, the blood sugar can be brought down to the normal level. So far we have been unable to demonstrate any very potent influence of insulin on the blood sugar of normal fish, although there is some indication that it can retard the development of asphyxial hyperglycaemia.

Taking all the evidence into consideration the conclusion seems justified that the only source from which physiologically effective insulin can be secreted within the animal body is the islet tissue.

And finally permit me to say something concerning the behaviour of depancreatized animals kept alive by means of insulin. By such studies it is possible that we may be able to determine whether the lost power to utilize carbohydrate can be reacquired in any measure, and also whether the secretion of pancreatic juice and of insulin include all the functions of the gland. With the collaboration of Frank N. Allan, I. L. Chaikoff, J. Markowitz, and W. W. Simpson, several completely depancreatized dogs have now been kept alive, by daily injections of insulin, for many months, the operation on

---

* There was one fish in which the blood sugar remained at a high level even eight days after asphyxia.

one of them, which is at present under observation, having been performed over eighteen months ago. But this result was not immediately achieved.

In the earlier observations it was observed that, notwithstanding the fact that the animals ate large amounts of meat, the body weight steadily fell, no doubt because of inadequate intestinal digestion and absorption. The addition of cane sugar, in amounts sufficient to cause a mild degree of glycosuria (50 to 100 g daily), had the immediate effect of preventing the loss of body weight, and in most animals, of causing it to become increased, especially when large amounts were given. Four of these animals lived in excellent nutritive condition for periods varying between one and seven months, when each in turn developed symptoms of acute jaundice (bile pigment in urine, yellowing of sclera and skin) accompanied by rise in rectal temperature, anuria and progressive bodily weakness, ending fatally in from two to three days after the onset. The *post mortem* examination revealed, in each case, an extremely fatty liver with no significant pathological changes elsewhere in the body. Under the microscope it was difficult to see any liver cells that were not completely filled with fat, except for a few towards the centres of the lobules which were only moderately invaded. In some way or other, absence of the pancreas leads to a fatal breakdown of the hepatic function. Two possibilities may be considered: the one, that the pancreas secretes internally, besides insulin, some other hormone which is necessary for the functional integrity of the liver, perhaps a hormone having to do with its action on fat metabolism; and the other, that in the absence of the pancreatic ferments, the process of intestinal digestion becomes of such a (bacterial) type that substances having a toxic effect on the liver cells are absorbed into the portal blood. It was therefore decided to add raw ox pancreas (50 g) to the daily diet, and it is as an outcome of this addition that the animals have thrived without showing any symptoms of hepatic breakdown. This favourable result may be dependent either on the restoration of the pancreatic enzymes, thus preventing the development of toxic substances, or because some hormone which withstands digestive action is absorbed from the ingested gland. We are at present observing the effect of adding trypsin to the food, instead of raw pancreas, but although the animal thus treated is in excellent nutritive condition we cannot as yet say whether it may not ultimately develop the hepatic symptoms. It may be added that the toxic theory is supported by the observation that in the absence of raw pancreas, or trypsin, not more than fifty per cent of the ingested meat is assimilated, whereas over eighty per cent is assimilated when either of these is present.

The carbohydrate balance is being determined at intervals in several diabetic animals, in order to see whether any of the lost power to secrete insulin may be reacquired. This is done by determining the proportion of the ingested sugar which reappears in the urine daily while the animals are under the same dose of insulin, but so far no change has been detected. While it is certain that any considerable reacquirement of the power to secrete insulin would be revealed by this method, it is possible that a very scanty secretion might be masked on account of the relatively large amounts administered daily from without, for it has been shown, by Frank N. Allan, that the glucose equivalent of each unit of insulin is very much higher when the total number of units administered is small than when it is large. Another method for investigating this problem remains available, namely to observe whether the diabetic symptoms which supervene when insulin is discontinued are less severe after several months treatment than they are soon after the removal of the pancreas. Our attempts to make this observation have, so far, been frustrated by the very rapid downward progress of the animals after discontinuing insulin. Unless they are given large quantities of meat they die in a few days of symptoms not unlike those of diabetic coma.

It has been stated by Carlson and Drennan that the diabetic symptoms are very much less than usual when pancreatectomy is performed on pregnant animals near full time, and this they have attributed to the secretion of insulin from the foetal pancreas. We could obtain no evidence in support of this hypothesis in the present investigations. Thus, one of the depancreatized dogs gave birth to five pups without any change whatsoever in her sugar balance throughout the pregnancy, although, on the day after the pups were born, severe symptoms of hypoglycaemia developed, no doubt because of the removal of glucose from the body to form the lactose of the milk. There was therefore no evidence that the developing foetuses contributed any significant amount of insulin to the maternal organism. In the face of the relatively large amounts of insulin injected into the mother, however, it is possible, in view of Allan's results, that the small contribution from the foetuses could have no measurable influence on the maternal sugar balance.

I have attempted to review but a small part of the work relating to insulin and have only cursorily referred to the perplexing problem of the mechanism of its action in the animal body. Facts of importance in this regard come almost daily to light and it is to be anticipated that, as these accumulate, a great advance will become possible in our knowledge of the history of carbohydrates in the animal body.

Biography

John James Richard Macleod was born on September 6, 1876 at Cluny, near Dunkeld, Perthshire, Scotland. He was the son of the Rev. Robert Macleod. When later the family moved to Aberdeen, Macleod went to the Grammar School there and later entered the Marischal College of the University of Aberdeen to study medicine.

In 1898 he took his medical degree with honours and was awarded the Anderson Travelling Fellowship, which enabled him to work for a year at the Institute for Physiology at the University of Leipzig.

In 1899 he was appointed Demonstrator of Physiology at the London Hospital Medical School under Professor Leonard Hill and in 1902 he was appointed Lecturer in Biochemistry at the same College. In that year he was awarded the McKinnon Research Studentship of the Royal Society, which he held until 1903, when he was appointed Professor of Physiology at the Western Reserve University at Cleveland, Ohio, U.S.A.

During his tenure of this post he was occupied by various war duties and acted, for part of the winter session of 1916, as Professor of Physiology at McGill University, Montreal.

In 1918 he was elected Professor of Physiology at the University of Toronto, Canada. Here he was Director of the Physiological Laboratory and Associate Dean of the Faculty of Medicine.

In 1928 he was appointed Regius Professor of Physiology at the University of Aberdeen, a post which he held, together with that of Consultant Physiologist to the Rowett Institute for Animal Nutrition, in spite of failing health, until his early death.

Macleod's name will always be associated with his work on carbohydrate metabolism and especially with his collaboration with Frederick Banting and Charles Best in the discovery of insulin. For this work on the discovery of insulin, in 1921, Banting and Macleod were jointly awarded the Nobel Prize for Physiology or Medicine for 1923.

Macleod had, before this discovery, been interested in carbohydrate metabolism and especially in diabetes since 1905 and he had published some 37 papers on carbohydrate metabolism and 12 papers on experimentally produced glycosuria. Previously he had followed the earlier great work of von Mering and Minkowski, which has been published in 1889, and although he believed that the pancreas was the organ involved, he had not been able to prove exactly what part it played. Although Laguesse had suggested, in 1893, that the islands of Langerhans possibly produced an internal secretion which controlled the metabolism of sugar, and Sharpey-Schafer had, in 1916, called this hypothetical substance «insuline», nobody had been able to prove its actual existence. Others had made extracts of the pancreas, some of which had proved to be active in affecting the metabolism of sugar, but none of these products had been found reliable, until Banting and Best, jointly with Macleod, could announce their great discovery in February 1922. The process of manufacturing the pancreatic extract which could be used for the treatment of human patients was patented; the financial proceeds of the patent were given to the British Medical Research Council for the Encouragement of Research, the discoverers receiving no payment at all. Subsequently, the active principle of these earlier pancreatic extracts, insulin, was isolated in pure form by John Jacob Abel in 1926, and eventually it became available as a manufactured product.

Earlier, in 1908, Macleod had done experimental work on the possible part played by the central nervous system in the causation of hyperglycaemia and in 1932 he returned to this subject, basing his work on the experiments done by Claude Bernard on puncture diabetes, and Macleod then concluded, from experiments done on rabbits, that stimulation of gluconeogenesis in the liver occurred by way of the parasympathetic nervous system.

Macleod also did much work in fields other than carbohydrate metabolism. His first paper, published in 1899, when he was working at the London Hospital, had been on the phosphorus content of muscle and he also worked on air sickness, electric shock, purine bases, the chemistry of the tubercle bacillus and the carbamates.

In addition he wrote 11 books and monographs, among which were his *Recent Advances in Physiology* (with Sir Leonard Hill) (1905); *Physiology and Biochemistry of Modern Medicine*, which had reached its 9th edition in 1941; *Diabetes: its Pathological Physiology* (1925); *Carbohydrate Metabolism and Insulin* (1926); and his Vanuxem lectures, published in 1928 as the *Fuel of Life*.

In 1919 Macleod was elected a Fellow of the Royal Society of Canada, in 1923 of the Royal Society, London, in 1930 of the Royal College of Physicians, London, and in 1932 of the Royal Society of Edinburgh. During

1921–1923 he was President of the American Physiological Society, and during 1925–1926 of the Royal Canadian Institute. He held honorary doctorates of the Universities of Toronto, Cambridge, Aberdeen and Pennsylvania, the Western Reserve University and the Jefferson Medical College. He was an honorary fellow of the Accademia Medica, Rome, and also a corresponding member of the Medical and Surgical Society, Bologna, the Societá Medica Chirurgica, Rome, and the Deutsche Akademie der Naturforscher Leopoldina, Halle, and Foreign Associate Fellow of the College of Physicians, Philadelphia.

Macleod was a very successful teacher and director of research. His lucid lectures were delivered in an attractive manner and his pupils and research associates found him a sympathetic and stimulating worker, who demanded exact work and the humility that was a feature of his character. He would not tolerate careless work. He was much interested in the development of medical education and especially in the introduction of scientific methods of investigation into clinical work.

Outside the laboratory he was keenly interested in golf and gardening and the arts, especially painting. A sensitive, loyal and affectionate man of engaging personality, his serene spirit met with courage and optimism the painful and crippling disabilities which troubled the final years of his busy life.

Macleod was married to Mary McWalter. He died on March 16, 1935.

Physiology or Medicine 1924

WILLEM EINTHOVEN

«for his discovery of the mechanism of the electrocardiogram»

Physiology or Medicine 1924

In regard to Einthoven's work, Professor J. E. Johansson, Chairman of the Nobel Committee for Physiology or Medicine of the Royal Caroline Institute, made the following statement*

The Staff of Professors of the Royal Caroline Institute has on 23rd October, 1924, decided to confer this year's Nobel Prize in Physiology or Medicine to the Professor of Physiology at the University of Leiden, Willem Einthoven, for his discovery of the *mechanism of the electrocardiogram*.

Einthoven's name is linked partly with the design of a physical instrument, the *string galvanometer*, partly with the so-called *electrocardiogram*, a record of the electrical potential fluctuations at the surface of the body, which accompany the heart beat. The heart beat, like the piston movement of a steam engine, is a cyclic process. Behind this process lies, in the first place, a similarly cyclic process in the heart muscle.

For the present this process is called the « muscular process», in analogy with «neural process» and «glandular process». All these processes, which with regard to energy must be considered as a conversion of chemical energy into forms of energy other than heat, are accompanied by a fluctuation in the electrical potential – the action current – which as a rule is extremely weak and which does not play any role in the life of the individual, but which from the viewpoint of experimental technique, however, is of the greatest interest, in so far as it allows the registration of the frequency of the functional process and its propagation through the individual organs.

The potential fluctuations concerned are measured in millivolts and in hundredths of seconds. To construct a self-registering measuring instrument which records directly and truly the potential variations of this order of magnitude was a problem which Einthoven has solved with his *string galvanometer* (1903). In constructing this, he started from the well-known Deprez-d'Arsonval «moving-coil galvanometer» and had herein replaced the moving parts – coil and mirror – with a fine, silver-plated quartz wire, which was stretched in the field between the poles of the magnet and at the same time between an optical illumination system, and another one for

* Professor Einthoven being on a lecture tour in the United states, and the other laureate of the year also being unable to come to Stockholm, the usual ceremony on December 10 was cancelled.

projection. The reduction in mass of the moving parts, achieved in this way, allows at the same time high sensitivity and short adjustment time.

After testing the practicability of the instrument for various purposes, and after a thorough analysis (1906) of the dependence of the string galvanometer curve on the mass and tension of the string, and on the damping of the deflection, the latter by electromagnetic means and by the effect of the air resistance, Einthoven published in 1909 the first detailed description of the instrument. Interest in the string galvanometer spread very rapidly, and string galvanometers of various types after Einthoven's specifications were supplied by several famous instrument firms.

Using strings of ultramicroscopic size in a vacuum between the poles of a magnet, Einthoven recently succeeded in registering potential fluctuations of a frequency far beyond the limits of known physiological phenomena. In this connection it may also be mentioned that he has registered sound waves with a frequency of more than 10,000 vibrations per second, by means of strings of the dimensions previously mentioned in association with suitable optical systems.

The construction of the string galvanometer was a purely physical problem. The interest shown by physiologists and physicians in this achievement, is caused, as already mentioned, by the possibilities of analysing, by means of the registration of the so-called action currents, some phenomena in the living organism. The string galvanometer has therefore been widely used for various purposes in physiology. To give an idea of this, some phenomena may be mentioned, which by means of the string galvanometer have been investigated by Einthoven himself: the retina current (1908, 1909), the action currents in nervus vagus (1908, 1909) and in the sympathetic chain (1923), the psychogalvanic reflex (1921), the Gaskell effect (1916), the muscular tone (1918). With regard to the action current of the muscle Einthoven demonstrated (1921) in a convincing manner that this occurs exclusively as a phenomenon accompanying the mechanical effect known for a long time – a fact very important to the concept of the action current.

The achievement for which the Staff of Professors of the Royal Caroline Institute awarded Einthoven the Nobel Prize, is in the field of the heart physiology. Einthoven's interest in the *action current* of the heart dated from 1891; at that time, as a result of the investigations of Burdon-Sanderson (1879) and Augustus Waller (1887, 1889) attention was focussed on this phenomenon.

Both scientists used the well-known Lippmann capillary electrometer, which registers potential variations; but the adjustment time is rather long, and the capillary electrometer curve, therefore, does not reflect in a direct manner the actual time process of the potential changes in the heart muscle during heart beat. Einthoven developed a rather simple method of correction (1894) and could with this derive the actual *electrocardiogram* from the capillary electrometer curve (1895). The details herein he denoted as P, Q, R, S, T: terms which are preserved to this day. This method, however, would never have any practical significance in reproducing the electrocardiogram of man. It is much too laborious for this. Einthoven saw the importance of an instrument which *directly* renders the potential variations with time during these processes, and the result was the string galvanometer described above (1903). The curve recorded by this instrument during the registration of the action current of the heart showed perfect agreement with the electrocardiogram derived from the capillary electrometer curve, and this agreement between the results of the two registration methods, fundamentally so different from one another, proved beyond all doubt that the actual time process of the potential variation accompanying the heart beat had been obtained. Einthoven can thus *with full justification be named the discoverer of the real electrocardiogram.*

One of the first results of this discovery was the demonstration that each *individual has his own characteristic electrocardiogram*, but that the electrocardiogram of all individuals in the main conform to a general type. In a publication « Le télécardiogramme » (1906) Einthoven returns to the same subject, revealing, however, at the same time a fact which has acquired the greatest clinical significance: that *different forms of heart disease reveal themselves characteristically in the electrocardiogram.* He gives examples of the electrocardiograms of patients with hypertrophia of the right ventricle during mitral insufficiency, hypertrophia of the left ventricle during aorta insufficiency, hypertrophia of the left auricle during mitral stenosis, of patients with degeneration of the heart muscle, also of electrocardiograms during various degrees of heart block, during extrasystoles, true « atypical heart systoles » of two different types, as well as during what is now called « ataxia cordis ». In a subsequent work « Weiteres über das Elektrokardiogramm » (More about the electrocardiogram) in 1908, he communicates other cases. Einthoven's interest for the electrocardiogram from clinical point of view is also evident from a proposal, put forward by himself (1906), namely, to establish so-called telecardiograms, i.e. to have electrocardiograms produced by a string galvanometer in a physiological laboratory from patients lying in a hospital several kilometers away. Nowadays, since a string galvanometer is available in almost any large hospital, this detail is only of historical importance.

It can be said that *this new method of investigation fulfilled a need in clinical medicine*. One needs only to remember the curves of venous and arterial pulses, and cardiograms at disposal up till then – all of them difficult to interpret – whenever a case of arrhythmia had to be cleared up. Moreover, some «stroke of luck» was indispensable, even if one is a well-trained experimentator, to obtain a «mechanical» cardiogram from a person, which entirely corresponds with one taken some hours before. The string galvanometer, on the other hand, once set up and adjusted, operates ideally, «accident»-free.

What did the electrocardiogram mean at that time? Einthoven said in his work in 1895 that the efforts to fully interpret the electrocardiogram should be abandoned for the moment, and in a survey of the relevant literature up to the first half of 1912, the author[*] put emphasis on the uncertainty of the efforts to interpret the cardiogram. It can therefore be said that Einthoven had in 1895 discovered some sort of writing the contents of which for many years after remained in virtual obscurity.

However, in his work in 1908 Einthoven gave an *interpretation of the electrocardiogram*. He starts from the fact that the stimulus (of the contraction process, the «negativity») is propagated as a wave in the muscular system of the heart. The string of the galvanometer, connected with the heart in a closed circuit in one of the usual ways, remains in the original position not only when the heart is at rest, but also when the «negativity» of the assemblage of points of the heart wall show the same value. A deflection is therefore in the first place to be expected at the beginning and at the end of a systole, and it presupposes that the condition of activity does not occur, respectively cease, simultaneously in all elements of the muscle. Further: if the contraction process (the stimulus) is propagated symmetrically in relation to the points connected to the galvanometer, then no deflection would take place either. Under such circumstances the electrocardiogram must be determined partly by the starting-point of the stimulus to the heart beat, partly by the conduction system within the heart. The point of departure for the normal heart beat has been sufficiently well known since the middle of the 1890's, the bundle of His also since that time, and Tawara's description of the ramification of the conduction system inside the ventricles known since 1906. According to Einthoven the P-peak is an expression of the propagation of the stimulus wave in the muscular system of the auricle. The negativity wave, corresponding to the stimulus wave in the His-Tawara system, is considered

[*] P. H. Kahn, «Das Elektrokardiogramm», *Ergeb. Physiol.*, 14 (1914) 1.

too weak by Einthoven to cause any deflection in the galvanometer. The QRS-complex is determined by the propagation of the stimulus wave in the muscular system of the two ventricles, proceeding in unsymmetrical fashion to the points of lead, starting at different moments at the transition of the tree-like ramified Purkinje's fibres into the various parts of the proper muscular system of the heart. When the contraction process has reached its maximum in all the points of the ventricular wall, the string returns to its original position. When the contraction ceases in the various parts at different moments, a T-peak is obtained.

It is unnecessary in this connection to consider the interpretations proposed by other investigators, as *Einthoven's concept is the only one which has proved to be tenable*. The interpretation that the P-peak belongs to the auricular systole is mainly based on his observation of electrocardiograms in cases of heart block in patients or during vagus stimulation in dogs. With regard to the interpretation of the QRS-complex Einthoven was evidently the first who has clearly recognized the *significance of the conduction system* in this connection. The train of thought in the interpretation of the T-peak can already be detected in Burdon-Sanderson's previously mentioned work.

Already Waller (1887) had observed that the deflections of the capillary electrometer vary accordingly as the lead is taken from both hands or from one hand and one foot, etc., and based hereupon his well-known scheme of the potential distribution in the body in relation with the heart's action current–a scheme later adopted in textbooks and handbooks. The scheme has principally been used to demonstrate that the amount of deflection, i.e. the «peaks» in the electrocardiogram, must vary in accordance with the manner in which the electrodes have been applied in relation to the heart axis. Einthoven pointed out, however, that not only the amount of the deflection but also *the shape of the entire electrocardiogram is changed when one manner of leading is replaced by another* (1908). One of the spikes may be accentuated while another may be suppressed, etc. One and the same spike resulting from different leads does not always correspond with the same phase of the heart period. Einthoven therefore found it essential to always indicate the manner of leading, and in connection with this he proposed (1908) the *now generally accepted standardization: Lead I, II, and III.*

In a publication (1913) Einthoven has shown how the *direction and definite amount of the resulting potential difference* at corresponding moments can be calculated from the simultaneous deflections at the three leads indicated. The direction of the resulting potential variation corresponds in a certain way to the electrical axis in Waller's scheme, and several authors use the term «electric axis» instead of Einthoven's designation. Waller made this axis to coincide with the anatomical axis of the heart, an obvious procedure, since at that time it was generally believed that the heart – as to the propagation of the stimulus wave – could be identified with a muscle with fibres running parallel, stimulated at one end. In fact, the resultant potential variation (the «electric axis»), as shown by Einthoven, changes its direction from one moment to the other during the heart period. The rotation of the electric axis during the heart period is nothing else than an expression of the course of the stimulus wave through the heart muscle, as is evident from the electrocardiogram at the three leads indicated. Already in his papers in 1906 and 1909 Einthoven pointed out on the basis of the shape of the electrocardiogram that the starting-point for the so-called atypical ventricular systoles must be other than for the normal, and showed that a combination of electrocardiograms at different leads supplies a possibility of deciding where this starting-point is located. The calculation of this direction of the resultant potential variation is a refinement of this method which can be used when an evaluation of the electrocardiogram by visual inspection is not sufficient.

Such a calculation is very simple. The difficulty consists in establishing the corresponding phases in the combination of electrocardiograms at the three leads. Hereby, as pointed out by Einthoven, we can make use of the electrophonocardiogram. However, the safest way is to register *simultaneously the electrocardiogram at the three leads*, or at least at two of them. Einthoven has given a particularly elegant design for such an instrument (1915, 1916)–two galvanometers one after the other, each transferring its string registration on the same plate. The firm of Carl Zeiss has carried out such a detailed construction.

Thus Einthoven has added the discovery of *the mechanism of the electrocardiogram* to the discovery of the true, individual electrocardiogram. Sir Thomas Lewis was the first who realized the importance of Einthoven's discovery and who followed his line of thought. His elegant demonstration (1916) of the QRS-complex in the electrocardiogram by means of an algebraic summation of dextro- and laevogram confirmed the correctness of Einthoven's interpretation, just as his demonstration of the «circus movements» of the stimulation wave (1921) in cases of auricular fibrillation proved conclusively the practical importance of Einthoven's calculation of the «direction and definite magnitude of the resultant potential variation». The examination of literature in this field fully justifies the statement that the importance of Einthoven's discovery of the mechanism of the electrocardiogram has only been conclusively proved by Sir Thomas Lewis's works previously mentioned.

Since Einthoven first described the details of the electrocardiogram and more so after the publication in 1906 of its appearance during the various heart diseases, a vast literature in this field has accumulated over the years. All these researches aim fundamentally to reveal the mechanism which underlies the electrocardiogram. The question then arises: Which facets of this mechanism have been brought to light? Let us imagine that a heart model made from fresh heart muscles is placed in a homogeneously conducting medium with leads connected with a string galvanometer, and let us ask ourselves the question: What should be put into this model so that it will give the customary electrocardiogram? The answer now would be: (1) the conduction system; (2) a conduction velocity in this system which is several times greater than that in the heart muscle. Einthoven was the first to point out the importance of the conduction system. The importance of the conduction velocity has been shown by Lewis.

The same mechanism governing the characteristics of the electrocardiogram, also governs the characteristics of the mechanical process during the heart beat. We should remember in this connection that the mechanical process not only consists of the succession of the stimulation of the separate parts of the heart compartments, but also of the cooperation of the individual parts of the heart wall which form the essential condition for the mechanical effect in the individual ventricle or in the individual auricle. A deficiency in this cooperation can, with regard to the mechanical effect, be as fatal as a valvular insufficiency. Today, the importance of the mechanism discovered by Einthoven can easily be realized.

WILLEM EINTHOVEN

The string galvanometer and the measurement of the action currents of the heart

Nobel Lecture, December 11, 1925

May I be permitted to communicate something about the string galvano-meter, its latest improvements and its use in electrocardiography.

The string galvanometer consists of a thin thread conducting the electric current which is stretched as a string in a magnetic field. The thread, as soon as the current passes through it, is displaced from its position of equilibrium in a direction at right angles to the direction of the lines of magnetic force. The amount of displacement is proportional to the strength of the current passing through the thread, so that this current can be easily and accurately measured.

The less the tension in the thread, the weaker are the currents which are adequate to cause a visible displacement. But, if the current sensitivity is in this way increased, the string movements are at the same time slowed down, whilst in most uses of the instrument, one has to deal with rapid variations in the current. To assess the usefulness of the instrument, we therefore introduce instead of current sensitivity the concept of normal sensitivity and define this as the sensitivity at a certain arbitrarily established, constant duration of the deviation of the string. In more precise form we write

$$G = \frac{u}{T^2\sqrt{A}}$$

in which G is the normal sensitivity and A that strength of current which causes the deviation u . T is the natural frequency of the freely oscillating string.

We assume that the movements of the string are damped down and u is the permanent deviation for a constant current. But T is the frequency which would be present if the damping down were removed.

In the string galvanometer the normal sensitivity is conditioned by the magnification v, the field strength H, the mass M of the string and the char-acteristics of the material of which the string is made, thus

$$G' = \frac{vH}{\sqrt{mwM}} \times \text{constant}$$

The quality of the material is rendered by mw, if $m =$ the mass and w the resistance of the string per cm of length. If H is expressed in gauss, m and M in grams, and w in ohms, the constant equals $1/320$.

The normal sensitivity of a string galvanometer can be made 1000 times greater than that of the most sensitive mirror galvanometer.

In order to increase the normal sensitivity one must select a favourable material for the manufacture of the string, use a high magnification and a strong field, and finally reduce the mass M of the string.

The possible improvement of the material soon reaches a limit and this also applies to the increase of the magnification and of the magnetic field used. A suitable electromagnet with a field of about 20,000 gauss can be constructed without special difficulty; it will however hardly be deemed possible to build a galvanometer with a field ten times bigger than this. The cost would be in any event disproportionately high.

If we now consider the mass M of the string and restrict ourselves to strings of a certain constant length, we then see from the formula that the normal sensitivity is inversely proportional to the diameter of the string. We must therefore make the string as thin as possible, only reaching the limit when it is so thin that the observation and reproduction of it are connected with difficulties.

While in many of the instruments in use for the most diverse purposes the string is 2 or 3 μ thick it is however possible to use strings of 0.1 μ diameter and still thinner ones with good results. The figure below (Fig. 1) shows a thin string which was photographed at a magnification of 1800. Its diameter could not be accurately measured. But there were good grounds to estimate this as 0.04 μ, i.e. about 15 times smaller than the wavelength of yellow light and nevertheless it was possible to obtain a photograph which, although it was not rich in contrasts, was still reasonably satisfactory.

A thin string can be induced to move rapidly and under suitable conditions to achieve almost aperiodic displacement, about 0.1 σ. Under these conditions the string is strongly stretched and its current sensitivity is only small. For most purposes a greater current sensitivity is needed, so that it is necessary to slacken the string. But then one encounters the difficulty of air-damping which in such a thin string is troublesome and inhibits the movement of the string excessively.

Engineer W. F. Einthoven has therefore made a model in which the string lies in a vacuum and this is specially suitable for the use of very thin strings. We obtained with this model with a string about 2 cm long and 0.1–0.2 μ thick and a magnification of the image of the string of 1800 times, a displacement of 1 mm in 0.01 sec. for a current of 10^{-11} A. This corresponds to an unparalleled normal sensitivity and one could think that these conditions would be uncommonly suitable for the study of the small variations of current which play so great a role in physiology and medicine.

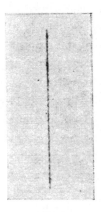

Fig. 1. A string the diameter of which is estimated
to be of the order of 0.04 μ.

There is however an insurmountable difficulty. The string does not remain still. Molecular movements due to the temperature cause the whole string to remain in continuous movement. These movements could be named Brownian string vibrations. If one puts the mean energy of a molecule at a certain temperature at $\frac{1}{2} m v^2$, the string movement which corresponds to one degree of freedom at the same temperature has a mean energy of $\frac{1}{6} m v^2$. In fact, one obtains about this amount if one calculates the mean energy from the mass of the string and the movement actually observed.

In a high vacuum the string oscillates very regularly with its natural frequency; if a little air enters the movements become irregular. It is self-evident that the air cannot damp down the movements of the string because the movement of the air molecules is itself the cause of the movements of the string. If the string is strongly stretched, its natural frequency is high but the amplitude of the oscillations is small. If one reduces the tension of the string

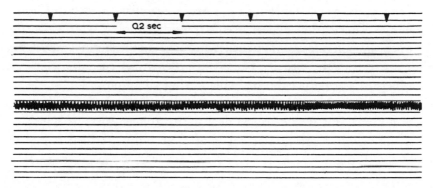

Fig. 2. Brownian string vibrations magnified about 2,000 times. The string is 18 mm long and has a diameter of less than $0.2\,\mu$ and a resistance of 250,000 ohm. High vacuum. No field–magnet current. The string–circuit is open.

its natural frequency will be lower, but at the same time the amplitude of the oscillations is greater to such an extent that the mean energy of the string's movement remains unchanged.

If the string is in a strong magnetic field and in a closed electrical circuit the usual string movements will be electromagnetically damped down. They induce currents in the circuit outside the galvanometer and must transmit heat there.

Since one might at first sight perhaps think that the Brownian string vibrations had the same result it will appear that here we have a case of « perpetuum mobile » of the second kind. Then, if the circuit is closed by an external resistance, this would be warmed even if its temperature were warmer than that of the galvanometer.

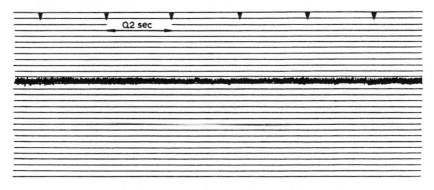

Fig. 3. Air pressure $100\,\mu$ Hg. Otherwise as in Fig. 2.

The Brownian string vibrations are made irregular by electromagnetic influences, but in fact they experience no trace of damping. The explanation of the phenomenon is found if one takes into account the thermal movements of the electrons in the circuit. These cause Brownian currents which can neither increase nor reduce the Brownian string vibrations.

We thus see that measurement of weak currents with thin strings is in practice limited. But if one has to deal with stronger currents which can cause a sufficiently large displacement of a strongly stretched string, the thinnest string is the best because one reaches with it the greatest normal sensitivity.

The string galvanometer is within the limits mentioned useful for numerous purposes. For wireless telegraphy a model has been built which has rendered very good service as a receiving apparatus. A short string *in vacuo* is tuned to the frequency of the ether oscillations which are sent from some transmitting station. From the ether oscillations high-frequency alternating currents are obtained by one of the usual methods which one sends through the accurately tuned string. As soon as a signal is sent the string is set in motion and its movements are photographed on a moving strip of paper, so that the dots and dashes can be recorded. Thus one has a recording apparatus which is very sensitive, and its decrement can be made precisely controllable and extraordinarily small.

A string 1 mm long can be so stretched that its natural frequency is 300 000 cycles per second which corresponds to a wavelength of 1 km. It is difficult to obtain such rapid oscillations of a material in any other way.

Another technical use of the string galvanometer is the recording of sound

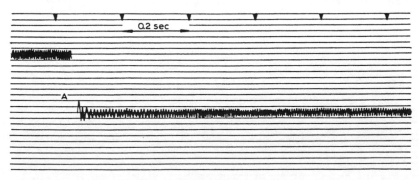

Fig. 4. High vacuum as in Fig. 2. The field-magnet current passes through and the galvanometer circuit is closed. At *A* a constant potential difference is applied to the galvanometer. Otherwise as in the two preceding figures.

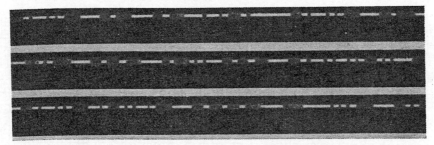

Fig. 5. String-galvanometer recording made in Leiden of wireless signals from Bandoeng, Netherlands Indies. Distance 12,000 km. The string is 6 mm long and is tuned accurately to the incoming waves of $\lambda = 7.5$ km. It thus has a natural frequency of 40,000 whole oscillations per sec. The dots and dashes of the signals are clearly seen in the figure.

waves which can by means of the telephone or microphone set the string in oscillations. One has made full use of the instrument in locating the source of the sound.

In physiology the string galvanometer has been used with success for recording muscle, nerve and skin currents and also those of sense organs.

We mention also the psychogalvanic reflex phenomenon, the heart sounds and the action currents of the heart, and must discuss the latter in further detail.

Just as each muscle when it contracts generates an electric current, so the heart develops electricity at each systole. This was first described by Kölliker and Müller. The English physiologist Augustus D. Waller then showed that the differences in potential which arise in the heart are transmitted to various parts of the body and that with a sensitive measuring instrument, the capillary electrometer, variations in potential in the human heart can be observed. One needs only lead the current from the hands and feet to the measuring instrument in order to see the variations of the current which show the same rhythm as that shown by the heart's action.

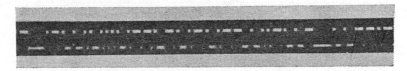

Fig. 6. Reception of wireless signals from an arc transmitter with two galvanometers. The string of one is tuned to the working wavelength and that of the other to the wavelength used when not transmitting signals. Whilst one string oscillates the other remains quiescent and vice versa.

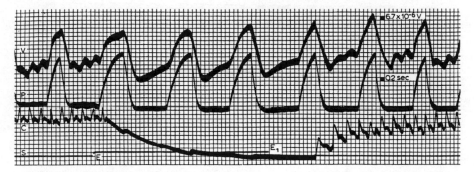

Fig. 7. Vagus currents of the dog. Recording of the currents from the peripheral stump of the left N. vagus. *V*, vagus currents; *P*, pneumogram of the own respiratory movements of the animal; *C*, blood-pressure curve; *S*, signal. Between *E* and E_1 the peripheral stump of the right vagus nerve was stimulated by induction currents. The heart became arrested. The heart waves disappear in the left electrovagogram, whilst the respiratory waves persist.

If one records the displacements in the measuring instrument, one obtains a curve called the electrocardiogram (ECG). But because of the imperfections of the capillary electrometer the directly recorded curve does not give an accurate picture of the actual variations in the potential. In order to obtain a fairly accurate picture one must construct a new curve based upon the properties of the instrument used and the data of the recorded curve and this takes much time. This stood in the way of the practical use of electrocardiography for the investigation of heart diseases, and general interest in the ECG only developed later, after the string galvanometer had made it possible to record easily and quickly the required shape with satisfactory accuracy.

There was a certain satisfaction in learning that the picture constructed by means of the time-consuming measurements and calculations from the data of a capillary-electrometer curve corresponded in shape and measurements

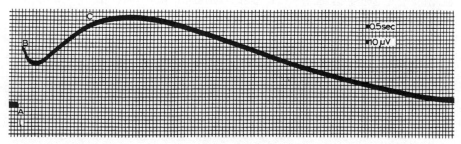

Fig. 8. Retinal currents. Combined reaction of the three substances *A*, *B*, and *C*. Eyes in darkness. At *l* instantaneous illumination with strong light for 0.01 sec.

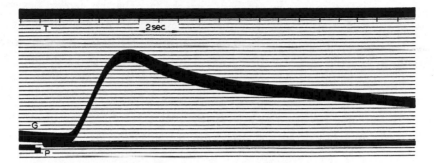

Fig. 9. Psychogalvanic reflex phenomenon. Current lead from the thumbs and fingers of a hand. At *P* acoustic stimulus is applied. The curve *G* gives the resultant change in the resistance of the skin of the person investigated and an upward deflection of one scale division corresponds to a reduction in the resistance of 20 ohm.

with the electrocardiogram directly recorded with the string galvanometer. Because this correspondence signified firstly that the earlier calculations were correct, and secondly that the new galvanometer fulfilled its purpose.

An advantage of electrocardiography over other graphic methods for the study of the heart and pulse is that the ECG can record in absolute units, and the shape of the curve no longer depends on the properties of the instrument used. Provided a string galvanometer is correctly built and is suited to its purpose, i.e. is sensitive and rapid enough, each curve, whenever and where-ever in the world it may be recorded, is directly comparable to any other curve.

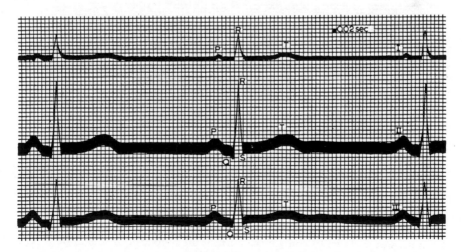

Fig. 10. Normal electrocardiogram taken with three leads simultaneously.

The investigator cannot be satisfied with the fact that he determines the shape of the potential variations at the hands and feet. This measurement is rather the means of bringing to light the functions of the heart. It is therefore desirable to take the ECG by three leads: I from the right and left hand; II from the right hand and the left foot; III from the left hand and left foot. Curves taken by these three leads have a certain relation to each other, i.e. the degree of potential variation in Lead III is equal to the difference in the potential variations recorded at the same time in Leads I and II.

$$\text{Lead III} = \text{Lead II} - \text{Lead I}$$

By means of these multiple leads it is possible to measure the direction and manifest magnitude of the potential variations in the heart itself. The technique of electrocardiography has developed so far that it is now possible to record curves with three strings simultaneously. This enables us to control the quality of the instruments and the accuracy of the methods used. For when one can confirm the formula practically by means of the curves, one demonstrates in this way that these curves have been correctly obtained.

What must we regard as the cause which determines a specific shape of the ECG or better: What determines the direction and the degree of the potential difference present in the heart at a given moment?

The process of contraction of the heart is associated, just as is contraction

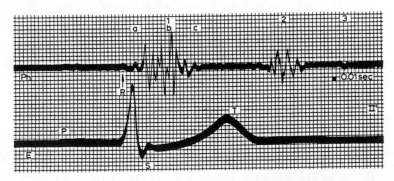

Fig. 11. Heart sounds and electrocardiogram. The sounds are taken from the apex of the heart and are indicated by *1*, *2*, and *3*. The third sound is only weak. In the phonocardiogram of the first sound, one can differentiate between the initial oscillations *a*, the main oscillations *b*, and the subsequent oscillations *c*. The arrow indicates both the beginning of the ventricular complex and that of the systolic heart sound. Both begin simultaneously.

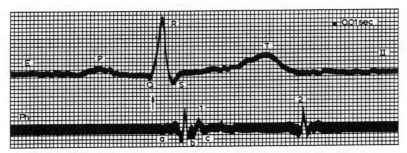

Fig. 12. Similar to Fig. 11. The ventricular complex and the systolic heart sound begin almost simultaneously.

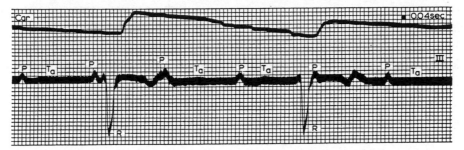

Fig. 13. Stokes-Adams disease with hypertrophy of the left ventricle. A positive T_a-wave.

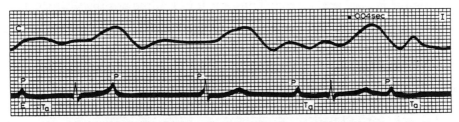

Fig. 14. Stokes-Adams disease. The T_a-wave is negative.

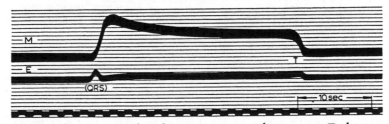

Fig. 15. A frog's heart poisoned with veratrin. M, mechanogram; E, electrogram of the ventricle. The mechanical systole and the electrogram are slowed to the same degree. They lasted almost half a minute.

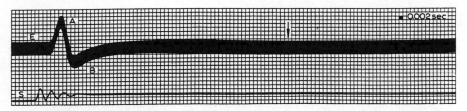

Fig. 16. Diphasic action current of a striated muscle. Stimulation at *S*. After going through the two phases *A* and *B* there is still a late, shallow depression of the curve visible. This is comparable to the *T*-wave of an electrocardiogram.

of a skeletal muscle, with the development of an electrical wave. Still the heart as a whole may not simply be compared with a skeletal muscle. It has a different structure. The skeletal muscle consists of fibres which are often of the same length as the muscle itself. When a contraction wave passes over a skeletal muscle, the locality at which the summit of the wave is found is electronegative with regard to the other parts of the muscle.

The heart muscle, however, consists of microscopically small segments, which are separated from one another by certain septa. So far as the production of electricity is concerned, each segment has to be considered an individuality. When a contraction wave passes over a segment of heart muscle, the point at which the summit of the wave occurs is electronegative towards all other points *in that same segment*. Thus it is for example possible that a contraction wave passing over the heart at a given moment may have its summit

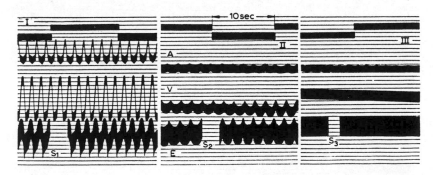

Fig. 17. Frog's heart poisoned with KCl. *I*, before the use of the KCl solution. *II* and *III*, during continued poisoning. *A*, mechanogram of the auricles; *V*, mechanogram of the ventricle; *E*, electrogram of the ventricle; S_1, S_2, S_3, calibration curves. It is evident from the curves that the mechanical and electrical phenomena decrease to the same extent. In *III*, the summits of the ventricular systole, as well as those of the electrogram are just visible.

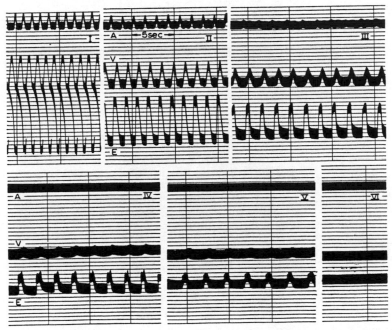

Fig. 18. Similar to Fig. 17, only the poisoning was with chloroform in stead of KCl.

in the base of the heart whilst nevertheless the direction of potential difference measured at the hands and feet indicates a negativity of the apex to the base. For, the direction of the created potential difference is determined not by the position of the wave in the heart, but by the position of the wave *in each individual segment.*

One has also to take into account the fact that the potential difference observed by indirect leads is always the resultant of all the potential differences present in all the segments of the heart.

There is strict relationship between the electrical and the mechanical phenomena of the heart systole. The accurate measurement of the moments at which both begin is beset with difficulties, but in so far as the measurements are made by a suitable technique and an appropriate method they indicate that at the moment at which the electrical current begins the first mechanical results of the contraction are also visible.

Also the duration of the mechanical contraction corresponds in general to the duration of the production of electricity.

And finally there is also a relationship between the force of the contraction and the magnitude of the potential difference produced.

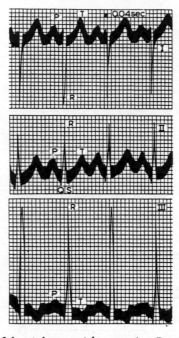

Fig. 19. Preponderance of the right ventricle, negative R_I-wave and very high R_{III}-wave. The three curves were taken successively.

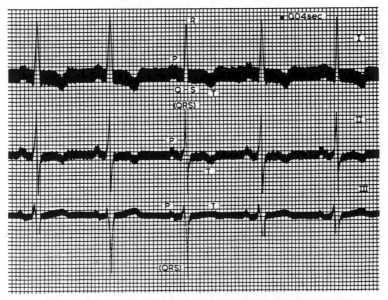

Fig. 20. Preponderance of the left ventricle. Strongly negative R_{III}-deflection. The record was taken simultaneously at the three usual leads.

In pathology and clinical work the ECG has attained great significance because it often facilitates the diagnosis of a disease. This is true for example in cases of unilateral hypertrophy of the heart.

In other cases new light has been thrown on known phenomena. As an example we can mention occasional cases in which a pulse becomes small or disappears in a series of otherwise regular successive pulses all of the same magnitude. Formerly one could think that the cause of this intermittence lies in a heart systole which because it appeared prematurely was so weak that it could not cause a forceful pulse. Now we know that also the premature systoles are of full strength and that the weakening or disappearance of the pulse is due to the fact that in the premature contraction the heart lacks time to fill its chambers with a sufficient amount of blood. A ventricle empty of blood cannot, when it contracts, even if its contraction occurs with full force, send blood into the arteries and thus cannot cause a pulse.

Fig. 21 shows a strong premature extrasystole with absence of the pulse, while some seconds later in the same patient and on the same photographic

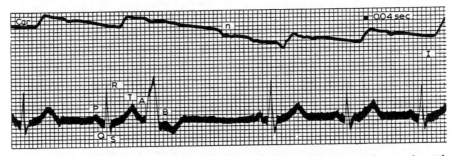

Fig. 21. Extrasystole which because of its premature occurrence caused no pulse. The heart was empty of blood during the extrasystole.

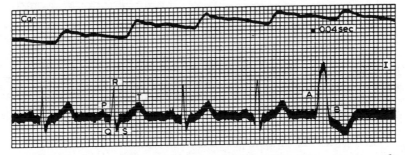

Fig. 22. An extrasystole similar to the one in the preceding figure. Because of its later appearance it caused a strong pulse. The heart had had time to fill with blood.

plate a similar but much less premature extrasystole appears (see Fig. 22). Although in both cases the kind and manner of the heart's contraction was the same, in the second case a forceful pulse was caused. This shows that what was lacking in the first case was not force but blood.

In Fig. 23 we see in succession an extrasystole *with* a pulse and one *without* it. The latter which is very premature is however very similar to the former. The atypical electrocardiograms shown in Figs. 21, 22, and 23 are all laevo-grams. In the main they are diphasic and in Lead I the first phase is positive as in Figs. 21 and 22; in Leads II and III it is negative as in Fig. 23.

The name laevogram indicates that the curve corresponds to a systole in which the contraction begins in the left side of the heart and passes from this to the right heart. A dextrogram gives the reverse picture.

It has been doubted whether in man a diphasic action current of the heart with downwardly directed phase I in Leads II and III corresponds to a laevo-gram and it has been believed on theoretical grounds that the position may be the reverse. But we have by chance had the opportunity to apply a direct control in man.

There was a hitherto quite healthy man whose sternum had been surgically removed. His heart was covered only by a thin layer of skin. Light tapping on the area of the heart caused an extrasystole. Because the right half of the heart lies in man on the anterior side and thus must be stimulated mechanically by the tapping this must cause a dextrogram. Fig. 24 confirmed this prediction.

It may also be stated that this finding is in full agreement with the theory of the heart muscle segments mentioned above.

In many cases of disease the diagnosis is possible only by electrocardio-graphy. This is especially true of certain diseases of the auricles and of anom-

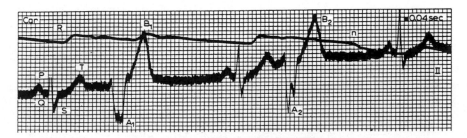

Fig. 23. Two extrasystoles. $A_1 B_1$ appeared late and caused a fairly strong pulse. $A_2 B_2$ appeared earlier. Instead of the pulse corresponding to them the record shows at n a depression.

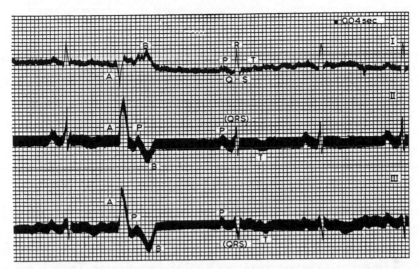

Fig. 24. Man without a sternum. The heart was as good as naked and covered only by a thin layer of skin. A light tapping on the heart area caused an extrasystole. This is a pure dextrogram which is in agreement with the theory that the segments of the heart muscle are electrically to be regarded as being individual entities.

alies in various parts of the conduction system which can nowadays be accurately localized.

I may be permitted to return in a few words to the capillary electrometer.

In 1894 we recorded with this instrument an ECG which had a shape markedly diverging from curves up till then recorded, which puzzled us. The oscillations of the mercury meniscus were so slow that the recorded curve must in itself already reproduce *approximately* the variations in potential actually present. Thus a detailed calculation and construction were not carried out.

In the course of 1925 we had the opportunity to investigate the same person who is now 79 years old with the string galvanometer. In Figs. 25 and 26 one sees how closely the capillary electrometer curve taken with Lead I 31 years earlier corresponds to the galvanometric curve taken with the same lead.

The typical heart anomaly has remained unchanged all this long time but what was then a puzzle is now explained. The patient had a block in the left branch of the bundle of His. The ventricles execute abnormal contractions and produce dextrograms. Clinically it is important that a patient can have this anomaly for more than 31 years without experiencing excessive discomfort and this case is for that reason alone well worth mentioning because un-

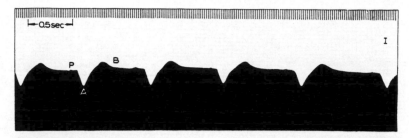

Fig. 25. Record of an electrocardiogram by means of a capillary electrometer taken on June 15, 1894, i.e. before the construction of the string galvanometer. Lead I. Blocking of conduction in the left branch of the bundle of His. Aberrant dextrogram.

til now clinical work on the whole has had at its disposal only experience in electrocardiography of limited duration.

The insight into the true nature of the deficiency has been given to us by later investigators. Among them the English investigator Thomas Lewis, who has played a great part in the development of electrocardiography deserves special mention. It is my conviction that the general interest in ECG would certainly not have risen so high, nowadays, if we had had to do without his work, and I doubt whether without his valuable contributions I should have the privilege of standing before you today.

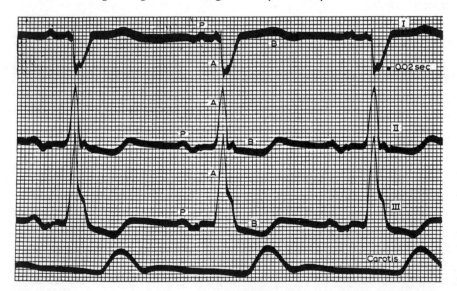

Fig. 26. The same patient as in Fig. 25 but investigated 31 years later with three leads simultaneously. The disease has remained unchanged. One notes the similarity of the curves taken with the capillary electrometer and the galvanometer with Lead I.

But in addition, innumerable other workers in the field of electrocardio-graphy have gained great merit. We cannot now name them all but we conclude with a reference to the happy circumstance that investigators of the whole world have worked together. A new chapter has been opened in the study of heart diseases, not by the work of a single investigator, but by that of many talented men, who have not been influenced in their work by polit-ical boundaries and, distributed over the whole surface of the earth, have devoted their powers to an ideal purpose, the advance of knowledge by which, finally, suffering mankind is helped.

Biography

Willem Einthoven was born on May 21, 1860, in Semarang on the island of Java, in the former Dutch East Indies (now Indonesia). His father was Jacob Einthoven, born and educated in Groningen, The Netherlands, an army medical officer in the Indies, who later became parish doctor in Semarang. His mother was Louise M. M. C. de Vogel, daughter of the then Director of Finance in the Indies. Willem was the eldest son, and the third child in a family of three daughters and three sons.

At the age of ten, Einthoven lost his father, and his mother decided to return with her six children to Holland, where the family settled in Utrecht.

After having passed the « Hogere Burgerschool » (secondary school), he in 1878 entered the University of Utrecht as a medical student, intending to follow in his father's footsteps. His exceptional abilities, however, began to develop in quite a different direction. After being assistant to the ophthalmologist H. Snellen Sr. in the renowned eye-hospital « Gasthuis voor Ooglijders », he made two investigations, both of which attracted widespread interest. The first was carried out after Einthoven had gained his « candidaat » diploma (approximately equivalent to the B.Sc. degree), under the direction of the anatomist W. Koster, and was entitled « Quelques remarques sur le mécanisme de l'articulation du coude » (Some remarks on the elbow joint). Later he worked in close association with the great physiologist F. C. Donders, under whose guidance he undertook his second study, which was published in 1885 as his doctor's thesis: « Stereoscopie door kleurverschil » (Stereoscopy by means of colour variation) – one of Einthoven's teachers was the physicist C. H. D. Buys Ballot, who discovered the well-known law in meteorology.

That same year, 1885, he was appointed successor to A. Heynsius, Professor of Physiology at the University of Leiden, which he took up after having qualified as general practitioner in January, 1886. His inaugural address was entitled « De leer der specifieke energieën » (The theory of specific energies). His first important research in Leiden was published in 1892: « Über die Wirkung der Bronchialmuskeln nach einer neuen Methode untersucht, und über

Asthma nervosum» (On the function of the bronchial muscles investigated by a new method, and on nervous asthma), a study of great merit, mentioned as «a great work» in Nagel's «Handbuch der Physiologie». At that time he also began research into optics, the study of which occupied him ever since. Some publications in this field were: « Eine einfache physiologische Erklärung für verschiedene geometrisch-optische Täuschungen» (A simple physiological explanation for various geometric-optical illusions) in 1898; « Die Accomodation des menschlichen Auges» (The accomodation of the human eye) in 1902; «The form and magnitude of the electric response of the eye to stimulation by light at various intensities», with W. A. Jolly in 1908.

Up till now, his talents had not yet been developed to the full. This opportunity came when he began the task of registering accurately the heart sounds, using a capillary electrometer. With this in view, he investigated the theoretical principles of this instrument, and devised methods of obtaining the necessary stability, and of correcting mathematically the errors in the photographically registered results due to the inertia of the instrument. Having found these methods he decided to carry out a thorough analysis of A. D. Waller's electrocardiogram – a study which has remained classic in its field.

This investigation led Einthoven to intensify his research. To avoid complex mathematical corrections, he finally devised the string galvanometer which did not involve these calculations. Although the principle in itself was obvious, and practical applications of it were made in other fields of study, the instrument had to be precisioned and refined to make it usable for physiologists, and this took three years of laborious work. As a result of this, a galvanometer was produced which could be used in medical science as well as in technology; an instrument which was incomparable in its adaptability and speed of adjustment.

He then, with P. Battaerd, took up the study of the heart sounds, followed by research into the retina currents with W. A. Jolly (begun earlier with H. K. de Haas). The electrocardiogram itself he studied in all its aspects with numerous pupils and with visiting scientists. It was this last research which earned him the Nobel Prize in Physiology or Medicine for 1924. In addition to this the string galvanometer has proved of the highest value for the study of the periphery and sympathetic nerves.

In the remaining years of his life, problems of acoustics and capacity studies came within the sphere of his interests. The construction of the string phonograph (1923) could be considered as a consequence of this.

Einthoven possessed the gift of being able to devote himself entirely to a particular field of study. (His genius was actually more orientated towards physics than physiology.) As a result he was able to make penetrating inquiries into almost any subject which came within the scope of his interests, and to carry out his work to its logical conclusion.

Einthoven was a great believer in physical education. In his student days he was a keen sportsman, repeatedly urging his comrades « not to let the body perish ». (He was President of the Gymnastics and Fencing Union, and was one of the founders of the Utrecht Student Rowing Club.) His first study on the elbow joint resulted from a broken wrist suffered while pursuing one of his favourite sports, and during the somewhat involuntary confinement his interest was awakened in the pro- and supination movements of the hand and the functions of the shoulder and elbow joints.

The string galvanometer has led countless investigators to study the functions and diseases of the heart muscle. The laboratory at Leiden became a place of pilgrimage, visited by scientists from all over the world. For this, suffering mankind has much to owe to Einthoven. In electrocardiography the string galvanometer is the most reliable tool. Although it has been superseded by portable types and by models utilizing amplification techniques used in radio communication (Einthoven has always mistrusted the use of condensers, fearing the distortion of curves), cardiograms from the string galvanometer have remained the standard of reference in numerous cases to this day.

Einthoven was a member of the Dutch Royal Academy of Sciences, the meetings of which he hardly ever missed. He frequently took part in the debates himself, and his sharp criticism frequently found weaknesses in many a lecture.

Einthoven married in 1886 Frédérique Jeanne Louise de Vogel, a cousin, and sister of Dr. W. Th. de Vogel, former Director of the Dienst der Volksgezondheid (Public Health Service) in the Dutch East Indies. There were four children: Augusta (b. 1887), who was married to R. Clevering, an engineer; Louise (b. 1889), married to J. A. R. Terlet, pastor emeritus; Willem (1893–1944) – a brilliant electro-technical engineer who was responsible for the development of the vacuum model of the string galvanometer and for its use in wireless communication, and who was Director of the Radio Laboratory in Bandung, Java; and Johanna (b. 1897), a physician.

He died on the 28th of September, 1927, after long suffering.

Physiology or Medicine 1925

Prize not awarded.

Physiology or Medicine 1926

JOHANNES ANDREAS GRIB FIBIGER

«for his discovery of the Spiroptera carcinoma»

Physiology or Medicine 1926

*Presentation Speech by Professor W. Wernstedt, Dean of
the Royal Caroline Institute*

Your Majesty, Your Royal Highnesses, Ladies and Gentlemen.

Few diseases have the power of inspiring fear to the same degree as cancer.
However, who would be surprised at that? How many times is this affliction
not synonymous with a long, painful and grievous illness, how many times
is it not equivalent to incurable suffering? It is therefore natural that we
should strive to throw light upon its nature; but the road to this discovery
is both long and difficult. Cancer always, in fact, presents the investigator
with a number of obscure and unsolved problems. Thus the cause of cancer
has for a long time baffled the penetrating studies of the most tireless research workers. Fibiger was the first of these to succeed in lifting with a sure
hand a corner of the veil which hid from us the etiology of the disease; the
first also, to enable us to replace with precise and demonstrable theories the
hypotheses with which we had had to content ourselves.

For example, it had been thought for a long time that a causal connection
existed between cancer and a prolonged irritation of some sort, mechanical,
thermal, chemical, radiant, etc.; this supposition was supported by the incidence, sometimes verified, of cancer as an occupational disease. Cancer
occurring in radiologists, chimney sweepers, workers in the manufacture of
chemical products, establish so many examples of cancerous infection that
one might believe they were provoked by radioactive or chemical irritation.
However, each time experiment was resorted to in an attempt to provoke
cancer in animals by irritants of this nature, it failed, and the animals refused
to contract the disease.

Others, with all the more reason, sought to find in cancer the work of
microparasites, for true neoplastic epizootics were thought sometimes to
have been established in the animal world. But research into the pathogenic
agent, the «cancer bacillus», and the experiments attempting to inoculate
the disease had remained fruitless. Cancer has been equally attributed to other
parasites, and notably to the worm. But, just as the attempts to provoke
cancer, whether by inoculation or by irritation remained unproductive, in
the same way it proved impossible to demonstrate experimentally that the

disease was attributable to worms. These authorities who continued to support this thesis were, moreover, frequently considered to be fantasts. *Because of the failure of attempts to establish, by experiment, the accuracy of any theory, there was no clear idea concerning the cause of cancer, and such in general was the position of this question. Then it was, in 1913, that Fibiger discovered that cancer could be produced experimentally.*

It is of the greatest interest to follow Fibiger along the laborious path of his research. The first idea of his discovery, which was to make his name celebrated the world over came to him in 1907: he recorded in three mice in his laboratory (originating from Dorpat), a tumour, unknown until that time in the stomach; in the centre of the neoplasm he noted the presence of a worm belonging to the family of Spiroptera.

Fibiger did not succeed at first in proving a relationship existing between the formation of the neoplasm and the worm. The attempts to provoke a cancer in healthy mice by making them ingest neoplastic tissue from diseased mice, and containing worms or eggs, failed completely. Fibiger then had the idea that perhaps this worm, like many others, underwent part of its evolution from an egg to an adult individual in another animal, which served as an intermediate host. After numerous and vain attempts to find again mice attacked by the tumours seen in 1907 – he unsuccessfully examined more than 1000 animals – Fibiger eventually discovered in a sugar refinery in Copenhagen mice who exhibited in considerable numbers the type of tumour he was seeking; in these tumours he found once again the worm he had observed in 1907. The factory was at this time infested with cockroaches, and Fibiger was then able to establish that the worm in its evolution used these cockroaches as intermediate hosts. The cockroaches ingested the excreta of the mice, and with them the eggs of the worm. These developed in the alimentary tract of the cockroaches into larvae, which, like the trichina, were distributed into the muscles of the insects where they become encapsulated. The cockroaches were in their turn eaten by the mice and in the stomach the larvae transformed into the adult form.

By feeding healthy mice with cockroaches containing the larvae of the spiroptera, Fibiger succeeded in producing cancerous growths in the stomachs of a large number of animals. It was therefore possible, for the first time, to change by experiment normal cells into cells having all the terrible properties of cancer. It was thus shown authoritatively not that cancer is always caused by a worm, but that it can be provoked by an external stimulus. *For this reason alone the discovery was of incalculable importance.*

But Fibiger's discovery had a still greater significance. The possibility of experimentally producing cancer gave to the particular research into this illness an invaluable and badly needed method, lacking until this time, allowing the elucidation of some of the obscure points in the problem of cancer. Fibiger's discovery also gave remarkable impetus to research. Whereas research had, in many respects, entered upon a period of stagnation, Fibiger's discovery marked the beginning of a new era, of a new epoch in the history of cancer, to which the fruitful research made by him gave fresh vigour. From his discoveries we have continued to march forward and have gained valuable ideas as to the nature of this illness.

It is thus that Fibiger has been and will remain a pioneer in the difficult field of cancer research. «To my mind», says the famous English expert on cancer, Archibald Leitch, to name only one of the numerous critical commentators on Fibiger's research, «Fibiger's work has been the greatest contribution to experimental medicine in our generation. He has built into the growing structure of truth something outstanding, something immortal, *quod non imber edax possit diruere*.» It is for this immortal research work that Fibiger is today awarded the Nobel Prize for Medicine for 1926.

Johannes Fibiger, honoured colleague. You have used the skill of your mature years to search for the cause of cancer. Thanks to penetrating observation, to conscientious and indefatigable work, you have succeeded in giving us convincing facts in place of unauthenticated hypotheses. You have thus, in a field of singular importance, enriched with new knowledge the sphere of medical research. You have at the same time given to the study of cancer a method for resolving points which are still obscure. You have stimulated this study as few others have done; you have drawn to its structure new workers who build on your foundations. We may perhaps hope that the day will come when we shall understand the problem of cancer in its entirety. If, on that day, we look back along the hard path we have travelled, your name will shine among the greatest, and you will remain a pioneer and a forerunner. The Staff of Professors of the Caroline Institute has decided to award you the Nobel Prize for Medicine for 1926 for your share in the annals of medicine. In expressing the warm congratulations of the Institute, I have the great honour of inviting you to present yourself before the King to receive your prize from his hands.

JOHANNES FIBIGER

Investigations on *Spiroptera carcinoma* and the experimental induction of cancer

Nobel Lecture, December 12, 1927

Ladies and Gentlemen. Now that it is my honour to address the Swedish Society of Medicine, I should like to begin by expressing the debt of gratitude I owe to this Society.

In 1913, some six months after I had published my first work on *Spiroptera carcinoma*, the Swedish Society of Medicine did me the honour of making me its member. This was the first recognition of this kind which came to me after the appearance of my work, and I would like to mention again today, here in the home of the Society, the great pleasure which I felt then and which I still recall with sincere gratitude.

Before I go on to relate the principal results of my investigations on *Spiroptera carcinoma* and the experimental induction of cancer, I would like to be allowed to offer the Nobel Foundation and the Staff of Professors of the Caroline Medio-Surgical Institute my most sincere and humble thanks for the great honour they have done me by finding my experiments worthy of the Nobel Prize in Physiology or Medicine.

It is possible to trace back attempts at supplementing clinical and anatomical studies of cancer by means of experimental work on the disease's origins, development and dissemination in the organism over a great number of years; not only the first, primitive experiments made over 150 years ago, but many more recent ones met with only negative results up to a short time ago.

Reports by Hanau and Morau, published during the years 1889–1894, contained the earliest description of successful transplantation of cancer from rats and mice to animals of the same species, but it was only at the dawn of this century that work done by Loeb, Jensen, Borrel, Bashford, Ehrlich, Haaland, Murray, and others (especially the classic experiments carried out by C. O. Jensen) ushered in the first experimental phase in cancer research, the era of transplantation experiments. From this time on the study of the cancer diseases could justifiably be said to belong to the field of experimental pathology.

Although a great deal of new and useful experimental work on transplantation has been done and is still being done, it must be admitted that on the whole it has not come up to early expectations. The development of cancerous tumours in one animal to which particles of a cancer which has occurred spontaneously in another have been transferred, is no more than the continued development of diseased tissue already in being. In fact, the study of transplanted neoplasms really only provides us with a possibility of closer investigation of the development, and conditions for development, of existing cancer tissue; it does not enable us to undertake any of the most important tasks in cancer research – an explanation of the original causes and the conditions for the onset of cancer, and of the processes surrounding the earliest beginnings of the disease. Moreover, the significance of transplantation experiments has decreased since the successful results of tests on immunization against transplanted cancer have shown themselves to be invalid for spontaneously developed neoplasms.

It follows that experimental studies must, for our purposes, necessarily be carried out on cancer growths which are not transplanted fully established from one animal to another, but which are induced experimentally and then develop in normal tissue in previously healthy animals.

In other words the problem of inducing cancer would have to be solved before the disease could be made the subject of the kind of experimental work which has provided important results in the study of the pathology of other diseases.

Recent attempts at solving this problem, which has occupied numerous workers in vain for a great number of years, have made use of methods based on the three famous theories of the origins of cancer: Virchow's irritant theory, Cohnheim's embryo theory, and the theory which ascribes cancer to *parasites*.

But up to recent times neither the constant application of chemical or physical irritants, nor the introduction of embryonic tissue, nor the implantation of various kinds of microbes into healthy animals produced any result at all; it was only around 1910 that there were reports of experiments in which the development of cancer was observed in isolated animals. Experiments carried out by Clunet, Marie and Raulot Lapointe, for example, produced sarcoma in two rats which had been exposed to X-radiation; Askanazy had, in isolated cases, observed the development of cancer in rats inoculated with embryonic tissue; and finally we must mention the sarcomatous fowl tumours first described by Fujinami and Inamoto, and by Peyton Rous, which could be produced by filtrates from tumour tissue, and which have generally come to be regarded as due to an invisible virus. There was doubt, however, as to whether these neoplasms could be considered as true cancers, and were really identical to normal sarcomatous tumours – and this doubt has still not been finally removed.

The first method which succeeded experimentally and systematically in bringing about a true epithelial carcinoma in healthy animals was the introduction of a Nematode, *Spiroptera neoplastica* or *Gongylonema neoplasticum*, into piebald rats, and it is as I mentioned the principal results of these experiments which I now have the honour of describing to you.

The starting-point for these studies was my finding in 1907, in the stomach fundus of three wild rats captured originally in Dorpat, of extensive papillomatous tumours which virtually completely filled the stomach and emerged from the cardiac region which was lined with pavement epithelium (Fig. 1; *cf.* normal rat's stomach, Fig. 2). Under the microscope the epithelium showed peculiar formations which were reminiscent of a section through egg-containing parasites, particularly Nematodes (Fig. 3), and after reconstruction and serial section it became possible to prove that the epithelium did in fact contain such parasites. This was later confirmed definitely by the separation

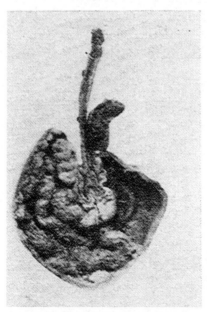

Fig. 1. Massive papillomata caused by *Spiroptera neoplastica* in the stomach fundus of a wild rat from Dorpat (natural size).

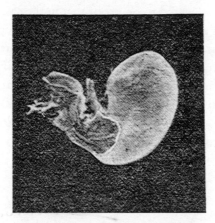

Fig. 2. Normal rat's stomach (natural size).

of isolated worms. My thoughts naturally turned to the possibility of these parasites having been the cause of the neoplasm developing – a supposition which was all the more justified since Borrel had already asserted in 1906, and subsequently, basing himself upon the presence of Helminthes and other types of animal parasites found in tumours in rats and mice, that such macro-parasites should be considered as having considerable significance in the genesis of the tumours. Haaland, too, had reported observations which could be used in support of this assumption; but the strongest argument for the pathological role of Helminthes in the development of cancer lay in the frequent appearance of cancerous neoplasms during schistosomiasis in the

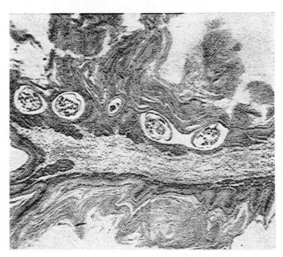

Fig. 3. Spiroptera embedded in the fundus epithelium of a brindled rat.

human bladder. There appeared to be proof that in the case of these tumours the presence of worms in the cancerous tissue was neither coincidental nor ascribable to a secondary invasion. In order to investigate whether the tumours in the stomach fundus of my three rats were actually due to a Nematode, and to contribute to a discussion whether the transmission of such parasites could result in the experimental inception of the development of a cancer, I set up a series of experiments of which I am only able here to give you the main features.

An attempt to demonstrate papillomata or worms of the type sought in

Fig. 4. *Periplaneta orientalis.*

Fig. 5. *Periplaneta americana.*

other rats came to naught, since investigation of the fundus in nearly 1,200 wild rats and laboratory animals gave only negative results. Negative results were also obtained when I tried to induce neoplasms by means of papilloma tissue introduced into healthy rats. Not even feeding the rats on tumour tissue produced positive results, and no tumours developed in rats kept for protracted periods (up to one year) in uncleaned cages previously occupied by diseased animals. There seemed, therefore, to be hardly any question of the direct transmission of Nematodes. This left the possibility of the Nematodes being transmitted through an intermediate host, in which the embryo-containing egg could undergo further development.

The literature was now found to countain a reference in 1878 to Galeb having found a Nematode, *Filaria rhytipleuritis*, in the stomach of rats. This had previously been discovered as a parasite in the fatty bodies of the common cockroach, *Periplaneta orientalis* (Fig. 4), and after three rats had been fed on cockroaches infected with this Nematode, Galeb was able to find worms in their stomachs, although no pathological change was evident.

Accordingly, I started on experiments in which rats were fed on cockroaches of the same species as those used by Galeb; these, too, were unsuccessful, the stomachs of rats captured in the same locality as the cockroaches containing neither worms nor papillomata. Results were only obtained when I changed over to experiments on rats which had lived in a sugar refinery where there were large numbers of *Periplaneta americana*, a species cockroach very seldom found in Europe (Fig. 5). Nematodes of the type under consideration were found in the stomachs of no less than 40 out of 61 wild brown rats from this refinery, and in 18 of these there were signs of pathological changes in the stomach fundus. In nine cases these were in the form of very extensive papillomatous neoplasms.

I began on fresh feeding experiments, now using cockroaches of the *Periplaneta americana* type from the sugar refinery, and these gave positive results. Nematodes of the type I was seeking appeared in 54 out of 57 rats fed on this species of cockroach, and of these 37 had stomachs exhibiting epithelial proliferation and papillomatous changes. Seven of the rats had large, papillomatous tumours.

I had thus been successful in tracing Nematodes and papillomatous tumours of the type under consideration in rats, and this had been achieved by feeding the animals on cockroaches from a given locality so as to infect healthy rats with Nematodes and to bring about the formation of neoplasms in their stomachs.

Fig. 6. Mature eggs of *Spiroptera neoplastica* (× 280 approx.).

Fig. 7. Coiled Spiroptera larvae in cockroach muscles.

Further research now showed that the larvae of the Nematodes were not, like the Filaria of Galeb, contained in the fatty bodies of the cockroaches but in the cross-striated muscle, and that their development was on the following pattern: after the infected cockroaches (or only their muscles) are eaten by the rats, the larvae are liberated and these invade the upper parts of the rat's alimentary canal, which is lined with pavement epithelium – the mouth lining, tongue, oesophagus and stomach fundus. Here they develop further, reach maturity and after about 45 days start to lay eggs, which are surrounded by a double membrane and contain embryos (Fig. 6). The eggs are evacuated with the rat's faeces. When a cockroach feeds on rat faeces containing these eggs, or on the Nematode eggs by themselves, it liberates the embryos which then penetrate into the cockroach's muscles and continue to develop. After 5–6 weeks they appear as Trichina-like, spirally-coiled larvae encysted in a thin capsule (Fig. 7). This completes the Nematode's development in the host and intermediate host.

Research work carried out by Hjalmar Ditlevsen, a member of the Staff of the University Zoological Museum, indicated that these Nematodes did not match the Galeb worm, but had rather to be added to the Spiroptera as a new species which had not so far been described. It was given the name *Spiroptera neoplastica*, which has since been changed to *Gongylonema neoplasticum*. In its fully developed form the male is $\frac{1}{2}$–1 cm long, with a diameter of 0.1–0.16 mm (Fig. 8). The female is 4–5 cm long with a diameter of 0.2–0.25 mm (Fig. 9).

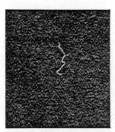

Fig. 8. *Spiroptera neoplastica* – male (natural size).

Fig. 9. *Spiroptera neoplastica* – female (natural size).

The presence of the Nematode in cockroaches and rats from the sugar refinery which I have mentioned, which had earlier received consignments of raw material from the West Indies, suggested that it must be a species coming originally from the tropics; when tests carried out on rats and cockroaches gathered in what used to be the Danish West Indies succeeded in showing the Nematodes, and when research done by others had the same result, it strengthened me in my belief that their proper habitat is in tropical countries from which they are brought into Europe with rats or cockroaches.

A fully developed papilloma in the stomach fundus was not only observed, as I have described, in wild rats from the sugar refinery, but also in piebald laboratory animals which were fed with Spiroptera-containing cockroaches in order to induce tumours experimentally. Five of these rats developed, in addition to papillomata, quite typical invasive squamous-celled carcinoma (Figs. 10, 11, and 12) which in two of them was accompanied by metastases; in one animal this affected the lung (Fig. 13), while in the other a lymph gland was involved (Fig. 14).

The way was now open for further research into the possibility of the planned, systematic inducement and close study of neoplasm and the development of cancer, and experiments along these lines were begun in the fol-

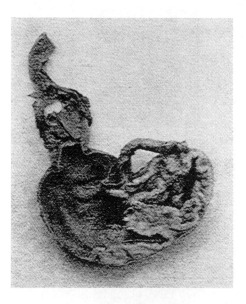

Fig. 10. *Spiroptera carcinoma* in stomach fundus of brindled rat fed on *Periplaneta americana* (natural size).

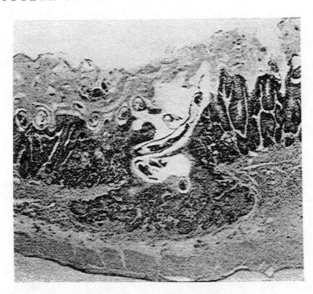

Fig. 11. Part of stomach fundus wall of brindled rat fed on Spiroptera, showing Spiroptera and heterotopic and incipient invasive epithelial growth (highly magnified).

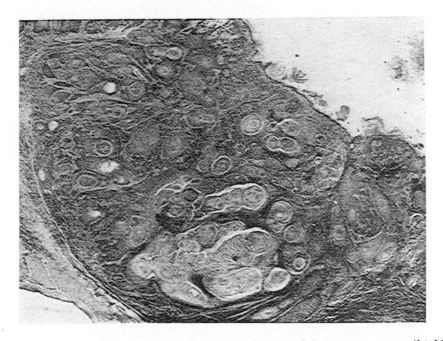

Fig. 12. *Spiroptera carcinoma* in fundus of brindled rat fed on *P. americana* (highly magnified).

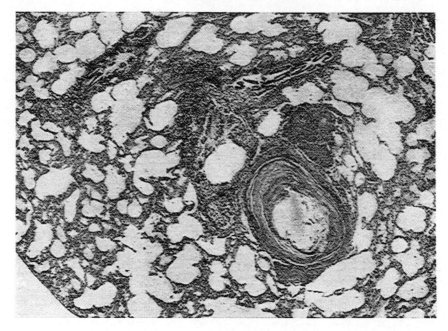

Fig. 13. Lung metastasis from *Spiroptera carcinoma* of stomach in brindled rat (highly
magnified).

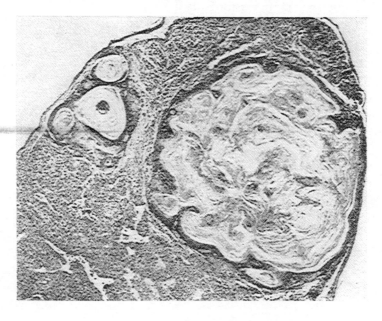

Fig. 14. Metastasis in retroperitoneal lymph gland from *Spiroptera carcinoma* of stomach
in brindled rat (highly magnified).

lowing year. Various species of rats and mice were used, most of them being piebald laboratory animals. The Gongylonema were transmitted partly by feeding the rats on cockroaches which had been infected by feeding on rat faeces containing the eggs, partly by feeding them only cockroach muscle containing larvae, and finally by means of direct injection into the rat's stomach of Gongylonema larvae prepared from muscle.

It was observed that the Gongylonema could not only live parasitically in rats and mice, but could also be transferred to the pavement epithelium-lined section of the upper alimentary canal in guinea pigs, rabbits, hedge-hogs, and squirrels. The common cockroach and the flour-mite, *Tenebrio molitor*, may serve as intermediate host as well as *P. americana*. My tests on the Gongylonema were carried out mainly on their natural host, *P. americana*, which were bred at the Institute for this purpose.

The experiments showed that in the stomach of a rat infected with a single Gongylonema, or with a small number, only epithelial hyperplasia or a slight inflammation occurred. If, on the other hand, the infection in-volved a large number of Gongylonema, then there was pronounced prolif-eration of the epithelium and of the connective tissue of the mucous mem-branes, accompanied by heterotopic growth at depth and the formation of papillomata. In the most pronounced cases there were deep-seated epi-thelial crypts, which might affect the entire stomach wall, and massive papil-lomata which could completely block the stomach lumen. In a number of animals squamous-celled carcinoma developed as well, typified partly by pronounced cornification with atypical cell formation and bulbous epithelial formations, partly (and particularly) by invasive growth penetrating the mucous membranes and muscularis mucosae and reaching down into – or through – the submucosa.

Desquamative inflammation and epithelial hyperplasia occurred in the mouth, oesophagus, and on the tongue. Characteristic carcinoma may be found on the tongue (Fig.15, normal tongue; Figs.16–18, *Spiroptera carcinoma* of the tongue).

The *Spiroptera carcinoma* is accompanied by metastases, localized mainly in the lungs; here they are, however, frequently only to be found by means of microscopic examination of serial sections. Metastases were discovered in six out of 33 brindled rats whose lungs were examined in this manner; in only one instance have I observed a gland metastasis. The structure of the metastases resembles exactly that of the primary growth, they never contain either Spiroptera or eggs, and it is only the proliferation of the cells without any contribution by the parasite which indicates that metastases (in the usual sense of the term) are involved.

By and large, the stage of development of the carcinoma is proportional to the length of life of the rats after the transmission of the Spiroptera. Among piebald laboratory rats the earliest observation of carcinoma was 45 days later.

Fig. 15. Normal rat's tongue.

Fig. 16. *Spiroptera carcinoma* in rat's tongue.

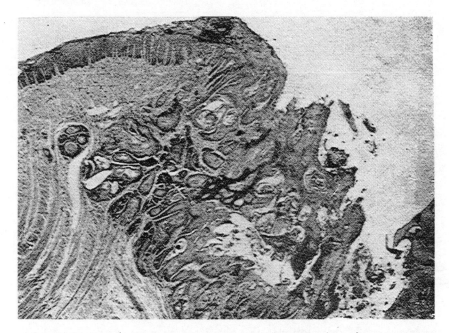

Fig. 17. *Spiroptera carcinoma* in rat's tongue (× 24).

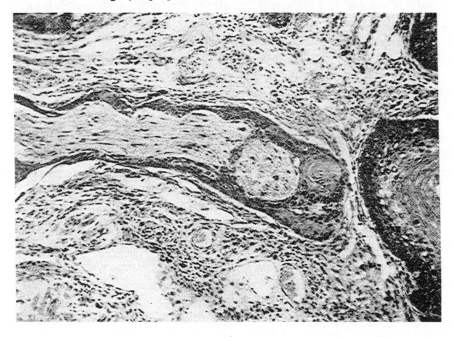

Fig. 18. *Spiroptera carcinoma* in rat's tongue, showing encroachment of the carcinoma into the lymphatic space surrounding a nerve (× 130).

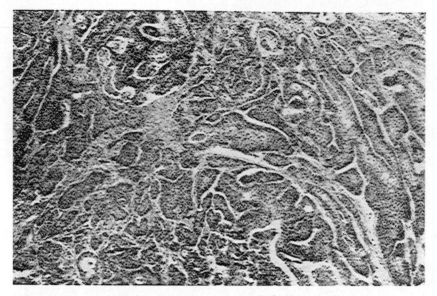

Fig. 19. Primary *Spiroptera carcinoma* in stomach fundus of white mouse (× 50).

The transfer of Gongylonema also produced changes in wild rats (*M. de-cumanes*), black rats (*M. rattus*), white mice, grey mice (*M. musculus*), and field mice (*M. sylvaticus*), and these were in most instances of the same type. In white mice I found instances where the carcinoma had completely pene-trated the stomach wall (Fig. 19) and in one case had led to its perforation. In one mouse the lungs contained a metastasis the size of an orange pip, while in another enormous metastases were found in the peritoneum and in the abdominal glands (Fig. 20).

I should finally like to emphasize that *Spiroptera carcinoma* is transplantable, this having been done during the study of the disease in a mouse (Figs. 21 and 22). It was possible to transplant the tumour through four generations in the course of a year.

We may thus summarize the results of the experiments I have described in the following terms: the carcinoma produced by *Gongylonema neoplasti-cum* is, in its structure, a typical epithelioma, or squamous-celled carcinoma; it is invasive and destructive in its growth, produces metastases and may be transplanted. Carcinoma of the stomach was produced experimentally in brindled rats in more than 100 cases, and carcinoma of the tongue in seven cases. As an illustration of the incidence of carcinoma of the stomach in the piebald rats, I might mention that out of 102 of these animals which sur-vived the transmission of Spiroptera by 45 days or longer, more or less pronounced stomach carcinoma developed in 54 – that is to say, rather more than half.

I shall return to the appearance of carcinoma in other rats and in mice at a later stage.

Before I go on to deal with the results of these experiments in greater detail, it ought first to be mentioned that this research gave experimental proof of the correctness of Borrel's theory that worms, like certain other animal parasites, possess the ability to produce neoplasms. Research similar to mine later provided further proof of this. In 1921, for example, the work of Bullock and Curtis (whose experiments covered a very wide scope) showed that the sarcoma which Borrel and others had described as occurring in the liver of rats when it has been a site for the cysticercus of *Tænia crassicollis* may be induced experimentally by feeding rats with the eggs of this Tænia. It was these experiments which provided us with the possibility of a systematic, experimental inducement of *sarcoma*.

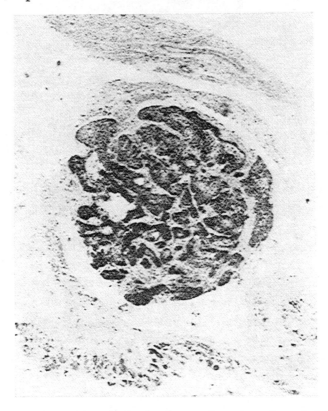

Fig. 20. Peritoneal lymph gland metastasis from the tumour in Fig. 19.

Fig. 21. Mouse with subcutaneously transplanted *Spiroptera carcinoma*, 84 days after transplantation – 1st generation.

In very recent times (1925) Yokogawa has published details of experiments in Formosa which have led to the discovery of a new type of Gongylonema known as *G. orientale*, which from the morphology viewpoint closely resembles *G. neoplasticum*; this, again, lives as a parasite in the stomach of rats, and makes use of cockroaches (*P. americana*, *P. australasiæ*) as an intermediate host during its development. Feeding rats on cockroaches whose muscles contained the larvae of the parasite showed that these Gongylonema, too, can bring about stomach carcinomas.

This therefore removes any shadow of a doubt that the Helminthes must be included among the causative agents of cancer, and the quite considerable number of studies and observations in which Nematodes and other Helminthes were shown to be present in various kinds of benign and malignant neoplasms must now be considered in a rather different light from formerly; although the possibility of fortuitous coincidence cannot, of course, be entirely ruled out, there are nevertheless grounds for suspecting that in such cases the parasites can have a pathogenetic significance. It would take too long to go here into all the observations of this kind which have been made, and I shall therefore mention only the Helminthes which must be assumed to play a greater or lesser role in the development of tumours and cancers in human beings.

As we have already mentioned, *Schizostomum hæmatobium*'s aetiological importance in the development of cancer of the bladder must be considered as proven. Nor can it be doubted that other Trematodes, such as *Opistorchis felineus*, *Schizostomum japonicum* and *Clonorchis sinensis* can, in certain cases, bring about primary carcinoma of the liver, and that *Schizostomum Mansoni* can be the cause of polyps and carcinoma in the colon.

In a number of cases, too, it has been found that the presence of *Echinococci* is accompanied by a primary carcinoma of the liver.

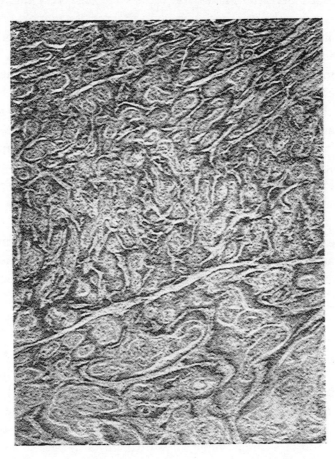

Fig. 22. Transplanted *Spiroptera carcinoma*–2nd generation (× 50).

In connection with the significance of the Nematodes, we must mention the statement by the American cancer research worker Ewing, that he has fairly frequently found Trichina in cases of carcinoma of the tongue; in Europe, too, there have been a whole series of reports of carcinoma in organs lying close to trichinous muscles in humans suffering from chronic trichinosis, in particular the breasts.

So far as can be ascertained, *Gongylonema neoplasticum* is never found in humans. On the other hand, four reports from America and one, or possibly two, from Italy indicate that another type of Gongylonema, probably G. *pulchrum*, can live in the mucous membrane of the human lips and tongue; this type is normally parasitic in the tongue and oesophagus of the pig and uses dung-beetles (*Aphodius* and others) as an intermediate host. In all the patients under observation, however, there were only very slight changes in the mucous membrane; there were no instances of carcinoma or similar processes, and there was likewise no cancer of the oesophagus in pigs infected with this Gongylonema.

This seems therefore to suggest that the Helminthes which must be assumed to result in the development of cancer are quite few in number.

If we wish to come to a conclusion on this question we must, however, keep in mind the fact that experiments on *Gongylonema neoplasticum* have (as I shall explain later) proved that this parasite may leave the cancer tissue after producing the carcinoma, which then still continues to grow. In cases where tumours have been brought about by Helminthes it consequently cannot always be assumed that these will be present; in such cases the original aetiological role of the Helminthes will be masked. This poses the possibility of a greater number of malignant neoplasms being due to such parasites than can be proved to be the case, and the part they play in the onset of tumours may in fact be greater than appears from our observations.

Even bearing this possibility in mind, we are nevertheless not justified in ascribing to the Helminthes any extensive aetiological role in the pathology of cancer in Man.

Nor can the endemic appearance of *Spiroptera carcinoma* among rats coupled to our knowledge of the presence and effects of the Gongylonema in human beings serve as a foundation for the theory, put forward in recent times, that the comparatively high incidence of cancer in the inhabitants of districts of buildings infested with rats and cockroaches may be caused by the transmission of a carcinogenic virus through *Gongylonema neoplasticum* or other Nematodes. Research on Spiroptera, again, does not give unqualified support for the theories that cancer is in general ascribable to macro-parasites or microbes.

All things considered, the present state of our knowledge allows us to allot the Helminthes only a modest place among the causes of neoplasms among humans, a fact that I stressed in my first work back in 1913.

In the course of the research work on *Spiroptera carcinoma*, it became possible for the first time to induce typical, metastasizing carcinomas systematically and at will. This provided *experimental proof* that the start of a cancer can, in agreement with the theory of Virchow, be brought about by external, exogenic influences, and lent support to experiments on the effects of long-term irritants of other kinds.

Experiments of this nature were first undertaken by Yamagiwa and Ichikawa, who chose coal-tar as the irritant; the carcinogenic effects of this had been observed in clinical practice, but in previous experiments had yielded only negative results. Works by Yamagiwa and Ichikawa over the years 1915–1918 however, proved, as a principal result, that a monthly application of coal-tar to the ears of rabbits was able to produce a skin cancer. Tsutsui reported, in 1918, that painting tar on the skin of the mouse would produce the same effect, and these experiments were taken further by Fibiger and Bang who were able, in 1920, to corroborate and supplement Tsutsui's results. This laid the foundation for a method of inducing cancer at will which is now in use in laboratories all over the world.

As I continue my account of the information which the study of *Spiroptera carcinoma* has contributed to our knowledge of the pathology of cancer, I shall add and compare some of these to the results which came from research on experimental tar cancer.

Before doing so, however, I should mention that the effect of the Gongylonema is generally taken to be due to their constant production of toxic substances; I originally suspected this through the analogy with the well-known fact that a number of the Helminthes exude pathogenic secretions.

A continuous excretion of this sort is not, however, essential for a carcinoma, once it has developed, to continue to grow. As tests showed, the growth of the fully developed carcinoma continues irrespective of whether the Gongylonema leave the tumour tissue or not; this is exactly analogous both with the continued growth of experimental tar cancer after the cessation of painting-on the tar, and with clinical observations that the development of chimney-sweep's cancer and cancers caused by aniline and X-rays does not come to a halt because the patients are removed from the influences which have given rise to the disease.

The fact that in a number of cases the *Spiroptera carcinoma* could be produced both on the tongue and in the stomach provided experimental proof that multiple carcinomas in the alimentary canal can be due to the formation of a primary cancer in different sites, stemming from the same cause but otherwise entirely independent. The experiments furthermore made it possible to confirm that in its earliest stages a carcinoma develops pluricentrically, starting from separate and well-defined small groups of cells or perhaps even from individual cells, and that its continued development is through *expansive*, *non-appositional* growth, in line with the view put forward earlier by Ribbert and others to explain the development of a carcinoma.

No proportional relationship was found between the appearance of carcinoma and the degree of development of epithelial hyperplasia and heterotopic growth at depth. In rats which had survived the transfer of Spiroptera by a long period the wall of the stomach could be the site of deeply penetrating cell masses or large retort-shaped epithelial cysts without a carcinoma having developed, and contrariwise carcinoma might develop in other animals starting from epithelium exhibiting only moderate and non-heterotopic hyperplasia. Again, there was no correlation to be found between the formation of a carcinoma, the development of inflammation and papillomata. Violent development of papillomata was not necessarily accompanied by carcinoma, and carcinoma could develop without papillomata.

Accordingly neither heterotopic benign epithelial proliferation nor papillomatosis need necessarily be preliminary stages of carcinoma; development of carcinoma follows as a *characteristic process* upon the hyperplastic growth of epithelium, irrespective of whether this is slight or pronounced, heterotopic or not, and whether at the same time as the epithelial hyperplasia there is substantial inflammation, proliferation of connective tissue and papillomatous changes, or whether there is no papilloma and only slight inflammation. This is not so say that there is never any relation between inflammation and carcinoma, but only that inflammation is not an essential factor in the genesis of the carcinoma and cannot unequivocally be regarded as a process which is absolutely essential for the onset of a cancer. If inflammation is present, it can most probably be seen as an accompanying phenomenon in the development of the cancer or as a secondary process which may even, particularly when it is far-developed, be regarded as having a defensive character and in its most pronounced forms (e.g. in mixed infection) can bring the cancer to decompose almost completely; this may be observed in cases of experimental tar cancer in the mouse. As you will know, Murphy and other workers even regard the lymphocyte reaction during cancer as being a significant part of the processes involved in general in the resistance to cancer. This is, however, a problem which can hardly be considered as finally resolved.

The disproportion between heterotopic epithelial hyperplasia, inflammation and papilloma formation on the one hand and the development of carcinoma on the other appeared especially clearly in the effects of the Gongylonema upon wild rats (*M. decumanus* and *M. rattus*), white mice, domestic mice (*M. musculus*) and field mice (*M. sylvaticus*). These can all survive the transfer of Spiroptera for as long a period as the brindled laboratory rats, and frequently even longer. The development and biological state of the Gongylonema, furthermore, seem to be entirely unaltered during their parasitic existence in these various rodents, in whose stomachs they give rise to substantial changes of exactly the same sort as those observed in brindled rats, and sometimes more extensive still.

Nevertheless *Spiroptera carcinoma* develops much less frequently in these animals than in the brindled rats, among whom – as I have said – stomach cancers developed in more than half of those who survived the transfer of Spiroptera by 45 days or longer. During experiments undertaken in collaboration with C. Krebs carcinoma was found in the stomach on only one out of 38 black rats (*M. rattus*) and only 11 out of 34 brown rats (*M. decumanus*). Among 56 white mice who, like all these rats, survived the infection with the parasites for at least 75 days and often very much longer, only three were found to have carcinomas. Despite numerous attempts, it was not possible to induce *Spiroptera carcinoma* in field mice and domestic mice.

Thus the development of a cancer following the same exogenic influence, and under the same conditions, does not always occur within the same period of time in all animals of the same species; what is more, it occurs with varying frequency among animals of different, though closely related, species.

This gave an experimental indication of a varying individual-and species-predisposition towards cancer, which was corroborated by later research on *Cysticercus sarcoma* and experimentally induced tar cancer. I will content myself here with mentioning that (so far as I know) it has only been possible to produce tar cancer on the skin of the rat in one single case, despite numerous attempts in various laboratories.

Tests on this made by Poul Møller in my Institute yielded only negative results, although the tar used for these would produce cancer in all the mice surviving by one month a 4-month course of painting. On the other hand, all the six most long-lived rats (who died only some 10–15 months after the course of painting with tar) developed a *primary carcinoma of the lung*.

This notion of a predisposition to cancer therefore not only applies to a differing individual-, race- or species-predisposition to develop the same form of cancer after the same carcinogenic effect; we must assume that it can vary within the same species as regards differing carcinogenic factors. Tar cancer develops quite easily among mice who will only contract *Spiroptera carcinoma* with the greatest difficulty, although both tar and Gongylonema produce a carcinoma of identical structure. The experiments of *Spiroptera carcinoma* also provided an indication of the existence of a varying predisposition among the organs, since, although I was never able to observe an incipient or extant carcinoma in the oesophagus of either rats or mice, the epithelium of this organ does not differ from that of the stomach fundus, and in the majority of the animals it contained numerous Gongylonema and was often the site of a pronounced and sometimes heterotopic epithelial hyperplasia. Yokogawa obtained exactly the same results with rats infected with *Gongylonema orientale*.

On the whole it seems, as Bashford had suspected as far back as 1912, that there may very likely be special predispositions, but no common general predisposition to all forms of cancer, to cancer in all organs or to all carcinogenic influences.

At the present time it is impossible either to make an overall, confident assessment of the influences to which these predispositions must be attributed, or to weigh up the importance of the special factors, the effects of which upon susceptibility towards cancer are discussed and studied in recent works: endocrine secretion, the significance of the spleen, the composition and salt, lipoid or vitamin content of the diet, etc., the effects of pregnancy, various diseases, the iso-haemagglutinin content of the blood, inherited and constitutional characteristics of various kinds, and so on.

I will discuss only one factor: old age, the considerable importance of which as a generally predisposing characteristic towards all forms of cancer has long been a universally accepted doctrine. This doctrine has not, however, found confirmation in a series of experimental works. Even my very early experiments indicated that *Spiroptera carcinoma* appeared with the same facility in both young and old rats, and later work done in my Institute by

Fridtjof Bang (as well as experiments carried out elsewhere) proved that in like conditions young mice contracted tar cancer just as frequently and just as rapidly as did old ones. It is possible to point to clinical experience, too, which bears out these experiments showing that it is the time at which the effect of the carcinogenic influence begins (and the nature, duration, and intensity of this effect) which determines the time of life at which cancer develops, rather than any particular age-predisposition.

As you will know, clinical experience of chimney-sweep's cancer and aniline cancer has made it clear that the appearance of a demonstrable and morphologically typical cancer need not be the immediate consequence of a carcinogenic influence – it may only appear after this has ceased to have any effect and after a certain period (which may be quite lengthy) has elapsed; conclusive proof of the existence of such a period of latency was first provided by research work on tar cancer, and in particular that of Fridtjof Bang and of A. Leitch.

I have attempted briefly to outline the most important results obtained from experiments on *Spiroptera carcinoma*.

Of the later methods for the experimental induction and study of cancer, that of painting mice with tar has so far shown itself to be the most suitable. We still do not know the specially active ingredient (or ingredients) in this highly complex substance, but we have incontrovertible proof that all types of coal-tar possess this property of inducing cancer in rabbits and mice. Tar has also shown itself, however, to be capable of causing malignant neoplasms in the lungs, breasts, testes, stomach, and other organs, and it will thus be possible to make use of the experimental induction of tar cancer in the pathological study of a number of forms of the disease.

In research on therapy, too, experimental work will provide an invaluable, and indeed indispensable, foundation. As an example of this I might mention a series of experiments carried out by Poul Møller and myself on the prevention of metastasis in cancer cases.

These were done on mice suffering from cancer of the skin, induced by tar painting. As an immunizing measure we employed subcutaneous injections of an emulsion made from sterile, living skin from a mouse foetus; earlier transplantation experiments had shown this to be effective in preventing the development of transplanted tumour tissue. The tests covered in all 293 mice, kept under identical conditions, who had a small area of the skin of the back treated every other day with the same coal-tar and in the same manner over a period of four months. The skin emulsion was injected into 156 of these animals from 2–7 times, while the remaining 137 mice acted as a control. Skin cancer was contracted by 127 (81%) of the treated animals and 102 (74.4%) of the controls, but metastasis occurred in only 38 (i.e. approx. 30%) of the treated mice as against 59 (about 58%) of the control animals.

This immunization treatment had thus, like earlier results, shown itself to be ineffectual in preventing primary cancer formation, although reducing the formation of metastases by about 50%.

There is consequently no doubt that the formation of metastases can be inhibited by this method, and I might add that after proper statistical checking these results have proved quite indisputable. I need not emphasize, therefore, that only further experiments covering a wider scope can show what methods can be evolved on the basis of our results, to serve as a therapeutic measure to combat cancer in Man. These methods will have to be free of the technical difficulties which affected ours; these included the fact that the tissue had not only to be homologous and sterile but also living, in order to be certain of avoiding the danger of hypersensitivity to cancer development which it had been shown could result from the injection of dead tissue. A detailed explanation of the processes by which metastasis is inhibited has still not been possible.

Other influences than those I have mentioned here have formed the subject of research by other workers in recent times, and have proved capable of inducing cancer. A point of interest in the study of occupational cancers is that it has been proved that carcinoma of the skin in mice can be produced not only by tar, but also by other products of the distillation and combustion of coal – soot, pitch, the paraffin oils and similar substances. Long-standing observations of experimentally induced X-ray sarcoma have been supplemented by recent reports of carcinomas from the same cause. The effect of traumata has been proved by tests of various kinds; of special importance is research showing the development of cancer in the gall bladder of the guinea-pig following on the penetration of gall-stone fragments or other foreign bodies into this organ.

Of even greater interest is the experimental work done by Carrel. I can only mention here the most important results of this, which show that by means of the inoculation of embryonic fowl tissue into fowls accompanied by the injection of extremely dilute solutions of tar, arsenic or indole (benzopyrrole) it is possible to induce rapidly growing and highly malignant, metastasizing and transplantable neoplasms identical to the Rous sarcoma; like the latter, these can be transmitted to healthy fowls by the inoculation of cell-free filtrates of the tumour tissue. Moreover, the filtrate is able, *in vitro*, to bring about the conversion of the blood macrophages into malignant tumour cells, and it has furthermore been shown by Fischer that a similar conversion can be brought about by the cultivation of spleen macrophages in a nutrient medium to which a small quantity of arsenic acid has been added.

Carrel's work has thrown new light on a number of problems. First and foremost, it is now difficult to uphold the theory that the Rous sarcoma must be attributed to an invisible virus, since sarcomas of exactly the same type were seen to be produced during these experiments by the effects of chemically unorganized compounds upon embryo tissue. The carcinogenic property of the tissue filtrate must therefore be assumed to depend upon the effects of a *cellular product*, possibly an enzyme.

These tests also provided support for further studies on the formation of malignant tumours following the inoculation of embryo tissue, and can also be seen as corroborating Cohnheim's theory. The conversion of the macrophages into tumour cells also suggests that neoplasms can develop from non-differentiated cells as well as embryo cells, and that foreign substances introduced into the organism can produce tumours in embryonic tissue; finally, it indicates that the formation of neoplasms does not depend solely upon local factors, but also on others originating from the organism generally.

It is however difficult at present to know how much importance can be attributed as a whole to the results of Carrel's work on the aetiology and pathogenesis of tumours, partly because the neoplasms he induced developed only from embryo rather than adult tissue, and that exclusively in fowls, and partly because in all instances they belong to the much discussed Rous type of sarcoma; the classification of this type alongside the sarcomas affecting mammals and human beings is problematical, and it is not recognized by many pathologists.

Recent work done by Askanazy has shown that the effect of arsenic is also able in the rat to foster the development of teratoids from inoculated embryonic tissue, and that there can be a simultaneous formation of sarcomas. If further experiments on mammals, involving entirely typical sarcomas and carcinomas, make it possible to obtain results matching those reached by Carrel, then the experimental study of tumours will have entered a new and fruitful era.

I will now close. The limited time at my disposal has permitted me to discuss only a few of the significant, principal results of modern cancer research, without mentioning either the vast number of other experiments which have been carried out, or all of the scientists who are working in this field. But what I have been able to tell you will have made it clear that the experimental induction of tumours in modern time has made it possible to realize that the causes of cancer include animal parasites and physical and chemical influences of various kinds, and that endogenic as well as exogenic factors must be regarded as playing a part in causing the disease.

Although even recent works attribute importance to microparasites and invisible viruses as causes of cancer, there has been no convincing evidence offered for the microbial origin of cancer in Man, and all theories of this kind must be regarded as based upon insufficient grounds.

The term cancer covers diseases of widely differing origins and of many different kinds, sharing the common feature of an uncontrolled, apparently autonomous, atypical and invasive growth of the cancer cells, with an ability to find a fertile site for development after metastasis in widely varying tissues in the organism. The anatomical and biological changes undergone by the cells in the course of their conversion into tumour cells, which give mature cancer cells their characteristic stamp, are far from well enough known; but the substantial improvements which have been made in the past few years in the methods of cultivating tissue *in vitro* have opened up new possibilities for studies of this kind, and have – as will be evident from what I have been telling you here – already given promising results. Warburg's work on the metabolism of cancer tissue, too, has meant a useful increase in our knowledge; I would mention only the very great importance which his experiments have taught us to attribute to glycolysis in the metabolism of the cancer cell.

The study of the manifold problems presented by cancer has, in recent years, seemed to offer many more riddles than were previously thought to exist; but the history of medicine has never known a period in which problems could be attacked in so many different ways as those made accessible today by the working methods now at our command.

Publications on Spiroptera carcinoma by Johannes Fibiger

1. Recherches sur un nématode et sur sa faculté de provoquer des néoformations papillomateuses et carcinomateuses dans l'estomac du rat, Académie Royale des Sciences et des Lettres de Danemark, 1913.

2. «Über eine durch Nematoden (Spiroptera sp. n.) hervorgerufene papillomatöse und carcinomatöse Geschwulstbildung im Magen der Ratte», Klin. Wochschr., (1913); «Undersøgelser over en Nematode (Spiroptera sp. n.) og dens Evne til at fremkalde papillomatøse og carcinomatøse Svulster i Rottens Ventrikel», Hospitalstidende, (1913).

3. «Untersuchungen über eine Nematode (Spiroptera sp. n.) und deren Fähigkeit, papillomatöse und carcinomatöse Geschwülste im Magen der Ratte hervorzurufen», Z.Krebsforsch., 13 (1913).

4. «Sur le développement de tumeurs papillomateuses et carcinomateuses dans l'estomac du rat sous l'action d'un ver nématode», Troisième Conférence Internationale pour l'Étude du Cancer, Bruxelles, 1913.

5. In colloboration with Hj. Ditlevsen: «Contributions to the biology and morphology of Spiroptera (Gongylonema) neoplastica sp. n.», Mindeskrift for Japetus Steenstrup, (1914).

6. «Weitere Untersuchungen über das Spiropteracarcinom der Ratte», Z. Krebsforsch., 14 (1914); «Fortsatte Undersøgelser over Spiropteracarcinomet hos Rotten», Hospitalstidende, (1914).

7. «Über Disposition der Ratten und Mäuse für die Wirkung der Spiroptera neoplastica», Zentr. Allgem. Pathol. & Pathol. Anat., 27 (1916).

8. «Undersøgelser over Spiroptera neoplastica's Indvirkning paa Mus», Foredrag paa Naturforskermødet i Kristiania, 1916.

9. «Investigations on the Spiroptera Cancer, III: On the transmission of Spiroptera neoplastica (Gongylonema neoplasticum) to the rat as a method of producing cancer experimentally», Kgl. Danske Videnskab. Selskab, Biol. Medd., I, 9 (1918).

10. «Investigations on the Spiroptera Cancer. IV: Spiroptera cancer of the tongue», Kgl. Danske Videnskab. Selskab, Biol. Medd., I, 10 (1918).

11. «Investigations on the Spiroptera Cancer. V: On the growth of small carcinomata and on predisposition to Spiroptera Cancer in rats and mice», Kgl. Danske Videnskab. Selskab, Biol. Medd., I, 11 (1918).

12. «Investigations on the Spiroptera Cancer. VI: A transplantable Spiroptera carcinoma of the mouse», Kgl. Danske Videnskab. Selskab, Biol. Medd., I, 14 (1919). «Sur la transmission aux rats de la Spiroptera neoplastica (Gongylonema neoplasticum)», Compt. Rend. Soc. Biol., (1920).
«Carcinome spiroptérien de la langue du rat», Compt. Rend. Soc. Biol., (1920).
«Sur l'évolution et la croissance du carcinome spiroptérien», Compt. Rend. Soc. Biol., (1920).
«Recherches sur le carcinome spiroptérien de la souris blanche et sur sa transplantabilité», Compt. Rend. Soc. Biol., (1920).

13. «On Spiroptera carcinomata and their relation to true malignant tumours; with some remarks on cancer age», J. Cancer Research, 4 (1919).

14. «Untersuchungen über das Spiropterakarzinom der Ratte und der Maus», *Z. Krebsforsch.*, 17 (1919).

15. «Recherches sur la production expérimentale du cancer chez le rat et la souris», *Bull. Assoc. Franc. Etude Cancer*, (1921).

16. «Le cancer spiroptérien et les autres cancers à parasites animaux», *Bull. Assoc. Franc. Etude Cancer*, (1923).

17. «Nieuwere onderzoekingen over den kanker», *Ned. Tijdschr. Geneesk.*, Eerste Helft, No. 14 (1923).

Biography

Johannes Andreas Grib Fibiger was born at Silkeborg (Denmark) on April 23, 1867. His father, C. E. A. Fibiger, was a local medical practitioner and his mother, Elfride Muller, was a writer. Fibiger gained his bachelor's degree in 1883 and qualified as a doctor in 1890. After a period of working in hospitals and studying under Koch and Behring he was, from 1891 to 1894, assistant to Professor C. J. Salomonsen at the Department of Bacteriology of Copenhagen University. While serving as an Army reserve doctor at the Hospital for Infectious Diseases (Blegdam Hospital) in Copenhagen from 1894 to 1897 he completed his doctorate thesis on « Research into the bacteriology of diphtheria ». He received his doctorate of the University of Copenhagen in 1895, and was subsequently appointed prosector at the University's Institute of Pathological Anatomy (1897–1900), Principal of the Laboratory of Clinical Bacteriology of the Army (1890–1905), and (in 1905) Director of the Central Laboratory of the Army and Consultant Physician to the Army Medical Service. After studying for some time under Orth and Weichselbaum, Fibiger was appointed Professor of Pathological Anatomy at Copenhagen University and Director of the Institute of Pathological Anatomy (1900).

Fibiger fulfilled a large number of official missions and took part in the direction of numerous institutions. He was First Secretary, and later President of the Danish Medical Society, Consultant to the Council of Forensic Medicine, member of the Planning Commission for the Construction of the Medical Institutes of the National Hospital; Vice-President, and later President of the Danish Medical Association's Cancer Commission, member of the National Radium Committee, member of the Administrative Council of the Rask-Ørsted Foundation, of the Northern Society to Promote a Biological Station in the Tropics, of the Pasteur Society; he was a founder-member and joint-editor of the *Acta Pathologica et Microbiologica Scandinavica*, co-editor of *Ziegler's Beiträge zur pathologischen Anatomie und zur allgemeinen Pathologie*, member of the International Commission for Intellectual Cooperation with Other Countries, representing his country at numerous congresses and meetings, and member of a great many academies and societies, both Danish and foreign. Fibiger was also Vice-President, and afterwards President, of «Die internationale Vereinigung für Krebsforschung», member of the Royal Academy of Science and Literature of Denmark, of the Swedish Medical Association, of the Finnish Medical Association, corresponding member of the «Association française pour l'Étude du Cancer», of the «Société de Biologie» of Paris, of the Helmintological Society of Washington, founder-member of «Van Leeuwenhoekvereeniging» for cancer study by experiment, honorary member of the Royal Academy of Medicine of Belgium and of the «Wiener dermatologischen Gesellschaft», member of the Royal Society of Physiography of Lund and of the Royal Society of Science of Uppsala, honorary doctor of the Universities of Paris and Louvain, etc. Fibiger was the winner of numerous prizes, among which should be mentioned the Nordhoff-Jung Cancer Prize and the Nobel Prize for Physiology or Medicine, 1927, for his work on cancer.

Fibiger died on January 30, 1928, at Copenhagen after a short illness (cardiac failure with multiple emboli and massive pulmonary infarcts; cancer of the colon: caecostomy), survived by his wife Mathilde, *née* Fibiger, whom he married in 1894.

Physiology or Medicine 1927

JULIUS WAGNER-JAUREGG

«for his discovery of the therapeutic value of malaria inoculation in the treatment of dementia paralytica»

Physiology or Medicine 1927

*Presentation Speech by Professor W. Wernstedt, Dean of
the Royal Caroline Institute*

Your Majesty, Your Royal Highnesses, Ladies and Gentlemen.

Fibiger won his laurels in the field of theoretical medicine, researching into the cause of a specific disease. Turning to the work that led Wagner-Jauregg to the list of Nobel Prize winners, we enter the field of practical medicine, or more exactly, the wide field where the means of healing diseases are sought. The disease in the treatment of which Wagner-Jauregg acquired such great merit is general paralysis, a mental disease which, on a syphilitic background, leads to a fatal idiocy and paralysis. It is therefore a very serious and moreover not uncommon disease.

Up to the arrival of Wagner-Jauregg, we were practically without any means of healing general paralysis, or even of influencing its course and outcome substantially and with any certainty. The inaccessibility of the paralysis to treatment, and its development leading as a rule directly to death within a few years even came to be regarded as a criterion whether the diagnosis of paralysis had been correct in cases where at first doubt had prevailed in this respect. It should be clear that whoever is successful in finding a means to eliminate such a disease, has thereby made an achievement of the greatest benefit to mankind. Wagner-Jauregg has performed such a deed and it is for this that he is to be rewarded today with the Nobel Prize for Medicine for the year 1927.

How then did Wagner-Jauregg proceed to heal the unfortunate victims of this terrible disease? There is a saying «one must expel evil with evil» that might aptly have been coined as a motto for his treatment of paralysis. He healed the mental patients by infecting them with another disease – malaria.

For a long time, ever since Hippocrates, it has been observed that every now and then, mental patients were healed or favourably influenced, when they were attacked by a fever. It was this ancient observation, which Wagner-Jauregg also made himself, that excited the idea in him, whether one could not obtain an effective method of treatment for chronic mental patients by infecting them with a febrile disease.

Even forty years ago, as a young lecturer at the University of Vienna, Wagner-Jauregg put forward his ideas in the professional press. At that time his proposals seemed to have attracted no attention, and he himself was for a long time prevented from putting them into practice. In 1917 the opportunity was first presented to him of realizing his ideas, in that in this year he injected nine persons suffering from paralysis with the infectious blood of malaria patients.

Wagner-Jauregg had not been deluding himself in his expectations. The infected patients developed malaria, their mental illness was favourably affected, and in three of the nine recovery was practically complete. The choice of the infecting disease which he had hit upon was also fortunate. The form of malaria (tertian fever) which he used is, if correctly treated, a relatively innocuous disease, which can always be cured by means of quinine treatment. It therefore requires no further motivation that under such conditions the method must be eminently well adapted, and that its practical application would be desirable in the highest degree.

The successful experiments of Wagner-Jauregg have been repeated throughout the whole world. Several thousand unfortunate people in various clinics and asylums in Europe and elsewhere – as well as in our own country – have received the benefits of this treatment during the last few years. Reports vary somewhat, but on one point they are unanimous, namely that never before have such remarkable results in the treatment of general paralysis been obtained. On the one hand, before Wagner-Jauregg it was possible to observe that about 1% of patients showed a « full remission » – that is to say, they recovered for a shorter or longer time, whether on account of treatment applied or by spontaneous remission, may be left an open question. With Wagner-Jauregg's malaria treatment on the other hand, it became apparent that on the average a complete cure from a practical point of view, and the ability to work, were obtained in no less than 30% of cases, and the best statistics even speak of nearly 50%. Approximately a third of all paralytics, formerly virtually condemned without exception for the rest of their lives to fall a burden on their relatives or society as useless beings, can, thanks to and as a consequence of the malaria treatment, count on being restored to a full life, fulfilling like others their duties in society.

For how long? On this it is impossible to speak with complete certainty, but the statistics are promising. It will suffice here to mention one set. I choose Wagner-Jauregg's own most recent compilation, as the cases observed for the longest time are found there. Wagner-Jauregg who in the course of years has treated over a thousand cases with malaria, took into consideration in these statistics only those cases, 400 in all, where at least two years had elapsed since the treatment. In spite of the length of time for which they were observed, varying between two and ten years, Wagner-Jauregg finds that about 30% – among them all three patients who had already recovered in 1917 (that is, ten years ago) – have enjoyed constant good health. This is quite remarkable, because previously, as stated, among the 1% of complete remissions observed, this lasted, as a rule, only a few months.

It is now quite clear from this that Wagner-Jauregg has given us a means to a really effective treatment of a terrible disease which was hitherto regarded as resistant to all forms of treatment, and incurable.

If it be considered that paralysis is, moreover, a disease which in general attacks persons between 32 and 45 years of age, and as a rule men – men, that is, in the best years of their lives and at an age when they are usually family providers and, as a rule, fathers of minors – it will be understood what a catastrophe for the whole family an attack by such a disease generally means. At the present time the great value of Wagner-Jauregg's achievement surely stands painted before our mind's eye in clearer colours than dry numbers can paint it. It is to such a one, who must be counted as one of the great discoverers and benefactors of mankind, that Alfred Nobel wished his prize to be awarded.

Julius Wagner-Jauregg, my most honoured colleague. As a young doctor the idea was born in your mind that by injecting the chronically insane with a febrile infectious disease it might be possible to cure the sick mind. After a long period of waiting came the moment when you were able to realize this idea. You injected malaria into human beings who were suffering from one of the most terrible mental diseases, one which up till now was thought to be incurable, and you led many, who were otherwise irretrievably lost, back to life and fit for work. Certainly, for you, the best reward for your life's work is the knowledge that you have given an unusually blessed gift to mankind, and the sense of the gratitude of the wretches whom you have made happy, as also of their families.

Recognition by the profession, the scientific world, is certainly, however, not a thing to be despised. The Caroline Institute has extended to you, in acknowledgment of your achievement just mentioned, the highest distinction that it has at its disposal, in that it has awarded you the Nobel Prize. I have the great honour to invite you to step before the King, and, accompanied by the heartfelt good wishes of the Institute and the gratitude and admiration of thousands, to receive your prize from the hands of His Majesty.

JULIUS WAGNER-JAUREGG

The treatment of dementia paralytica by malaria inoculation

Nobel Lecture, December 13, 1927

Two paths could lead to a cure for progressive paralysis: the rational and the empirical. The rational path appeared to be practical, as since Esmarch and Jessen, in 1858, attention had been drawn to a connection between progressive paralysis and syphilis. If incontestible proof that progressive paralysis was a syphilitic brain disease was first given much later (I mention in this connection the names Wassermann and Noguchi), therapeutic attempts to apply anti-syphilitic treatments were nevertheless instituted much earlier.

Established psychiatry, it is true, soon turned away from the specific therapy. In all the textbooks it was stated that the mercury cure was of no use against paralysis and was usually harmful.

However, systematic research on this point was not undertaken. It appears that the demand made by Jadassohn in 1912, that «one must compare, one with the other, a large series of cases untreated and of cases energetically treated with mercury», was nowhere put into effect.

When rather belated Runge-Kiel in 1914 established that among 555 paralytics collected between 1901 and 1912 there were remissions in 3.9% of the untreated cases, 9.3% of those treated with iodine, and 11.4% of those treated with mercury it was still not possible to say that the iodine-mercury treatment was completely ineffective. However, its effect was unsatisfactory and, with rare exceptions, not lasting.

The discovery of arsphenamine (Salvarsan) by Ehrlich brought new hope. The disappointment which soon followed was due to quite insufficient dosages. As one reads the reports of writers who have given arsphenamine in large doses and in rapidly repeated courses of treatment, and when one hears of the remissions obtained, in number and in duration far superior to the number of remissions observable in untreated paralysis, it cannot confidently be maintained that arsphenamine is ineffective against progressive paralysis. Yet it seems indeed, disregarding rare exceptions, that sooner or later a point is reached where arsphenamine treatment is unable to halt the fatal progression. The augmentation of the treatment by the employment of bismuth preparations could not change this.

There are, however, still always writers who expect the cure of paralytics from specific treatment alone.

But the question is not one of prestige between specific and non-specific treatment, but of what is the most far-reaching therapeutic effect on the disease obtainable.

And thus we have arrived at the empirical method.

Progressive paralysis has always been regarded as an incurable disease leading within a few years to insanity and death.

Nevertheless there were records of cured cases of progressive paralysis; cases in which there was such a complete retrogression of all the symptoms of the disease that it was possible for the person concerned to go about his life and business independently for many years. And even though such cases were extraordinarily rare, there were still relatively frequent remissions of a considerable duration in which the symptoms of the disease already developed retrogressed to a greater or less extent. Thus, in principle at least, progressive paralysis was necessarily a curable disease. And Francis Bacon, Lord Verulam, had already pronounced that it must be of the greatest interest for the physician to study healed cases of incurable diseases.

Now, the observation has been made that, in the rare cases of cure and in the frequent remissions of progressive paralysis, a febrile infectious disease or protracted suppuration had often preceded the improvement in the state of the disease.

In that lay a pointer. These cures following febrile infectious diseases, of which I had experienced striking instances myself, led me to propose as early as 1887 that this natural experiment should be imitated by a deliberate introduction of infectious diseases, and I suggested at that time malaria and erysipelas as suitable diseases. I singled out as a particular advantage of malaria that there is the possibility of interrupting the disease at will by the use of quinine, but I did not then anticipate to what degree these expectations from induced malaria would be fulfilled.

At that time I did not proceed to the direct application of these proposals, apart from an unfortunate experiment with erysipelas, and I also hardly had the authority then to carry on with them.

On the other hand I attempted to imitate the action of a febrile infectious disease by the use since 1890 of tuberculin which Koch had just introduced. At first this was used not only in progressive paralysis, but also in other mental disturbances, not infrequently with beneficial consequences. (This was to some extent a forerunner of the use of protein therapy, which later attained a great advance.) As there were among these, some cases of progressive paralysis, my interest soon concentrated on this disease because a favourable result cannot be so easily regarded as fortuitous as in other psychoses.

It was ascertained by means of a preliminary experiment of a large number of paralytics that those treated with tuberculin (with a maximum dose at that time of 0.1) showed more and longer-lasting remissions and a longer duration of life than an equal number of untreated paralytics. Afterwards, this treatment was carried out systematically and with an increasing dose of tuberculin (up to 1.0), and simultaneously a vigorous iodine-and-mercury treatment, later accompanied by arsphenamine injections, was also introduced.

In 1909, at the International Medical Congress in Budapest I gave some information on these methods of treatment, which were thus the first combined treatment – i.e. specific and non-specific – of a syphilitic disease.

Qualitatively the remissions which were obtained by means of the mercury-tuberculin treatment did not differ from those to be attained through induced malaria. The complete disappearance of the mental disturbances and the resumption of business activity, even in professions which make greater intellectual demands – such as civil servant, officer, barrister, solicitor, teacher, industrialist, actor, etc. – and the duration of the remissions was in individual cases quite remarkable; amounting to up to fifteen years.

But the number of relapses was great, the lasting remissions were in the minority.

I attempted to increase the effectiveness of the non-specific treatment by the utilization of various vaccines – staphylo-streptococcal vaccine, typhus vaccine – without altering the frequency of discouraging relapses in the slightest.

In the course of this experimentation with treatments I was able to observe repeatedly that particularly complete and long-lasting remissions presented themselves precisely in those cases in which an unintentional infectious disease, such as pneumonia or an abscess, appeared during the course of the treatment.

In 1917, therefore, I commenced to put into practice my proposal made in the year 1887, and I injected nine cases of progressive paralysis with tertian malaria.

The result was gratifying beyond expectation: six of these nine cases showed an extensive remission, and in three of these cases the remission proved enduring, so that I was able to present these cases of cured patients who have without interruption taken up again their former occupations, to this year's annual meeting of the German Psychiatric Society as having been able to follow them for ten years. After the result of this first experiment was pursued for two years, I went on, in the autumn of 1919, to continue this experimental treatment on a large scale, and I made a report on it in 1920 to the annual meeting of the German Psychiatric Society in Hamburg.

Whereas the earlier non-specific methods of treatment of progressive paralysis had met with little approval, it was otherwise with the malaria treatment. After Weygandt and Nonne, stimulated by Mühlens in Hamburg, had first tested the method of treatment on a large number of patients, it found quick acceptance in many psychiatric clinics and insane asylums, and is currently used, as far as I am informed, in all the countries of Europe, in North and South America, in South Africa, in the Dutch East Indies, and in Japan.

The overwhelming majority of writers agree that with this method remissions can be obtained which are on a scale far exceeding those attained by any other method.

Nevertheless, the malaria treatment should not simply replace specific treatment but should be used in conjunction with it. Some writers believed at first that they could dispense with the specific treatment, in that they obtained brilliant remissions by malaria alone. But the question is how to obtain the maximum therapeutic effect from the treatment. So, I undertook comparative investigations in which paralytics admitted to the clinic were treated alternately, the one with malaria only, the other with malaria followed by neoarsphenamine. The superiority of the combined specific-nonspecific treatment was clearly shown.

The cases which had been subsequently treated with neoarsphenamine had 48.5% full remissions, those with no subsequent treatment only 25%. On the other hand, the number of deaths in the latter group was higher – 18.7% against 12% – and likewise the number of rapidly deteriorating cases was 22% against 6.7%.

The malaria treatment is thus to be associated with a specific treatment. Insofar as neoarsphenamine is concerned, the drug should be given first after the fever has subsided, as otherwise the malaria is cut short. In my clinic now 5.00 grams of neoarsphenamine are given over six weeks after each malaria treatment.

Malaria treatment is the more effective the earlier in the course of the paralysis it is carried out. Therefore it is impossible to get a correct picture of its potential effectiveness by simply calculating that out of such and such a number of paralytics treated with malaria so many per cent obtained complete remission. It depends very much on how many among the material in question were in the initial stages of paralysis and how many were in the advanced stages.

We have therefore for some time singled out from the first, those cases of paralysis on their very first arrival at my clinic which from the degree and duration of their illness promised a favourable outcome and followed their progress separately. It was shown, that of these cases 84.8% obtained a full remission and 12.1% a partial remission, and that out of the total number of this series only one in thirty-eight had to be committed to the asylum.

Hence it was shown that progressive paralysis is, in principle, curable and that the practical success of the malaria treatment will be the greater the earlier the diagnosis of the illness is established – the more, that is, that the early stages of paralysis are recognized by physicians. It has become apparent that it is unwise to employ other methods of treatment against paralysis before the malaria treatment, as this means time wasted.

As the malaria treatment is the more effective the earlier it is employed, it would thus be best if it were carried out immediately on those luetics who are threatened with progressive paralysis. Which luetics are these? We know that they are those luetics in whom the cerebrospinal fluid in the advanced period of latency gives a positive reaction.

It is due to the late Kyrle of Vienna that the malaria treatment was extended to these luetics in that in these cases, which are not yet immediately threatened, he prescribed a course of arsphenamine to precede the malaria, and a second course to follow it. The results in respect of the readjustment of the cerebrospinal fluid, which in such cases with other methods of treatment is on the contrary frequently very refractory, were so gratifying that already a large number of syphilologists have become acquainted with these methods. And it is to be hoped that once these methods become public property, psychiatrists will have very much fewer paralytics to treat.

That the malaria treatment attained so great a dissemination is due to some favourable results, only apparent during its application, which could therefore not have been expected from the beginning.

It would have been difficult in many places to continue the malaria treatment if it had not been possible to maintain for an unlimited period a malaria strain by continual passage through human beings – that is in the asexual cycle. This was at first doubted, or at least the fear was stated, that such a strain would, in the course of its passage, change its properties, i.e. might become either no longer infectious or too virulent. Those fears have proved groundless. In my clinic there is a malaria strain in use that since September 1919 has made about two hundred passages through human beings, without its infectiousness, its virulence, and its therapeutic properties having been altered. Similar experiences have been had in many places.

The uninterrupted breeding of such a strain is, however, only possible where there is access to a sufficiently large number of paralytics needing treatment and possibly also of luetics in the advanced latency.

In places with little patient material, however, such a strain of induced malaria will always die out again, and it would involve great and often insurmountable difficulties to always procure a new case of natural malaria again to start a treatment, for the malaria virus will not breed in cultures.

Fortunately, however, the malaria parasites in human blood remain infectious for some time outside the human body, and this capacity can, by special methods of preservation be maintained for up to three days and in rare exceptions even longer, so that it is possible to send the virus over considerable distances by various means of transport.

It is therefore possible to supply with malaria virus an area of a very large radius from a centre, especially if use is made of the most modern form of transport, air mail. In this way we once successfully supplied malaria blood to Constantinople from Vienna.

It was finally a fortunate circumstance, which was not expected from the first, that tertian malaria brought on by injection proved to be so extraordinarily sensitive to quinine that a few grams of quinine suffice to cure the malaria completely and permanently, so that there is no fear of a relapse. It was through this that the great expansion which induced malaria has gained was first made possible.

When tertian malaria is acquired naturally the attacks of fever may also be cut short very effectively with quinine, but the patients remain carriers of the plasmodium and frequently relapse sooner or later. How would it have been possible to release so many paralytics and advanced syphilitics from the hospitals when outside they first of all ran the continual risk of a relapse and secondly, particularly where there were anopheles, were a danger to their environment?

The patient inoculated with malaria who has been adequately treated with quinine neither endangers himself further (in the sense of a malaria relapse), nor can he endanger his environment. However, he can from the moment of infection up to the elimination of the malaria present a danger to his environment, as malaria can be transferred from him to other persons through the sting of the anopheles, and that is then not induced malaria, but natural malaria, with its resistance to quinine.

This danger, which was assumed with the presence of anopheles in places of treatment, can be excluded with a fair degree of safety, if the patients are kept under mosquito-proof netting during the whole duration of the treatment. This has been done in several countries, such as England and Sweden.

The question is whether it is not possible to meet this danger in yet another way. An experiment was made in my clinic in 1924 with a large number of patients and mosquitoes to see if induced malaria could be transferred to other patients through anopheles; the experiment was without results. Such transfers have, however, been obtained by other writers, notably Shute and James, and also Warrington Yorke, in England, have carried out numerous successful transfers of induced malaria by means of anopheles. The Vienna strain has, however, been proved at that time to be free of gametes by an experiment of the Italian malariologist D. Vivaldi. The strains which were transferable through anopheles have all proved to be gamete producers; and the English writers mentioned in particular state that the transference by anopheles is the easier, the richer in gametes the donor's blood is.

Plehn and Schulze of Berlin, and Vonkenel of Munich, also reported on such gamete-free strains.

I have therefore in the preceding year made the demand that everyone who practises malaria therapy, should procure a gamete-free strain and thus eliminate the danger of a transfer by means of anopheles.

More recent investigations carried out in my clinic this year have, however, shown that this demand cannot be realized, as gamete-free malaria strains cannot be obtained by transferring preserved blood unaltered from one place to another. That is to say that it has been shown that from the moment malaria blood leaves the human body, the malaria parasites deviate from the normal course of development; they leave the red blood corpuscles and assume gametic forms.

We thus have in the preserved blood, not a gamete-free strain, but predominantly gamete-containing injection blood.

Thus the propagation of a gamete-free strain would not be possible with preserved blood but only by direct transfer from one patient to another. It would not be possible to effect this by transferring the blood but only by transferring the patients.

Induced malaria is, however, of itself a dangerous disease. The attacks of fever usually reach 40°C by the third attack. The temperature often remains above 40°C for many hours in the later attacks. It frequently reaches 41°C. The highest temperature that I have observed was 42°C. In addition, the attacks frequently assume the quotidian type or take that course right from the beginning. It appears, incidentally, that paralysis plays a role in the appearance of the quotidian type, as in luetics of advanced latency the malaria usually remains tertian. Perhaps in this respect, however, different strains of malaria behave differently, since Bravetta in Novara has at his disposal a strain of which he reports that it causes without exception attacks of the tertian type.

The high temperatures on the one hand and the brief pauses in the quotidian type on the other, make on the usually already weakened organism of the paralytic, especially on his heart, often too great demands; and thus we and others also have seen not by any means infrequent cases of death during the fever period or immediately afterwards.

However, by various measures, this danger has been decreased to such an extent that fatal cases are now almost never seen. We use several methods to this end. Something can frequently be effected by the mode of inoculation. That is to say that, if one inoculates intracutaneously with a small quantity of blood, about 0.1 cm³, the fever usually develops into the tertian type, especially when the blood groups of the donor and recipient correspond, and the avoidance of the quotidian type is already an alleviation.

In other cases we mitigate the fever with small doses of quinine (0.2–0.3) which must not, however, be given two days in succession, otherwise the fever ceases entirely. After a single administration of such a dose, the fever disappears for some days, during which the patient recovers; and when the fever sets in again it runs a milder course, as a rule. Alternatively one gives 0.1 quinine every two or three days from immediately after the injection, and in this way obtains a general alleviation of the course of the fever.

Finally, in cases which on account of their physical constitution or on account of their age – somewhere between 55 and 70 – appear particularly endangered, a division of the course into two parts has been proved particularly successful. In such patients the fever will be interrupted by quinine after two to at the most four attacks. This is followed by a six weeks' pause taken up with injections of neoarsphenamine, after which the patient will be infected a second time. He has meanwhile recovered, and now endures the continuation of the cure very well.

In this connection the question also arises, how many attacks of fever are necessary for a successful malaria cure? This can only be decided by experience, as we have no biological evidence as to when the optimum activity occurs. In my clinic the fever is, as a rule, terminated after eight attacks. English writers, by comparing therapeutic results after a shorter or longer duration of the fever period have likewise come to the conclusion that the optimum therapeutic effect lies at around eight attacks.

Some writers have let their patients have very much longer fevers. However, I believe that it is much better to give to a patient in whom a course of some eight attacks of fever has had an unsatisfactory result, a second course soon afterwards, than to endanger the reconstruction which should follow each malaria injection by weakening the patient too severely by continuing the course too long.

This reconstruction as an aftereffect of induced malaria, and the long duration of its aftereffect in general, is something that must be taken into account by every explanation of the mode of action of induced malaria.

The improvement in the physical and mental health of the patients is not as a rule demonstrable immediately after the last attack of fever and never to the full extent. On the contrary, it often happens that a paralytic who on completion of the treatment has been committed to the asylum as uncured, presents himself again after six or twelve months and states that he has taken up his occupation again.

The most convincing, because numerically demonstrable, expression of this delayed action of induced malaria is in the reactions of the serum and of the cerebrospinal fluid. The immediate effect of the malaria treatment on these reactions is negligible, and the changes do not run parallel with the clinical symptoms. It does change however, if these reactions are repeatedly investigated at intervals.

Kyrle has already noticed that in the malaria treatment of advanced syphilitics distinguished by a positive cerebrospinal fluid, the immediate effect of the treatment was relatively small. However, after the space of a year the cerebrospinal fluid was negative, although since the malaria, no further treatment had taken place, and in spite of the fact that before the malaria treatment the most vigorous specific therapy had been applied without any result.

The same thing happens with paralytics, only at an even slower tempo. In them the negativity of the cerebrospinal fluid reactions often first appears two, three, and even four years after the malaria treatment and still without any specific or non-specific treatment being introduced after the latter.

My assistant D. Dattner reported three years ago on the results of treatment from a particular aspect on a series of 129 paralytics treated with malaria in the period between the beginning of 1922 and the beginning of 1924; 66 of them underwent cerebrospinal fluid examinations at more frequent intervals up to the present day. They were thus cases in whom the malaria treatment lay about three to five years behind. Of these cases, repeated examinations of the cerebrospinal fluid in 1927 showed completely negative findings in 36, and nearly negative findings in 23. This favourable result, however, had first appeared in many of them two or more years after the cessation of the malaria treatment and without any further treatment having been administered in the meantime.

It has been shown by these investigations that the serum reaction is more refractory than the cerebrospinal fluid reaction.

The regularly repeated examination of the serum and cerebrospinal fluid also provides good evidence to establish a prognosis for the remissions achieved. Relapses, that is, do occur, but they form by far the minority beside cases which have attained a full remission. However, the cases in which this progressive improvement in the cerebrospinal fluid appears, do not relapse; but the contrary does not hold good. Curiously enough, this progressive improvement of the cerebrospinal fluid appears also in a number of the cases which do not improve clinically. It is thus of prognostic value only in conjunction with the clinical findings.

How is the action of induced malaria on the paralytic process to be explained?

It is certain that it is not the high temperature alone that is effective. The spirochaetes, it is true, disappear from the brain during the fever. When, however, the fever has passed, they are immediately to be found again in the brain, at least in cases where the course is not successful, as Forster has shown. Where are they in the meantime? Does malaria act against syphilis in general or predominantly against progressive paralysis? We know that syphilitic processes in the secondary period are also influenced by malaria, yet this action appears to be less permanent than the action on progressive paralysis. Vascular syphilis appears to be less favourably affected than progressive paralysis. Further, it has been experienced that soon after the malaria treatment gummata appear, even in cases in which the paralytic process has been favourably affected.

It appears then, that malaria besides a non-specific action against the syphilitic infection, also exerts a specific elective action on the cerebral process of progressive paralysis, including advanced infection of the cerebrospinal fluid.

It is also very likely that malaria creates favourable conditions for all reparatory processes because of its cyclic course, and because ultimately a rapid transition takes place from a serious state of illness to a full recovery. The superiority of induced malaria over the different types of stimulation therapy, e.g. by the injection of vaccines and proteins, has been shown by Schilling and his colleagues on the cytological blood-picture, and by Donath and Heilig on the chemical blood-picture. It is certain that induced malaria therapy will yet pose many worthwhile problems for research to explain.

Biography

Julius Wagner – his father Adolf Johann Wagner was granted the title « Ritter von Jauregg » only in 1883 – was born on March 7, 1857, in Wels, Austria. He attended the famous old Schottengymnasium in Vienna and started reading medicine at Vienna University in 1874.

From 1874 to 1880 he studied with Salomon Stricker, in the Institute of General and Experimental Pathology, obtaining his doctor's degree in 1880 with a thesis entitled « L'origine et la fonction du coeur accéléré » (Origin and function of the accelerated heart). He left the Institute in 1882. It was during this period that Wagner-Jauregg became acquainted with the use of laboratory animals in experimental work – a practice little followed at that time.

For a short period he worked in the Department for Internal Diseases under Bamberger, but gladly accepted the post of assistant to Leidesdorf in the Psychiatric Clinic in 1883, although he had never previously considered the possibility of becoming a psychiatrist and had practically no experience of this specialized field. Nevertheless, he was invited to lecture on the pathology of the nervous system already in 1885 and three years later this field was extended to include psychiatry. In 1887 his chief, Leidesdorf, fell ill and Wagner-Jauregg took charge of the clinic. In 1889 he was appointed Extraordinary Professor at the Medical Faculty of the University of Graz as successor to Krafft-Ebing and Director of the Neuro-Psychiatric Clinic. It was there that he started his investigations on the connections between goitre and cretinism; on his advice the Government, some time later, started selling salt to which iodine had been added, in the areas most affected by goitre.

In 1892 followed the appointment to the « Landesirrenanstalt » (State Lunatic Asylum) and in 1893 he became Extraordinary Professor of Psychiatry and Nervous Diseases, and Director of the Clinic for Psychiatry and Nervous Diseases in Vienna, as successor to Meynert. Ten years later, in 1902, Wagner-Jauregg moved to the psychiatric clinic at the « Allgemeines Krankenhaus » (General Hospital) as this offered more scope and a more varied activity. However, when in 1911 the « Landesirrenanstalt » was rebuilt and enlarged on the outskirts of Vienna at Steinhof, thus making the setting up of a larger psychiatric-neurological department, Wagner-Jauregg returned to his former post.

Wagner-Jauregg's initial study was concerned with the origin and function of the N. accelerantes, and this was followed by another on the respiratory function of the N. vagus.

The main work that concerned Wagner-Jauregg throughout his working life was the endeavour to cure mental disease by inducing a fever. Already in 1887 he systematically investigated the effects of febrile diseases on psychoses, later also making use of tuberculin (discovered in 1890 by Robert Koch). As this and similar methods of treatment did not yield satisfactory results, he turned in 1917 to malaria inoculation, which proved to be very successful in the case of dementia paralytica. This discovery earned him the Nobel Prize in 1927. His numerous other distinctions included the Cameron Prize (1935).

Among his numerous publications may be mentioned: *Myxödem und Kretinismus*, in the *Handbuch der Psychiatrie*, (1912); *Lehrbuch der Organotherapie* (Textbook of organotherapy), with G. Bayer, (1914); *Verhütung und Behandlung der progressiven Paralyse durch Impfmalaria* (Prevention and treatment of progressive paralysis by malaria inoculation) in the Memorial Volume of the *Handbuch der experimentellen Therapie*, (1931).

Wagner-Jauregg occupied himself also intensively with questions concerning forensic medicine and the legal aspects of insanity; he assisted in formulating the law regarding certification of the insane, which is still in force in Austria today. In recognition of his services to forensic medicine he was awarded the diploma of Doctor of Law.

Wagner-Jauregg was judged by his pupils and friends to be rather reserved, cool and aloof, but was generally respected, and all his students were proud to work under him. He worked very hard and conscientiously, and was well known for his sense of justice. Among his numerous pupils should be mentioned C. von Economo, who in 1917 isolated epidemic encephalitis (since then also called Economo's disease) – a discovery giving rise to the abolishment of certain classical views in neurology.

Professor Wagner-Jauregg married Anna Koch. There were two children from this marriage: Julia (b. 1900) now Mrs. Humann-Wagner Jauregg, and Theodor (b. 1903) now «Privatdozent» in Chemistry at the University of Vienna.

In 1928, Wagner-Jauregg retired from his post in Steinhof, but was by no means idle, publishing about 80 scientific papers after his retirement. He enjoyed good health and remained active until his death on September 27, 1940.

Physiology or Medicine 1928

CHARLES JULES HENRI NICOLLE

«for his work on typhus»

Physiology or Medicine 1928

Presentation Speech by Professor F. Henschen, member of the Staff of Professors of the Royal Caroline Institute

Your Majesty, Your Royal Highnesses, Ladies and Gentlemen.

In awarding the 1928 Nobel Prize for Medicine to Dr. Charles Nicolle, Director of the Pasteur Institute at Tunis, the Caroline Institute wished to pay tribute to a man who has realized one of the greatest conquests in the field of prophylactic medicine, i.e. the vanquishing of typhus.

Typhus is an acute infectious disease which, by its clinical evolution, its contagiousness and the conditions under which immunity is conferred shows considerable resemblance to ordinary measles. In severe cases, a state of stupor or even deep coma may occur. On account of the rash it causes the disease has been called exanthematous typhus; it has nothing in common, however, with what is properly called typhoid, i.e. enteric fever. In certain epidemics, particularly where children are concerned, epidemic typhus takes a relatively benign form; in adults, however, and if conditions are unfavourable, the mortality rate can reach the frightening proportion of between 50 and 60%.

The epidemiology of typhus presents a number of characteristics which, understandably, appeared most mysterious to physicians of past ages. It seemed, in fact, impossible to protect oneself against this disease which claimed many victims, even in the medical profession itself.

The general opinion used to be that typhus was transmitted more or less in the same way as measles or influenza, i.e. by direct contact, by dust, or by what is known as « droplet » infection. Around 1880 and 1890 when the part played by insects as carriers of infection was established, various people began to suspect that typhus could well be transmitted in the same way, in particular by parasites affecting man. This hypothesis, however, excited no special interest. The way in which the disease was actually disseminated was unknown and remained so, and there were, therefore, no effective measures available to combat the disease.

One of the peculiarities of typhus is the way in which it tends to cause serious epidemics, flaring up suddenly and coinciding with the previous occurrence of some serious public calamity. Populations in the throes of war or famine fell victim to the disease, which caused numerous deaths, some-times as many as hundreds of thousands. Thus originated certain expressive synonyms, such as « camp typhus », « famine typhus », « jail typhus » by which the disease has sometimes been known. As one author so truly says, the history of typhus is the history of human misfortune.

The disease has been known since the beginning of all time. The plague which devastated Attica, especially Athens in the year 430 B.C., and which Thucydides describes in his work on the Peloponnesian War, was most likely an epidemic of typhus. The picture that the great historian draws of the disease agrees in certain respects, down to the smallest details, with the clinical picture we were able to observe during the Great War. Epidemics followed one another without respite during the great wars of the sixteenth and seventeenth centuries. At the end of the Thirty Years' War, typhus raged over the whole of Central Europe. The Napoleonic Wars caused the disease to flare up again. In the general disorganization which followed the Grand Army's retreat from Russia, typhus claimed innumerable victims amongst the troops and amongst the civilian population. Further epidemics broke out during the Crimean War and the Russo-Turkish War, affecting both sides.

With the progress of civilization and during the period of peace and pros-perity which, in all, lasted from the end of the nineteenth century until 1914, typhus seemed of its own accord to have become restricted to certain remote regions of Europe and to certain extra-European countries where, from time immemorial, the disease had existed endemically.

At the beginning of this North Africa was among these non-European countries where the disease had been a veritable national scourge for several centuries. As soon as he took up his appointment as Director of the Pasteur Institute at Tunis, young Dr. Charles Nicolle was immediately brought into contact with the scientific and practical problems that typhus had created in this country. As soon as he took up his post, Nicolle immediately with extraordinary energy attacked these problems. He visited patients in their own homes, examining their beds and their sordid rags, whilst undertaking at the same time a strict enquiry within the hospitals. This work was to cost the life of two of his collaborators. He was led to an observation which could hardly have escaped the attention of earlier workers, i.e. that whilst typhus patients continued to spread infection up to the point when they entered the hospital waiting-room and amongst those who took charge of their clothing, they became completely inoffensive as soon as they had been bathed and dressed in the hospital uniform. At this point they could be admitted to the general wards without the slightest risk. Nicolle concluded that the pathogenic agent must necessarily be related to one factor, carried by the patient himself and transmissible to others, a factor which no longer acted once the patient had bathed and changed his clothing; this factor, therefore, could only be a parasite, the body louse, which lives on the patient's body and in his clothing. This simple observation contains in essence Nicolle's discovery.

In order to further his research Nicolle now made experiments on animals. Previously, some research workers had succeeded in inoculating healthy individuals with typhus by injecting blood from a patient, but all attempts to inoculate animals had failed up till then. After several unsuccessful attempts, Nicolle succeeded at the beginning of 1909 in inoculating chimpanzees with typhus and, from the chimpanzee, he was able to inoculate monkeys of a lower order by injection of blood. As early as September of the same year Nicolle and his collaborators were able to demonstrate that lice which had previously bitten contaminated monkeys transmitted the infection to healthy animals simply by biting them. The part played by the body louse as a transmission agent had thus been proven experimentally.

The secrets of this terrible disease were then laid bare one after another. The first problem was to define the conditions under which infection by the body louse took place. It was possible to establish that the blood of a typhus patient could transmit infection from some hours before the appearance of fever until the first days of recovery. The insect could therefore absorb the pathogenic agent during the whole course of the disease, even before the disease became apparent and after the fever had disappeared.

The parasite's bite is not however immediately dangerous; it only becomes virulent after about a week when the pathogenic agent has had time to multiply in the parasite's digestive tube. Contagion is also possible by means other than the parasite's bite: it is sufficient that the skin and clothing be soiled by the excreta of an infected parasite and that the patient infect himself by scratching, for the disease to develop. This form of transmission most probably plays at least as important a role as the direct bite.

Nicolle was not long in making another important discovery: he established that the germ of typhus is not transmitted to new generations of parasites. The epidemic dies out of its own accord when the contaminated adult insects die. All these observations are obviously of the greatest importance from the point of view of combating the disease.

Nicolle and his collaborators had established at an early stage of their re-searches that monkeys who had recovered from a first attack of the disease became resistant to further contamination. This observation led them to a series of important discoveries concerning the conditions necessary for im-munization against typhus, and led to a series of successful attempts to exert preventive and attenuating effects on the disease with serum taken from convalescents and by vaccination.

Nicolle's discovery that it is possible to inoculate the guinea pig with typhus was an important step forward in the study of this disease. By succes-sive inoculations from one animal to another, it became possible to preserve the agent of typhus in the laboratory for an unlimited period; it has not yet been possible to cultivate the virus on artificial substrates, and our know-ledge of its morphology and biology is still extremely limited.

The study of typhus in the guinea pig led Nicolle to another very impor-tant discovery: certain infected animals may be germ carriers even though they present no apparent symptoms. Not even a slight fever indicates that they are contagious. This form of the disease had been hitherto unknown. Nicolle calls this form of typhus «inapparent» typhus and considers it to be the prototype of a group of latent infectious diseases of the same type. Nicolle's discovery of the inapparent infection orientated the work of scien-tists towards a hitherto unexplored field of research. These new concepts are highly significant, even as regards direct action against infectious diseases.

It soon became apparent that the discovery of the part played by the body louse in the transmission of typhus was of the greatest practical importance: it now became possible to combat the terrible disease by rational methods. In fact, within two years Nicolle and his collaborators succeeded in ridding Tunis entirely of a disease which had raged there each winter from time immemoral.

But in 1910 who would have guessed that the results of Nicolle's research were about to be put to practical use on a vast scale?

When the Great War broke out and many Russian and Serbian prisoners were interned in German and Austrian prison camps, typhus, which until then had hardly attracted the attention of European doctors at all, was not long in making an appearance. In spite of the precautions ordinarily taken against epidemics, it was quickly transmitted from one man to another, from home to home, regardless of age and in defiance of all the laws of epidemiology. The armies were threatened with a veritable catastrophe. The epidemic broke out in the same way amongst the civilian population in regions of the eastern front devastated by war. The Balkan Peninsula was affected badly, but the disease spared no part along the whole front, from Finland to Mesopotamia. The value of Nicolle's discovery was once again made apparent. The Great War provided the opportunity for a clinico-experimental application of Nicolle's work on a large scale. As a French doctor said, one had to see the deserted Serbian towns to realize the desolation that can be caused by typhus, the satellite of war; one had to see for oneself the resurrection of whole areas, entirely due to the hygienic measures developed from Nicolle's discoveries, to appreciate fully the significance of these discoveries. Remembering the great losses incurred through another wartime epidemic, Spanish influenza, a far less serious disease in itself, one shudders to think of what might have happened in the Great War if we had been unable to combat typhus successfully.

Indubitably, the nature has not changed and we still know of no effective therapeutic action against this disease. Nevertheless, this terrible plague has become a mere contagious disease and is no longer considered in terms of devastating epidemics. Thanks mainly to the work of Charles Nicolle, we have now completely mastered the disease. The man who has vanquished typhus deserves the gratitude of the whole human race.

In the absence of Mr. Nicolle, the honour of whose presence is denied us today, I ask your Excellence, in your quality of Minister representing the French Republic, to accept on his behalf and to convey to him the prize and diploma. May I also ask you to convey to your illustrious compatriot the tribute and sincere congratulations of the Caroline Institute.

CHARLES NICOLLE

Investigations on typhus

*Nobel Lecture**

I am going to give you an account of how I arrived at the results for which
I have received the Nobel Prize for Medicine. I shall also summarize these
results.

It did not seem likely that I was destined to undertake research on typhus.
I was born, first studied medicine and undertook my first research work in a
French province from which typhus had disappeared since 1814. It is true
that I came across a few imported cases at Rouen in 1889. They made no
particular impression on me.

My arrival in Tunisia placed me immediately in contact with typhus. Ten
days after my arrival, at the beginning of January 1903, I saw a few cases in
natives living in a suburb of Tunis.

In those days the disease flared up each winter in the rural districts of
Tunisia. From these remote districts it spread to the doss houses, the prisons
and the outskirts of towns. The native districts of Tunis and the prisons were
regularly stricken. The epidemic receded in June; it drew back into the
remote country districts and was not heard of again until the end of the
year.

Of all the problems which were open to me for study, typhus was the
most urgent and the most unexplored. We knew nothing of the way in
which contagion spread. The field of experimental study was virgin ground.
We were scarcely able to conclude, from the results of debatable experiments, that it was possible to inoculate the disease from one man to another
by means of the blood.

In June of 1903 I was determined to carry out a preliminary study. At
that time typhus was raging in a native prison, 80 kilometres South of Tunis,
Djouggar. I requested the doctor in charge of this establishment to allow
me to accompany him on his weekly visits. We made an appointment. The
evening before I had a haemoptysis. If it had not been for this accident, my

* As Professor Nicolle has been prevented by ill health from coming to Stockholm
to deliver his Nobel Lecture, he has very kindly sent the text to the Editor of *Les
Prix Nobel* for publication.

first contact with typhus would undoubtedly have been my last. My colleague, Motheau, and his servant went to Djouggar; they spent the night there, contracted typhus and both died.

Most of the doctors in the Tunisian administration, especially those in country districts, contracted typhus and approximately one third of them died of it. The fact that I was fortunate enough to escape contagion, in spite of frequent, sometimes daily contacts with the disease, was because I soon guessed how it spread.

The native hospital in Tunis was the focal point of my research. Often, when going to the hospital, I had to step over the bodies of typhus patients who were awaiting admission to the hospital and had fallen exhausted at the door. We had observed a certain phenomenon at the hospital, of which no one recognized the significance, and which drew my attention. In those days typhus patients were accomodated in the open medical wards. Before reaching the door of the wards they spread contagion. They transmitted the disease to the families that sheltered them, and doctors visiting them were also infected. The administrative staff admitting the patients, the personnel responsible for taking their clothes and linen, and the laundry staff were also contaminated. In spite of this, once admitted to the general ward the typhus patient did not contaminate any of the other patients, the nurses or the doctors.

I took this observation as my guide. I asked myself what happened between the entrance to the hospital and the wards. This is what happened: the typhus patient was stripped of his clothes and linen, shaved and washed. The contagious agent was therefore something attached to his skin and clothing, something which soap and water could remove. It could only be the louse. It was the louse.

Even if it had not been possible to reproduce the disease in animals and consequently to verify the hypothesis, this simple observation would have been sufficient to demonstrate the way in which the disease was propagated.

Fortunately, it was also possible to provide experimental proof.

My first attempts to transmit typhus to laboratory animals, including the smaller species of monkeys, had failed, as had those of my predecessors, for reasons which I can easily supply today.

I asked my teacher, E. Roux, to get me a chimpanzee, thinking that an anthropoid might be more susceptible to infection than animals of other species. The day I received the chimpanzee I inoculated it with the blood of a patient from Dr. Broc's department at the hospital in Rabta. The chim-

panzee contracted a fever. I inoculated a macaco (*M. sinicus*) with blood
from the chimpanzee taken during the fever, and he also developed a fever.
I cultivated lice on the macaco, which I then transported to other macacos.
The latter became infected and subsequently proved to be vaccinated against
a test inoculation of the virus.

These decisive experiments did not take very long. I had reproduced
typhus in the chimpanzee in June, 1909; I demonstrated the role played by
the louse in August. I published these results in September with Charles
Comte and Ernest Conseil. This was the yield for the year 1909. During the
years that followed I undertook with Conseil and Alfred Conor, and later
with Georges Blanc, a more detailed experimental study of the disease and
the conditions of transmission.

At the beginning of my research the only animals I knew to be susceptible
to the disease were monkeys. All species of monkeys can be infected, provided that a sufficient quantity of virus is inoculated through the peritoneum.
The monkey was well suited for preserving the virus. But the monkey is an
expensive animal. During the first two years, therefore, I had to be satisfied
with studying typhus during the months of its seasonal expansion, experi-
menting from man to monkey and, exceptionally, from one monkey to
another. The discovery that I soon made that the guinea pig was also susceptible to infection made it possible for me, from the third year on, to
preserve the virus on this animal. It was then easy for me to conduct my
research independently of epidemics, that is, all the time. From the practical
point of view, the susceptibility to infection of the guinea pig proved to be
the most useful step forward. Today, all laboratories use this animal for
preserving the virus.

In man typhus is characterized by a triade of symptoms: fever, rash, nervous symptoms. In animals, fever is the only sign of infection. The fact that
the virus is localized more particularly in the brain explains the nervous
symptoms to be found in our species.

I demonstrated the characteristics of experimental fever. It appears after
an incubation period which is never less than five days. It follows the same
pattern as natural fever in man, but is of shorter duration and less pronounced. There follows a period of hypothermia, clearly apparent in the
monkey, but less obvious in the guinea pig. In the monkey there are minor
general symptoms and subsequent loss of weight is usual. In the guinea pig
the disease would be inapparent if the temperature were not noted. At the
time when I was conducting my research there was no known method for

taking the guinea pig's temperature. I demonstrated a technique which is now widely used.

I have already mentioned that the virus could be kept indefinitely by injecting guinea pigs. For this purpose, brain tissue gives more constant results than blood.

It can sometimes happen, especially when blood is used, that out of a group of guinea pigs inoculated with the same dose of the same product, certain of them show no signs of fever. At first I attributed this fact to a technical fault or to greater individual resistance. Repeated negative results precluded my continuing to accept these oversimplified explanations. Animals for which the virus is pathogenic present a whole scale of degrees of susceptibility – from the grave form, often fatal in adult Europeans, down to the fever, revealed only by the thermometer, in the guinea pig, passing through all the intermediate stages: typhus of the native adult, benign typhus in children, still more benign in monkeys. I wondered if there was not, below the very slight susceptibility of the guinea pig, an even lesser degree in which the only sign of infection was the virulence of the blood during the period when more susceptible animals showed signs of the characteristic fever. This was indeed the case. I was able to ascertain this fact in 1911. A little later I went back to study this problem with Charles Lebailly. It led me to the conception of what I have called *inapparent infections*. Apyretic typhus, without symptoms, the *inapparent typhus* presented by the guinea pig, is the first case described and the most well known.

Inapparent typhus in the guinea pig may be a primary infection typhus, as was the case I discovered. The cause in this case is the inoculation of a quantity of virus insufficient to produce fever. Blood from a guinea pig who has contracted inapparent typhus will always produce pyretic typhus in another guinea pig if the dose inoculated is sufficient. The positive result of this inoculation proves in fact that the first guinea pig had contracted typhus, in spite of the lack of fever.

Having ascertained inapparent typhus in cases of primary infection, it was then easy for me to demonstrate the existence of apyretic typhus in certain guinea pigs that had been reinoculated a fairly long time after contraction of a primary infection of pyretic typhus.

Lebailly and I then went on to demonstrate that inapparent primary infection typhus, which is exceptional in inoculated guinea pigs, is the only form which the disease takes experimentally in the rat and the mouse. This curious disease, which presents no symptoms, which has an incubation period and a period when the blood is virulent, and which confers a certain degree of immunity, can be transmitted from one rat to another. On two occasions I effected twelve such transmissions. At the twelfth, brain tissue from the rats induced pyretic typhus in the guinea pig.

This new concept of inapparent infections that I introduced to pathology is, without a doubt, the most important of the discoveries that I was able to make. In the absence of fever, experimental work on typhus must take into account the possible existence of inapparent forms. We shall see that this concept alone can explain the preservation of the virus in nature and the reappearance of seasonal epidemics. I applied this new concept to measles and certain spirochaete infections, and Georges Blanc has just discovered the existence of inapparent dengue in man, as well as in the monkey and the guinea pig.

Thus my research opened a new chapter, *sub-pathology*, which is doubtless a vast field where almost everything remains to be discovered.

We have known for a long time that primary infection of typhus confers immunity in man in almost all cases and that this immunity lasts a lifetime. I established that laboratory animals were subject to a similar immunity; but I also demonstrated that this was of shorter duration than was generally thought before the discovery of inapparent typhus in secondary infections. Although benign, inapparent typhus also confers a certain degree of immunity.

Lucien Raynaud (of Algiers) and E. Legrain (of Bougie) had used serum from convalescent typhus patients empirically in treating typhus. Conseil and I established that this method was not effective. On the other hand, we demonstrated that serum from convalescent patients and from animals that had recovered from the disease had preventive properties as regards subsequent inoculation of the virus. Following these discoveries, we were able to institute a preventive method against typhus by using serum from convalescents. This method gives sure protection to persons in contact with typhus patients, to doctors and to nurses. It proved to be particularly effective, in our hands, in protecting persons contaminated by lice from typhus patients and who would otherwise certainly have contracted the disease. The immunity conferred by the serum is of short duration, but inoculation may be repeated where necessary.

The knowledge of the preventive properties contained in serum from convalescent typhus patients led me, with Conseil, to use serum taken from children cured of measles to prevent this disease. This method spread from Tunis throughout the whole world and has saved thousands of lives.

I was less successful in my attempts to effect preventive vaccination against typhus by using the virus and in trying to produce large quantities of serum using large animals.

I did, certainly, succeed in vaccinating a number of people by injecting very small, repeated doses of virulent blood (with Conseil) and achieved rather better results with brain tissue from the guinea pig (with Hélène Sparrow). However, the method is unreliable and an element of danger precludes any recommendation to generalize it.

Prevention of typus by serum from convalescent patients presupposes the existence of such patients; moreover, the quantity of serum provided by a convalescent is very small. Although useful, therefore, this method can be considered only as a makeshift. If it were possible to produce these preventive properties in a large animal, the quantity of serum obtained would be unlimited and, by preserving the virus in guinea pigs, it would thus be possible to produce the serum as and when required.

I had realized that repeated virulent inoculations of the donkey did not cause appreciable preventive properties to be developed in the blood of this animal, but in attempting to find a solution to the problem, I tried to ascertain whether the donkey, apparently resistent to the disease, could not contract inapparent typhus and I found that this was sometimes the case. As inapparent typhus produced no preventive properties, I then endeavoured to produce pyretic typhus in the donkey. Working first alone and later with Hélène Sparrow, I succeeded by intracerebral inoculation of the virus, but only in very rare cases and without being able to determine the conditions necessary to produce frequent, if not constant, results. Serum from the donkey cured of pyretic typhus does in fact possess preventive properties. Up to the present time, this is as far as we have gone in our research on this point.

I was more successful in my study of the conditions in which transmission of typhus takes place. After clearly establishing the role of the louse, I demonstrated the detailed mechanism of transmission. Two factors are involved: man and the louse.

The patient's blood is virulent, not only during the whole duration of the fever, but also two or three days before it appears and two or three days after the temperature has fallen. During this whole period, therefore, the louse can be infected by the patient. Children play an important part in the propagation of typhus. Conseil and I have demonstrated that native children, particularly very young children, contract a benign form of typhus, sometimes so slight that it can only be detected by positive results obtained from inoculation of blood in the monkey or guinea pig.

The role of inapparent typhus in cases of re-infection is indubitably even more important. This alone explains the preservation of the virus in nature when there is no epidemic and the seasonal reappearance of the epidemics themselves. It would be impossible to understand how typhus could remain active if this depended on the louse continually finding different people to infect. And if this is self-evident in the case of typhus, it is even more so for measles. We have no doubt discovered here the most useful application of the concept of inapparent infections.

Just as the only reservoir for the typhus virus in nature is provided by man, so the only vector of infection is the louse. The bite of the louse is not virulent immediately after the infecting meal. It becomes so only towards the 7th day following infection. On the 9th and 10th days the bite is invariably virulent. It is therefore a necessary condition that the virus should multiply within the louse for it to become dangerous. I have shown that this multiplication takes place in the digestive tract and that louse faeces become virulent at the same period as does the bite. This observation was made in 1910. Developed by further, more detailed research undertaken in conjunction with Georges Blanc in 1914, it provided a guide to those who undertook to locate the pathogenic organism responsible for typhus in the intestine of the louse. Edmond Sergent (from Algiers) was the first to describe inclusion bodies, later called *rickettsia*, in the intestinal cells and faeces of the louse. Not being attracted to purely morphological research, I have not myself studied these inclusion bodies, the significance of which is not clearly demonstrated.

I demonstrated the more immediately important fact that typhus can be transmitted through louse faeces. By soiling the natives' skin, they are easily inoculated by scratching or by soiling the fingers, which are then brought into contact with the conjunctiva, which is an ideal means of entry for such an active virus.

I finally demonstrated that typhus infection is not hereditary in the louse.

These observations formed the basis of typhus prophylaxis. The pioneer in this field was Conseil, Director of the Bureau of Public Health in Tunis; in three years he eradicated typhus from a town where it had raged year after year since the beginning of history. Thanks to the endeavours of our medical administrators, and in particular Gobert, Cardaliaguet, and Henry, typhus has been eliminated from the mines and prisons and has retreated to remote rural districts, where it will remain a permanent menace until it is finally eradicated. From Tunisia the method has spread over the world.

From the very beginning of the Great War steps were taken, on my instruction, to institute medical control of troups in North Africa. No native left African soil for Europe without having been previously deloused. This was the precaution, taken by all nations in similar conditions, which saved the armies from typhus. Although lice appeared and multiplied in the trenches and became a plague in themselves, for the first time in the history of man typhus did not go hand in hand with a long war.

If in 1914 we had been unaware of the mode of transmission of typhus, and if infected lice had been imported into Europe, the war would not have ended by a bloody victory. It would have ended in an unparalleled catastrophe, the most terrible in human history. Soldiers at the front, reserves, prisoners, civilians, neutrals even, the whole of humanity would have collapsed. Men would have perished in millions, as unfortunately occurred in Russia.

And this is the ultimate lesson that our knowledge of the mode of transmission of typhus has taught us: Man carries on his skin a parasite, the louse. Civilization rids him of it. Should man regress, should he allow himself to resemble a primitive beast, the louse begins to multiply again and treats man as he deserves, as a brute beast.

This conclusion would have endeared itself to the warm heart of Alfred Nobel. My contribution to it makes me feel less unworthy of the honour which you have conferred upon me in his name.

Biography

Charles Jules Henry Nicolle was born in Rouen on September 21, 1866, where his father, Eugène Nicolle, was a doctor in a local hospital. Charles received, together with his brothers, early tuition in biology from his father and, after education at the Lycée Corneille de Rouen, he entered the local medical school where he studied for three years before following his elder brother, Maurice, who was working in Paris hospitals. (Maurice later became Director of the Bacteriological Institute of Constantinople and a Professor at the Pasteur Institute, Paris.) Meanwhile, Charles had studied under A. Gombault in the Faculty of Medicine and under Roux at the Pasteur Institute (serving at the same time as demonstrator in the microbiology course) to complete a thesis « Recherches sur la chancre mou » (Researches on the soft chancre), which gained him his M.D. degree in 1893. He returned to Rouen to become a member of the Medical Faculty and in 1896 he was appointed Director of the Bacteriological Laboratory. He continued in this capacity until 1903 when he was appointed Director of the Pasteur Institute in Tunis, a position he held until his death in 1936.

Early in his career, Nicolle worked on cancer, and at Rouen he investigated the preparation of diphtheria antiserum. In North Africa, under his influence, the Institute at Tunis quickly became a world-famous centre for bacteriological research and for the production of vaccines and serums to combat most of the prevalent infectious diseases. His discovery in 1909 that typhus fever is transmitted by the body louse helped to make a clear distinction between the classical louse-bound epidemic typhus and marine typhus, which is conveyed to man by the rat flea. He also made invaluable contributions to present-day knowledge of Malta fever, where he introduced preventive vaccination; tick fever, where he discovered the means of transmission; scarlet fever, by experimental reproduction with streptococci; rinderpest, measles, influenza, by his work on the nature of the virus; tuberculosis and trachoma. He was responsible for the introduction of many new techniques and innovations in bacteriology. Nicolle was one of the first to recognize the protective properties of the convalescence serum against typhus and measles; and succeeded in cultivating Leishmania donovani and Leishmania tropica on artificial culture media. His discovery of the mechanism of the transmission of typhus fever has created the basis for the preventive precautions against this disease, during the 1914–1918 and 1939–1945 Wars.

Nicolle wrote several important books including *Le Destin des Maladies infectieuses; La Nature, conception et morale biologiques; Responsabilités de la Médecine,* and *La Destinée humaine.*

Nicolle was an Associate of l'Académie de Médecine and he was awarded the Prix Montyon in 1909, 1912, and 1914; the Prix Osiris in 1927, and a special Gold Medal to commemorate his Silver Jubilee in Tunis in 1928. On this occasion he was also appointed member of the Académie des Sciences, Paris. In 1932, he was elected Professor in the College of France.

Charles Nicolle also enjoyed considerable reputation as a philosopher and as a writer of fanciful stories, such as *Le Pâtissier de Bellone*, *Les deux Larrons*, and *Les Contes de Marmouse*. He was said by Jean Rostand to be « a poet and realist, a man of dreams and a man of truth».

Nicolle married Alice Avice in 1895; two children came from this marriage, Marcelle (b. 1896) and Pierre (b. 1898).

He died on February 28, 1936.

Physiology or Medicine 1929

CHRISTIAAN EIJKMAN

«for his discovery of the antineuritic vitamin»

Sir FREDERICK GOWLAND HOPKINS

«for his discovery of the growth-stimulating vitamins»

Physiology or Medicine 1929

Presentation Speech by Professor G. Liljestrand, member of the Staff of Professors of the Royal Caroline Institute

Your Majesty, Your Royal Highnesses, Ladies and Gentlemen.

That the fruits of civilization are not solely beneficial is shown by, *inter alia*, the history of the art of medicine. Not a few illnesses and diseases follow close on the heels of, and are more or less directly caused by, civilization. This is the case with the widespread disease beriberi, first described more than 1,300 years ago from that ancient seat of civilization, China. In modern times, however, it was not until towards the end of the 17th and the beginning of the 18th century that the disease attracted more general attention. Subsequently it has, on different occasions and with varying degrees of violence, made its appearance in all five continents, but more particularly its haunts have been in Eastern and South-Eastern Asia. At times the disease has been a serious scourge there. Thus in 1871 and 1879, Tokio was visited by widespread epidemics, and during the Russo-Japanese War it is said that not less than one-sixth of the Japanese army was struck down.

Beriberi shows itself in paralysis accompanied by disturbances in the sensibility and atrophy of the muscles, besides symptoms from the heart and blood vessels, *inter alia*, tiredness and oedema. Decided lesions have been shown in the peripheral nerves which seem to explain the manifestations of the disease. Mortality has varied considerably, from one or two per cent to 80 per cent in certain epidemics.

A number of circumstances indicated a connection between food and beriberi: for example, it was suggested that the cause might be traced to bad rice or insufficiency in the food of proteins or fat.

The severe ravages of beriberi in the Dutch Indies led the Dutch Government to appoint a special commission to study the disease on the spot. At the time, bacteriology was in its hey-day, and it was then but natural that bacteria should be sought as the cause of the disease, and indeed it was thought that success had been attained. The researches were continued in Java by one of the commission's coadjutors, the Dutch doctor Christiaan Eijkman. As has so often been the case during the development of science, a chance observation proved to be of decisive importance. Eijkman observed a peculiar sickness among the hens belonging to the laboratory. They were attacked by an upward-moving paralysis, they began to walk unsteadily, found difficulty in perching, and later lay down on their sides. The issue of the disease was fatal unless they were specially treated. It has been said that the secret of success is to be prepared for one's opportunity when it presents itself, and indubitably Eijkman was prepared in an eminent degree. With his attention focussed on beriberi, he immediately found a striking similarity between that disease and the sickness that had attacked the hens. He also observed changes in numerous nerves similar to those met with in the case of beriberi. In common with beriberi, this ailment of the hens was to be described as a polyneuritis. In vain, however, did Eijkman try to establish micro-organisms as the cause of the disease.

On the other hand, he succeeded in establishing the fact that the condition of the hens was connected with a change in their food, in that for some time before they were attacked they had been given boiled polished rice instead of the usual raw husked rice. Direct experiments proved incontestably that the polyneuritis of the hens was caused by the consumption of rice that by so-called «polishing» had been deprived of the outer husk. Eijkman found that the same disease presented itself when the hens were fed exclusively on a number of other starch-rich products, such as sago and tapioca. He also proved that the disease could be checked by the addition to the food of rice bran, that is to say, the parts of the rice that had been removed by polishing, and he found that the protective constituent of the bran was soluble in water and alcohol.

Eijkman's work led Vorderman to carry out investigations on prisoners in the Dutch Indies (where the prisoner's food was prepared in different ways according to the varying customs of the inhabitants), with a view to discovering whether beriberi in man was connected with the nature of the rice food they consumed. It proved that in the prisons where the inmates were fed on polished rice, beriberi was about 300 times as prevalent as in the prisons where unpolished rice was used.

When making investigations to explain the results reached, Eijkman considered that protein or salt hunger could not be the cause of the disease. But he indicated that the protective property of the rice bran might possibly be connected with the introduction of some particular protein or some special salt. At the time it might have been readily imagined that the polyneuritis in the hens and beriberi were due to some poison, and Eijkman set this up as a working hypothesis, though his attempts to establish the poison were in vain.

In his view, however, such a poison was formed, but it was rendered innocuous by the protective substance in the bran. It was only Eijkman's successor in Java, Grijns, who made it clear that the substance in question was used directly in the body, and that our usual food, in addition to the previously known constituents, must contain certain other substances, if health is to be preserved. Funk introduced the designation *vitamins* for these substances, and since then the particular substance that serves as a protection against polyneuritis has been called the « antineuritic» vitamin.

It might have been expected that Eijkman's discovery would lead to an immediate and decided decline in beriberi – perhaps to the disappearance of the disease. But this was by no means the case, and not even in the Dutch Indies, where Eijkman and Grijns had worked, were the results particularly brilliant. The reasons for this were several: the reluctance of the inhabitants to substitute the less appetizing unpolished for polished rice, the opinion that polyneuritis in birds was not a similar condition to beriberi in man, and an inadequate appreciation of Eijkman's work. As a result of numerous experiments by different investigators on animals and human beings, who offered themselves for experimental work, it has gradually become clear that beriberi is a disease for the appearance of which lack of the vitamin found in rice bran – but also other circumstances – is of decisive importance. These experiences, in addition to successful experiments made in various places on the basis of Eijkman's observations, especially in British India, have gradually led to a general adoption of Eijkman's views. The successful attempts to combat beriberi which are now proceeding are the fruits of Eijkman's labours.

It was the analysis of the nature of the food used in cases of polyneuritis in hens that led Eijkman to his discovery. As a rule, analysis and synthesis complete each other, and indeed the employment of both these avenues of approach has been of decisive importance also for the development of the science of vitamins.

Although a number of experiments carried out about 50 years ago supported the assumption that, if our food is to have its full value, it must contain something more than the long-known basic constituents – proteins, fat, carbohydrates, water, and salts – yet it is not until our own days that complete certainty has been reached. One line of development has been sketched above. But numerous investigations have also been carried out by different experimentors with a view to testing the value of foods composed exclusively of the above-mentioned constituents in pure form. Sometimes it has proved to be a matter of some difficulty to get young animals to grow on such foods. One explanation put forward for this was the monotony of the food, and another was that the excessive purity resulted in the absence of certain substances giving the food taste which are necessary for appetite, and which must be present if the food is to be taken in sufficient amount. From other quarters, however, it was reported that even from the pure constituents, a food had been successfully produced which led to growth in the young organism.

When Hopkins joined the numbers of those who were trying to find a solution to this problem, he had the advantage of a far-reaching experience within similar fields of research, for he had done a great deal of detailed work on the presentation in pure form of certain proteins, and in connection therewith he had discovered the amino acid tryptophane as an element in certain proteins. As early as 1906 he had carried out careful feeding experiments on mice with different proteins, and by means of regular weighings it was observed whether the food was sufficient or not. It appeared from these experiments that the animal organism cannot itself build up tryptophane – proteins which do not contain it are not sufficient for the needs of the body. The simple methods employed by Hopkins came to play an important role later on.

When Hopkins continued his experiments, he fed young rats on a basic diet which, in addition to the necessary salts, contained a carefully purified mixture of lard, starch, and casein, i.e. the protein that is most abundantly found in milk. After some time the animals ceased to grow, which showed the insufficiency of this basic food in itself. By various experiments, however, Hopkins demonstrated that it was only necessary to add a very small daily amount of milk – two to three cubic centimeters for each animal – for growth to recommence. This amount of milk only corresponded to one or two per cent of the energy-content of the food, so that in this respect the addition of milk was insignificant. It was indeed found that incompletely purified casein, e.g. the ordinary casein of commerce, owing to the slight quantities of active substances present, was sufficient, with the other basic food, to maintain growth, even though it was considerably delayed. It was evident, as Hopkins was able to show more explicitly, that here was to some extent the explanation of the older and conflicting results.

Hopkins showed that there was a sufficiency of food consumed without the added milk, but it could be fully utilized by the body only when the growth-promoting influence of the milk was present. This effect was found not to be connected with any of the known constituent parts of milk. It was found also with yeast and the green parts of plants.

Hopkins communicated certain of his main results – but in an extremely brief form – as early as in 1906, and he returned to the subject in 1909 in a series of lectures, but it was not until three years later that his work was published in its entirety. By then Stepp had given accounts of experiments which, though they certainly seem less capable of one definite interpretation than those of Hopkins, yet point in the same direction, and the ground was also in other respects prepared, so that Hopkins' work was a great incentive to continued experiments in the young science of vitamins. Chiefly by American investigations it was shown that there are at least two vitamins necessary for growth, one soluble in fat, the other in water. It is still an open question whether the latter is identical with the antineuritic vitamin.

Just as at one time the newly acquired knowledge of bacteria as causes of illness opened the door to an entirely new province of research of extraordinary importance, so now the discovery of vitamins – even though to a lesser degree – has opened up new vistas to medicine, and we have advanced nearer to the understanding of numerous obscure maladies. Under the influence of Eijkman's discovery, Holst, with Frölich, exposed the nature and character of scurvy. Above all by the efforts of Hopkins' pupil Mellanby, it was found that rachitis was an illness due to lack of certain substances, and others have shown similar conditions for a large number of other maladies, the last one being pellagra, the similarity in principle of which to beriberi was already indicated by Eijkman in his classic work.

At the same time, extensive and important contributions have been made to the question of the nature of the physiological processes which are affected by vitamins.

Thus the discovery of vitamins, which is this year rewarded with the Nobel Prize, implies an advance of extraordinary significance, but there is still much of importance to be discovered that can at present be but dimly discerned or suspected.

Your Excellency, Baron Sweerts de Landas Wyborgh, Sir Frederick Gowland Hopkins. Many years have passed, since Eijkman found the antineuritic principle in food, but the great importance of this work has been appreciated but slowly. Today, however, the outstanding significance of the discovery is universally acknowledged not only for our understanding and our attempt to combat beriberi, but also because it has indicated a way of investigating and controlling many other deficiency diseases.

You, Sir Frederick, have demonstrated the physiological necessity of the vitamins for normal metabolism and growth, thus very considerably extending our knowledge of the importance of vitamins for life processes as a whole.

The discoveries of the antineuritic and the growth-promoting vitamins, for which the Caroline Institute has awarded the Nobel Prize in Physiology or Medicine this year, are foundation stones of the science of vitamins. Great as has been the progress in this field, yet we may still hope to reap rich harvests in the future.

On behalf of the Caroline Institute I express its hearty congratulations to the prizemen, and I beg Your Excellency to convey to your famous countryman its felicitations. With these words I have the great honour of asking you to accept the Nobel Prize for Physiology or Medicine from the hands of His Majesty the King.

CHRISTIAAN EIJKMAN

Antineuritic vitamin and beriberi

*Nobel Lecture**

Beriberi is a disease prevalent, epidemically, in tropical and subtropical regions of Eastern Asia, where rice is the staple food of the natives; it is found elsewhere among sago-eating peoples (Molucca Islands), as well as in South America, in places where rice or cassava meal is the staple diet, as in certain parts of Brazil. However, the disease also occurs sporadically – here and there even with abundant frequency – in the temperate zone and, in some circumstances, in the frigid zone.

The symptoms of the disease are paralysis and numbness starting from the lower limbs, as well as cardiac and respiratory disorders accompanied by dropsy. These latter symptoms rapidly come to the fore, and we then speak of « wet beriberi ». Soon the motility and sensitivity disturbances, eventually accompanied by severe muscular atrophy, become more pronounced (« dry beriberi »). Mixed and transitional forms, however, are not rare. Where the paralysis is fairly advanced, the peculiar gait of the patient is noticeable. As the extensors of the foot are paralysed, the patient has to raise the knee up and swing the foot forward in order to avoid stumbling over the downhanging toes.

It is mostly young men in full vigour who are stricken by the acute form of the disease; they not infrequently die suddenly, in terrible distress through inability to breathe. This fatal issue, even in the more chronic cases, is often due to intercurrent causes, e.g. physical strain, or diseases such as malaria, dysentery, etc.

Although beriberi was described as long ago as in the first half of the 17th century by the Dutch specialist in tropical medicine Bontius, it has only shown itself as a really devastating disease in the Malay Archipelago since about the sixties of last century, and then generally speaking not so much among the free population as among people living to some extent under constraint, such as soldiers, sailors, prisoners, imported coolies in the mines

* As Professor Eijkman has been prevented by ill health from coming to Stockholm to deliver his Nobel Lecture, he has very kindly sent the text to the Editor of *Les Prix Nobel* for publication.

and plantations, and so forth. It was particularly common in the native prisons; even detainees awaiting trial, sometimes died of the disease. And in the native hospitals it claimed many a victim among people admitted for a relatively minor ailment, such as gonorrhoea or a fractured bone.

During the protracted war – a true guerilla war – which we had to wage with the Achin sultanate, both army and navy suffered very severely from the disease. Newly drafted native troops were often unfit for service after about six weeks and the pitiful remnant had to be evacuated as quickly as possible. For instance, a man might apparently be in good health in the morning, and even gave proof of his skill during target practice, and then fell victim to the disease by evening. An army doctor of that time mentions that at his hospital 18 soldiers died of beriberi on one day.

European troops also suffered from the disease in Achin, though to a lesser extent than the native troops.

Faced with this serious situation, the Home Government decided in 1886 to send out a commission to investigate the nature of beriberi and its cause. Pekelharing, Professor of Pathology, and Winkler, reader in neurology, both of the Medical Faculty of the University of Utrecht, were in charge, and I was seconded to the commission as an assistant. Winkler immediately established that the disease was essentially a form of polyneuritis (more accurately: multiple neuratrophy), a finding which agreed with results obtained by Bälz and von Scheube from investigations conducted by them in Japan. The clinical symptoms were also consistent with this pathological-anatomical diagnosis, which moreover could be confirmed in the living subject by electrical examination (reaction of degeneration, etc.).

At that time there were naturally many different theories as to the cause of beriberi. Two of these were based on the fact that the disease was particularly prevalent among rice-eating populations, although – as already mentioned – this is not always the case. On the one hand, rice poisoning was suspected; on the other, a deficiency of the rice diet, but not in the same sense as we now understand it.

With regard to the first theory, the disease was found to occur all too often where the rice was of excellent quality from the culinary point of view. Neither was any actual proof of a pre-existing poison in the rice diet found, or even sought. This also applies to other poison-theories, e.g. that the disease was due to rotten fish or to « mephitic » gases emanating from the soil.

The second theory was formulated by Van Leent in 1879 in the light of his experience with the East Indies Navy. He considered that an one-sided rice diet resulted in malnutrition owing to its very low content of protein and fat, and that this condition would promote beriberi. He found that the incidence of the disease among native sailors fell considerably when they were put on to a European diet.

There were weighty arguments against this theory, in particular the fact that people with great muscular strength and a well-developed *panniculus adiposus* were often attacked by the disease; and also that European crews, on a diet containing sufficient protein and fat, were not immune, even when they were given virtually no rice at all. In the report made after its return to Holland, the commission rejected this theory because it could not be assumed that malnutrition in itself would result in destruction of peripheral nerves. For this a directly injurious effect on the nerves or their centres would be necessary. In the light of our knowledge today, however, we see that there was a very large grain of truth in Van Leent's theory, and it can scarcely be doubted that a lasting improvement in the health of the native crews could have been achieved if it were possible to introduce the Dutch navy food and persuade the men to take it. It is however a very delicate matter to alter the diet to which people have been accustomed from their youth, and consequently with few exceptions the native sailors soon forsook pea soup with sausage and bacon, potatoes, bread with cheese, etc. and went back to the monotonous and poor rice diet, which they were usually able to procure.

Takaki, who – for reasons similar to those of Van Leent – a few years later changed the diet of the Japanese Navy to the European style, thereby – it is reported – achieving lasting success in the fight against beriberi, also had to admit that this measure was difficult to put into effect. «By last year's experience,» he said, « we have found that most of the men dislike meat as well as bread and we do not know what we shall do next.»

However this may be, we failed to check the disease permanently, although at first it did seem to subside a little. But this diminution does not prove much since in any case beriberi was subject to considerable periodic fluctuations, apart from the fact that the statistics on the disease were not very reliable. For instance it was quite obvious that these statistics frequently included cases of palpitations of the heart, oedema of the feet, as well as a new cause of death described as « hydraemia perniciosa tropica » – conditions which should obviously have been ascribed to beriberi. In those days diagnosis of beriberi was clearly taboo.

In the limited time which the commission, for official reasons, had at its disposal it was unable to carry out experimental investigations in every conceivable direction, and it is not surprising that at a time when bacteriology was all-triumphant under the leadership of Pasteur and Koch, the commission should have decided first of all to apply the methods of these two men to the problem in hand. It had all the more reason to take this line since beriberi is a disease with pronounced local and periodic tendencies (as defined by von Pettenkofer). The periodic fluctuations, the prevalence under certain climatic and weather conditions, the epidemic occurrence in certain countries and places, the association with buildings, and finally the cases of transmission of the disease described in the literature – all this pointed to an infection. True, under these circumstances a parasitic disease might have been considered – in particular hookworm, widely spread in the tropics, has been accused of being the main cause of beriberi – but the commission was able to establish that anaemia of any severity is not one of the characteristics of this disease.

I shall not dwell on the results of the bacteriological investigations because they have long since ceased to have any significance, although at that time they were apparently of fundamental importance. Polymorphic bacteria were found in the blood of patients, although not regularly, and degeneration of the nerves could be induced in animals by repeated injections of a coccus culture from them. In its report, following its return to Holland, the commission suggested very tentatively that the cause of beriberi had thus been discovered. The commission succeeded in isolating a coccus, which it considered identical with the above-mentioned one, from the air in barracks which, from its point of view, could be regarded as infected.

Following the departure of the commission I was entrusted with the task of continuing its investigations, but at first I failed to get any further. The disinfection measures which, as is logical, were recommended had not the hoped-for success. Then a chance happening put me on the right track.

A disease, in many respects strikingly similar to beriberi in man, suddenly broke out in the chicken-house at the laboratory in Batavia, and this called for a thorough study. The symptoms of this disease are as follows: The initial stages, following defective evacuation of the crop some days earlier, are characterized by an unsteady gait. The bird has difficulty in perching and has to exert itself in order not to fall; the legs are spread through weakness, and the knee and ankle joint are bent. The bird frequently collapses and falls over when walking. Finally it remains lying on its side and paresis of the wing muscles now becomes obvious from its vain efforts to get up. Paralysis of the body muscles advances rapidly from below. Within a few days the condition of the bird has so deteriorated that it can no longer eat anything without assistance; although swallowing movements are still produced, the bird is unable to lift its head. Symptoms indicating the onset of paresis of the respiratory muscles then appear. Respiration is slowed down, the beak opens, comb and skin become cyanotic, the neck is bent back and the head drawn in. The bird now becomes more and more soporific, the eyes are covered by the nictitating membrane and the body temperature falls a degree or two centigrade below normal.

This was a case of polyneuritis, as indicated by the symptoms and course of the disease and proved beyond question by a microscopic examination.

With regard to the etiology, our original supposition was not confirmed, i.e. that in the case of the strikingly epizootic occurrence of the disease an infection was involved. Attempts to induce the infection with material from affected birds or from birds which had died of the disease were inconclusive since all the chickens, even those kept separate as controls, were affected. No specific micro-organism or any higher parasite was found.

Then suddenly the disease cleared up and we were unable to continue our investigations. The affected chickens recovered and there were no new cases. Fortunately suspicion fell on the food, and rightly so, as it very soon turned out.

The laboratory was still housed provisionally and in a very makeshift manner at the military hospital, although it was administered by the civilian authorities. The laboratory keeper – as I afterwards discovered – had for the sake of economy fed the chickens on cooked rice which he had obtained from the hospital kitchen. Then the cook was replaced and his successor refused to allow military rice to be taken for civilian chickens. Thus, the chickens were fed on polished rice from 17th June to 27th November only. And the disease broke out on 10th July and cleared up during the last days of November.

Deliberate feeding experiments were then conducted in order to check more thoroughly whether or not the propable connection between diet and the disease actually existed. It was found for certain that the polyneuritis was due to the diet of cooked rice. The chickens were attacked by the disease after 3–4 weeks, and in many cases somewhat later, whereas the controls which were fed on unpolished rice remained healthy. In many cases, birds suffering from the disease could be cured by a suitable alteration in diet.

This difference between polished and whole rice did not lie in the fact that the former had not kept so well during storage, since cooked rice which had been freshly prepared from the whole kernel would also cause the disease. In addition to this, rough rice, i.e. rice with only the coarse husk removed, which deteriorates much more easily (being attacked by mites, mould fungi, etc.), proved harmless during the feeding experiments. This rice, as obtained simply by stamping, still has its inner hull – known as its « silver skin» (pericarpium) – and germ wholly or largely intact. We were then able to conclude from a series of highly varied experiments that the antineuritic principle is situated mainly in these parts of the rice kernel, and indeed of any cereal grain. It can easily be extracted with water or strong alcohol and is dialysable. I also established that it can be used medicinally either through the mouth or parenterally.

Feeding with other Amylaceae, such as sago and tapioca, had exactly the same results as cooking rice. Since these contain only traces of primary nutrients other than starch, it was impossible to dismiss out of hand the suspicion that the disease was due to simple inanition, especially as it was accompanied by considerable emaciation. And at that time – when the effectiveness of minute doses of the antineuritic principle had not yet been established – the fact that the affected birds recovered when fed on a diet consisting only of meat, could be regarded as being tantamount to the result of an experimentum crucis. On the other hand, by adding meat to the starch-rich diet it was possible to prevent the emaciation but not outbreaks of the disease, even though these were somewhat delayed. Thus, inanition in itself could not be the main cause of the disease (any more than «protein» or «salt» deficiency), even though it promoted it. I also concluded from this that an antineuritic principle must have been present in the meat, i.e. in the animal organism, but that this principle was gradually used up while the chickens were being fed on a starch diet. And finally the outcome of these experiments was encouraging for me in that one apparent difference as compared with human beriberi had thus been eliminated, for, as already stated, beriberi not infrequently attacks well-fed, strong individuals. Starvation experiments which I conducted at that time also produced a negative result. This was confirmed many times over (Holst, Shiga, & Kusama; Fraser & Stanton, et al.) before I resumed these experiments (in Holland) in view of a communication from Chamberlain et al. (1911) that they had observed in 3 out of 8 cases, polyneuritis among chickens which had been given drinking water without any solid food. This time I obtained a positive result in no less than six out of eight cases. However, it could not be concluded from this alone that the cause of the disease was general inanition, for, apart from the fact that total starvation includes partial starvation, we were able to show that the starved birds which were suffering from the disease, just as birds which had been fed on an unbalanced diet, recovered again, despite continuing loss of weight, when given 8–10 g vitamin-rich yeast per day.

Investigations with other animal species on the East Indies showed that birds such as pigeons and the Indian rice bird can also be used with success. In contrast, results of experiments with mammals were almost entirely negative. This disappointment could, of course, not prevent me from testing whether and to what extent the discovery which had been made as a result of the study of polyneuritis in birds could be used in the fight against human beriberi, although at that time there was still reason enough to doubt the identity of the two diseases. I was therefore accused of not being logical, a reproach which I am happy to pass over without comment.

The obvious thing to do was to test the preventive and curative effect of rough rice in the case of beriberi by means of experiments and controls. Rough rice is the staple diet wherever the rural population in the Indies grows rice for its own use and shelled it by primitive methods. White, or polished, rice is processed mechanically and is one of the blessings of European civilization, of our improved techniques which, for instance, have also changed the colour of bread here in Europe from brown to white. Thought-provoking too was the fact, already referred to, that the free population, which fed itself, was much less subject to beriberi than the population whose freedom was restricted, and which was often dependent on imported and therefore mechanically processed, polished, rice. (As already noted, rough rice is unsuitable for prolonged storage.)

When the projected nutritional experiments on man had just begun, an observation was made by Vorderman which had almost the value of a check experiment. As civilian medical inspector for the island of Java he knew that, in accordance with local custom, the native prisoners were given polished rice in some areas, but rough rice in others. The question, therefore, was to ascertain whether there was a connection between the nature of the staple diet and the incidence of beriberi at the prisons. This on-the-spot investigation was carried out by Vorderman at my request, on behalf of the East Indies Government, and at the same time attention was paid to other factors – apart from diet – with a bearing on the subject. As a result of this research, conducted with admirable proficiency and perseverance, my theory based on the chicken experiments was proved correct. The enquiry covered no less than 101 prisons with almost 300,000 inmates. In brief, the proportion of cases of beriberi in the prisons where polished rice was used as staple diet was some 300 times greater than in those where rough rice was used.

Vorderman's remarkable results did not at once find recognition and were received with scepticism and adverse criticism in some quarters. For me, who had meanwhile been repatriated, the results agreed so unequivocally with those of the chicken experiments that the possibility of coincidence could not seriously be considered. However, the nutritional experiments on which I had already embarked were discontinued, and the alteration in the catering regulations for prisoners, suggested after my departure by Vorderman and Grijns, in the long run did not meet with the necessary cooperation of the authorities. Only when Grijns, who continued my investigations, discovered that a certain bean, the «katjang idju» (*Phaseolus radiatus*), has the protective and curative action in the case of chickens, successful experiments – with this bean – were made on man (Roelfsema, Hulshoff Pol, and Kiewiet de Jonge). It is, however, obvious that this bean is of far less value as a staple food than a suitable rice diet.

New encouragement then came from British India (Braddon, Ellis, Fletcher, and Fraser & Stanton), on the strength of which in 1910 the following resolution was drawn up by the Far Eastern Association of Tropical Medicine meeting in Manila:

«That in the opinion of this Association, sufficient evidence has now been produced in support of the view that beriberi is associated with the continuous consumption of white (polished) rice as the staple article of diet, and that the Association accordingly desires to bring this matter to the notice of the various Governments concerned.»

And in Hong Kong, 1912:

«That the accuracy of the opinion of the Association, recorded in 1910, has received further and more complete confirmation by investigators in Japan, China, French Indo-China, the Philippine Islands, Siam, Netherlands India, the Straits Settlements and the Federated Malay States, namely that beriberi...»

Experiments on animals were naturally also used in order to trace the distribution of the antineuritic principle in living organisms. Numerous animal and vegetable foodstuffs which are more or less rich in this principle are listed in the literature on the subject. In practice this is of the greatest importance, since here too it has been found that the vitamin-rich foodstuffs can be used successfully to combat human beriberi. Concentrated extracts of these have also been prepared and these likewise have a pronounced curative action both against polyneuritis in birds and against beriberi.

Until a short while ago, however, no one had succeeded in extracting the active principle reliably and in pure form from these initial materials, although indeed efforts to do so were not wanting. And I was particularly delighted when a few years ago Jansen and Donath were able to report from the laboratory in Batavia – now a splendid improvement on my old workshop – that they had isolated the antineuritic vitamin from rice bran. It is a crystalline substance, obtained in the form of a hydrochloride, an analysis of which gives approximately the formula $C_6H_{10}ON_2 \cdot HCl$.

With this substance they conducted experiments in preventive feeding with rice birds and pigeons, and they found that an addition of 2 mg per 1 kg polished rice, i.e. in the ratio of $1:500,000$, is sufficient to give protection against polyneuritis.

They sent me approximately 40 mg, sufficient to carry out not only preventive but also curative experiments on pigeons and young cocks. I was able to confirm that such a minute addition gives protection and – which, indeed, could not have been expected otherwise – also has curative power.

Accordingly the human requirement of antineuritic vitamin can be estimated at 1–2 mg per day.

Biography

Christiaan Eijkman was born on August 11, 1858, at Nijkerk in Gelderland (The Netherlands), the seventh child of Christiaan Eijkman, the headmaster of a local school, and Johanna Alida Pool.

A year later, in 1859, the Eijkman family moved to Zaandam, where his father was appointed head of a newly founded school for advanced elementary education. It was here that Christiaan and his brothers received their early education. In 1875, after taking his preliminary examinations, Eijkman became a student at the Military Medical School of the University of Amsterdam, where he was trained as a medical officer for the Netherlands Indies Army, passing through all his examinations with honours.

From 1879 to 1881, he was an assistant of T. Place, Professor of Physiology, during which time he wrote his thesis *On Polarization of the Nerves*, which gained him his doctor's degree, with honours, on July 13, 1883. That same year he left Holland for the Indies, where he was made medical officer of health first in Semarang later at Tjilatjap, a small village on the south coast of Java, and at Padang Sidempoean in W. Sumatra. It was at Tjilatjap that he caught malaria which later so impaired his health that he, in 1885, had to return to Europe on sick-leave.

For Eijkman this was to prove a lucky event, as it enabled him to work in E. Forster's laboratory in Amsterdam, and also in Robert Koch's bacteriological laboratory in Berlin; here he came into contact with A. C. Pekelharing and C. Winkler, who were visiting the German capital before their departure to the Indies. In this way medical officer Christiaan Eijkman was seconded as assistant to the Pekelharing-Winkler mission, together with his colleague M. B. Romeny. This mission had been sent out by the Dutch Government to conduct investigations into beriberi, a disease which at that time was causing havoc in that region.

In 1887, Pekelharing and Winkler were recalled, but before their departure Pekelharing proposed to the Governor General that the laboratory which had been temporarily set up for the Commission in the Military Hospital in Batavia should be made permanent. This proposal was readily accepted, and

Christiaan Eijkman was appointed its first Director, at the same time being made Director of the «Dokter Djawa School» (Javanese Medical School). Thus ended Eijkman's short military career – now he was able to devote himself entirely to science.

Eijkman was Director of the «Geneeskundig Laboratorium» (Medical Laboratory) from January 15, 1888 to March 4, 1896, and during that time he made a number of his most important researches. These dealt first of all with the physiology of people living in tropical regions. He was able to demonstrate that a number of theories had no factual basis. Firstly he proved that in the blood of Europeans living in the tropics the number of red corpuscles, the specific gravity, the serum, and the water content, undergo no change, at least when the blood is not affected by disease which will ultimately lead to anaemia. Comparing the metabolism of the European with that of the native, he found that in the tropics as well in the temperate zone, this is entirely governed by the work carried out. Neither could he find any disparity in respiratory metabolism, perspiration, and temperature regulation. Thus Eijkman put an end to a number of speculations on the acclimatization of Europeans in the tropics which had hitherto necessitated the taking of various precautions.

But Eijkman's greatest work was in an entirely different field. He discovered, after the departure of Pekelharing and Winkler, that the real cause of beriberi was the deficiency of some vital substance in the staple food of the natives, which is located in the so-called «silver skin» (pericarpium) of the rice. This discovery has led to the concept of vitamins. This important achievement earned him the Nobel Prize in Physiology or Medicine for 1929. This late recognition of his outstanding merits has ended all criticism of his work. In addition to his work on beriberi, he occupied himself with other problems such as arach fermentation, and indeed still had time to write two textbooks for his students at the Java Medical School, one on physiology and the other on organic chemistry.

In 1898 he became successor to G. Van Overbeek de Meyer, as Professor in Hygiene and Forensic Medicine at Utrecht. His inaugural speech was entitled *Over Gezondheid en Ziekten in Tropische Gewesten* (On health and diseases in tropical regions). At Utrecht, Eijkman turned to the study of bacteriology, and carried out his well-known *fermentation test*, by means of which it can be readily established if water has been polluted by human and animal defaecation containing coli bacilli. Another research was into the rate of mortality of bacteria as a result of various external factors, whereby he was able to show that this process could not be represented by a logarithmic curve. This was followed by his investigation of the phenomenon that the rate of growth of bacteria on solid substratum often decreases, finally coming to a halt. Beyerinck's auxanographic method was applied on several occasions by Eijkman, as for example during the secretion of enzymes which break down casein or bring about haemolysis, whereby he could demonstrate the hydrolysis of fats under the influence of lipases.

As a lecturer he was known for his clarity of speech and demonstration, his great practical knowledge standing him in good stead. He had a pre-eminently critical mind and he continuously warned his students against the acceptance of dogmas. But Eijkman did not confine himself to the University he also engaged himself in problems of water supply, housing, school hygiene, physical education; as a member of the Gezondheidsraad (Health Council) and the Gezondheidscommissie (Health Commission) he participated in the struggle against alcoholism and tuberculosis. He was the founder of the Vereeniging tot Bestrijding van de Tuberculose (Society for the struggle against tuberculosis).

His unassuming personality has contributed to the fact that his great merits were at first not really appreciated in his own country; but anyone who had the privilege of coming into close contact with him, quickly perceived his keen intellect and extensive knowledge.

In 1907, Eijkman was appointed Member of the Royal Academy of Sciences (The Netherlands), after having been Correspondent since 1895. The Dutch Government conferred upon him several orders of knighthood, whereas on the occasion of the 25th anniversary of his professorship a fund has been established to enable the awarding of the Eijkman Medal. But the crown of all his work was the award of the Nobel Prize in 1929.

Eijkman was holder of the John Scott Medal, Philadelphia, and Foreign Associate of the National Academy of Sciences in Washington. He was also Honorary Fellow of the Royal Sanitary Institute in London.

In 1883, before his departure to the Indies, Eijkman married Aaltje Wigeri van Edema, who died in 1886. In Batavia, Professor Eijkman married Bertha Julie Louise van der Kemp in 1888; a son, Pieter Hendrik, who became a physician, was born in 1890.

He died in Utrecht, on November 5, 1930, after a protracted illness.

F. GOWLAND HOPKINS

The earlier history of vitamin research

Nobel Lecture, December 11, 1929

When the present century began, animal nutrition was being viewed too exclusively from the standpoint of energy requirements. The fundamental pioneer work of Rubner and its later extension to human subjects in the remarkable enterprise of Atwater, Benedict, Rosa, and others in the United States could not fail to produce a deep impression upon the thought of the time. The quantitative character of the data obtained and the attractive circumstance that such data appeared to supply a real measure of nutritional needs, independent of, and apparently superior to, considerations based upon chemical details, induced a feeling that knowledge concerning these needs had become highly adequate and was approximating even to finality. As a matter of fact, however, these calorimetric studies, invaluable in themselves, were then leading to doctrinal teaching which contained inherent errors. So fundamental an aspect of the then dominant doctrine as, for instance, the law of isodynamic equivalence among foodstuffs, is at the most approximately true, and fails entirely when the equivalence is tested by physiological results rather than by purely physical data. The assumption indeed that carbohydrates and fats can replace each other indefinitely in a diet, so long as the total energy supplied remains the same, has led to serious errors in practical dietetics. More serious in leading to error was the assumption that all proteins were of equal nutritive value, and most serious of all in this respect was, as we now know, the belief that proteins, carbohydrate, fat, and suitable inorganic materials supplied in themselves all the needs of the organism. Inhibitory, moreover, was the odd assumption often to be detected in the writings of leading authorities that to view nutritional needs from the standpoint of energetics was not alone more convenient, but more scientific, and even more philosophical, than to discuss them in terms of the material supply.

It must, of course, be most fully recognized that the calorimetric studies so dominant twenty years ago and, no less, the quantitative studies of respiratory exchange which aided and extended them, provided knowledge which is of the utmost importance and a permanent asset of science. The prime demand of the active organism is for energy, and when all the more specific demands are met in a well-balanced food supply, the available energy becomes, of course, the limiting factor which determines the adequacy of that supply. The studies in question were, moreover, usually made upon subjects consuming natural foods, and the diet was therefore, in general, sufficiently well balanced. When it was not, the inevitable brevity of calorimetric observations failed to reveal effects of dietetic deficiencies which require time for their display. Such studies have therefore retained their value and will yet doubtless be extended: but had they continued to dominate the whole field of activity, had they not been supplemented by a more discriminative enquiry into the various factors which determine real adequacy in nutrition, our knowledge would have remained highly imperfect, and our views erroneous. For now that the organic material factors present in dietaries actually capable of maintaining normal nutrition for long periods have been more carefully explored, it has become evident that, so far from comprising protein, fat, and carbohydrate alone, they are many and diverse, and each is indispensable. We have even come to know that a diet which, apparently at least, will support the individual throughout life may yet lack a factor which is necessary to maintain an adequate capacity for parentage, so the diversity of materials indispensable for normality is seen to be yet greater. The complexity of these nutritional needs as we now view them is indeed astonishing. We find them displayed in the details of the protein supply, in the call for a right balance among the inorganic constituents of a diet, and, particularly, in the urgent call of the body for a number of organic substances specific in nature and function, in respect of which, however, the quantitative supply is, in accordance with the demand, so small as to contribute little or nothing to the energy factor in nutrition. These substances, following the suggestion of Casimir Funk, we have agreed to call *vitamins*.

Who was the « discoverer » of vitamins? This question has no clear answer. So often in the development of science, a fundamental idea is foreshadowed in many quarters but has long to wait before it emerges as a basis of accepted knowledge. As in other cases, so with recognition of vitamins as physiological necessities. Their existence was foreshadowed long ago, but a certain right moment in the history of the science of nutrition had to arrive before it could attain to universal recognition. Some workers had discovered suggestive facts, but failed to realize their full significance. On the other hand, the work and words of true pioneers lay forgotten because published when average minds were not ready to appraise them at their right value.

Some fifteen or sixteen years ago, the importance of vitamins became somewhat suddenly recognized. So enormous now is the existing literature concerning them, so complex and, sometimes, so uncertain, are the issues raised, that it is impossible to survey the subject adequately in a single lecture. The circumstances of my most enviable position here today will justify me in dealing rather with the earlier history of the subject, and I will venture in virtue of that position to put before you certain personal experiences which have no place in the proper history of the subject. They have not been, and will not be, published elsewhere.

No one can deny that the recorded experience of voyagers and explorers in the eighteenth century, and particularly perhaps the records of the British Navy which deal with the incidence and cure of scurvy, would have directed thought towards our modern conception of vitamins, had the times been ripe. The knowledge concerning nutrition was then, however, entirely vague, and the days of experiment in such matters had not yet come. The earliest experiments indeed supplied evidence which was of an independent kind reached from a different angle.

It is now generally agreed that the first clear evidence, based upon experiment, for the existence of dietary factors of the nature of vitamins came from the school of Bunge at Basel. In 1881 Lunin, one of the workers in that school, fed mice upon an artificial mixture of the separate constituents of milk; of all the constituents, that is, which were then known, namely the proteins, fats, carbohydrates, and salts. He found that upon such a mixture the animals failed to survive and was led to conclude that «a natural food such as milk must therefore contain besides these known principal ingredients small quantities of unknown substances essential to life». Such a statement, already half a century old, when allowed to stand out clear and apart from a context which tended to bury it, seems to contain the essentials of what is believed today.

It will not be uninstructive to seek for an explanation as to how such a significant remark could remain for years almost without notice, instead of being an immediate challenge to further investigation. It may be noted first of all that the title of Lunin's paper, « Ueber die Bedeutung der anorganischen Salze für die Ernährung des Thieres» was one prone to conceal the important but incidental suggestion it contained. The thought of the Bunge school was, as a matter of fact, both then and long afterwards concerned largely with the inorganic factors of nutrition. After an interval of ten years, during which, so far as one can discover, no effort had been made to follow up the pregnant suggestion of Lunin's work, another member of the school came again to Lunin's point of view. He, however, was once more concerned primarily with the study of an inorganic element in nutrition. Socin, the worker in question, though, as the title of his paper, « In welcher Form ist das Eisen resorbiert? », indicates, he had not set out to seek further for those wholly unknown substances essential to life, the existence of which Lunin had predicted, was nevertheless owing to the nature of his experiments inevitably driven to a belief in them. Socin, it is true, was led to think that the ineffectiveness of the synthetic diets employed by him might be due to an inadequacy in the quality of the protein contained in them, a view not wholly unjustified and foreshadowing knowledge which only long afterwards became definite and proven. He became clear, however, that other and unknown substances must be sought, substances which, he remarks, are certainly present in whole milk and egg-yolk. He remarks, moreover, with conviction that to discover them was the first task to be faced, before new feeding experiments are undertaken. Here was a strong enough challenge to investigate, yet we have again, I think, no knowledge of any attempt in Bunge's laboratory to follow up the challenge. Perhaps an explanation of this is to be found in the circumstances that Bunge, though in his well-known book he remarks that it would be worth-while to continue the experiments which had suggested the existence of unknown nutritional factors, was, as I happen to know, himself inclined to disbelieve in them. He thought that the real error in the synthetic diets used by his pupils (which was, so to speak, « dissected milk ») was that the method of its preparation had involved the separation of inorganic constituents from certain organic combinations in which latter form alone could they adequately subserve the purposes of metabolism. Other causes may have contributed to the long neglect, first of Lunin's and then of Socin's suggestions, and it must be admitted indeed that no *experimentum crucis* was carried out in Bunge's laboratory.

In Lunin's experiments the fate of six mice only (those placed by him upon a normal salt mixture) really suggested the existence of unknown factors; and no data are given as to their consumption of food. Neither Lunin nor Socin made any attempt to complete the evidence by making discriminative additions to the diets which had proved inadequate. Lastly, as I have already suggested, since the main intention of their work, and the titles of their publications, were remote from the special issue, their significant remarks might well appear as mere *obiter dicta* when read without the light of modern developments.

Yet the pregnant suggestions arising from the observations just discussed did ultimately, though not for fourteen years after the latest of them were published, awaken (as we may suppose) the interest of an investigator distinguished in many fields who was led to repeat and extend them. I allude to the late Professor Pekelharing, whose own observations (published in 1905) unhappily again remained unknown to the majority till very recently. It is indeed astonishing that the results of such significant work as his, though published in the Dutch language alone, should not have become rapidly broadcast. I cannot refrain from referring to the circumstance that the paper was not abstracted or mentioned in Maly's *Jahresbericht für Thierchemie*, so adequate, and in general so complete, in its dealings with current literature. Many of us were accustomed to rely upon it for references to work published in journals that we could not consult, or in a language that we could not read. Though other work by Pekelharing was duly recorded at this time, no mention was made of the extraordinarily interesting paper in question. My own experiments began soon after the paper was published, and as a proportion of my own work was very similar to that of Pekelharing, I shall never cease to regret that, in common with so many others, I was then completely ignorant of the latter. After speaking of experiments carried out on lines similar to some of those done in Bunge's laboratory, and indicating that they pointed to the existence of some unknown essential, Pekelharing goes on to say: « Till now my efforts, constantly repeated during the last few years, to separate this substance and get to know more about it, have not led to a satisfactory result, so I shall not say any more about them. My intention is only to point out that there is a still unknown substance in milk which even in very small quantities is of paramount importance to nutrition. If this substance is absent, the organism loses the power properly to assimilate the well-known principal parts of food, the appetite is lost and with apparent abundance the animals die of want. Undoubtedly this substance occurs not only in milk but in all sorts of foodstuffs both of vegetable and animal origin. » Here we have a clear statement of the vitamin doctrine already a quarter of a century old. It is noteworthy that Pekelharing records prolonged endeavours towards the isolation of a vitamin.

Between the publications of Lunin, Socin and that of Pekelharing appeared the extremely important papers of Professor Eijkman. There is no need for me to discuss those remarkable publications. A most adequate review of them is fresh in your minds. The fundamental importance of Professor Eijkman's work to the whole subject of *vitamins* and *deficiency diseases* is now universally recognized. It would be difficult to rejoice wholeheartedly in one's own good fortune had I not the happiness of sharing with him the Nobel Prize for Medicine.

In my own earlier thought about the subject, though in so far as I was familiar with Eijkman's observations I could not fail to recognize their importance, yet I thought of them in a wrong and too narrow category. Eijkman's own earlier teaching as based on his results was that the function of the substance in the cortex was to neutralize a nutritional error due to excess of carbohydrate in a diet of rice. A substance which functions in the neutralization of an error is not the same thing as a substance universally necessary, and it was to the existence of substances of the latter type that my own thoughts had turned. Eijkman did not at first visualize beriberi clearly as a deficiency disease; but the view that the cortical substance in rice supplied a need rather than neutralized a poison was soon after put forward by Grijns and ultimately accepted by Professor Eijkman himself. I was myself, however, quite ignorant of these later views when my experiments were begun twenty-four years ago.

It is abundantly clear that before the last century closed, there was already ample evidence available to show that the needs of nutrition could not be adequately defined in terms of calories, proteins, and salts alone. How came it then that this limited definition was then in vogue and so remained for the first ten years of the present century, while no effective attention was given to the facts which have been discussed? The main reason, I feel sure, is that the minds of the leaders of thought in *nutritional science* were obsessed by a sense of the overwhelming importance of calorimetric studies with their impressive technique. A few experiments done mostly on small animals, and not very quantitative or crucial in their nature, were at best unimpressive. The former studies were clearly scientific and sound; the results of the latter appeared to be on a less sound basis, while deficiency diseases, in so far as their existence was proven, seemed to stand in so special a category that their indications might be neglected when the needs of normal nutrition were being estimated.

No general or widespread belief in the view that an adequate diet must contain indispensable constituents other than adequate calories, a minimum of protein, and a proper mineral supply, could be said to exist till the years 1911–1912. Those years saw the appearance of my own publications. Thereafter started a period of great activity in the study of the facts: immediately in the United States of America, a little later, and still more after the war, in many centres.

I will now refer to my own experiments, and will also intrude upon your patience with a reference to those personal experiences of which I spoke at the beginning of this address.

Early in my career I became convinced that current teaching concerning nutrition was inadequate, and while still a student in hospital in the earlier eighteen nineties I made up my mind that the part played by nutritional errors in the causation of disease was underrated. The current treatment of scurvy and rickets seemed to me to ignore the significance of the old recorded observations. I had then a great ambition to study those diseases from a nutritional standpoint; but fate decreed that I was to lose contact with clinical material. I had to employ myself in the laboratory on more academic lines. I realized, however, as did many others at the last century's close, that for a full understanding of nutrition, no less than for an understanding of so many other aspects of biochemistry, further knowledge of proteins was then a prerequisite; and when I was first called to the University of Cambridge I did my best to contribute to that knowledge.

As an ultimate outcome of my experiments dealing with the relative metabolic importance of individual amino acids from protein, my attention was inevitably turned, without, I think, knowledge, or at any rate without memory, of the earlier work, to the necessity for supplying other factors than the then recognized basal elements of diet if the growth and health of an animal were to be maintained. This indeed must at any time come home to every observer who employs in feeding experiments a synthetic dietary composed of adequately purified materials. It was the experience of the workers in Bunge's laboratory long ago; it was, as we have lately learned, the experience of Pekelharing. A good many investigators using synthetic dietaries have, it is true, from time to time expressed doubts upon the point, but we now know that it was because the constituents they used were not pure and not free from adherent vitamins. In 1906–1907 I convinced myself by experiments, carried out, as were those of Lunin and Socin, upon mice, that those small animals at any rate could not survive upon a mixture of the basal foodstuffs alone. I was especially struck at this time, I remember, by striking differences in the apparent nutritive value of different supplies of casein in my possession. One sample used as a protein supply in a synthetic dietary might support moderate growth, while another failed even to maintain the animals. I found that a sample of the former sort, if thoroughly washed with water and alcohol, lost its power of support, while addition of the very small amount of extract restored this power and also, if added to the samples originally inadequate, made them to some degree efficient in maintaining growth. I found further at that time (1906–1907) that small amounts of a yeast extract were more efficient than the casein extracts. Similar experiences were encountered when otherwise adequate mixtures of amino acids were used to replace intact proteins. By sheer good fortune, as it afterwards turned out, I used butter as a fat supply in these early experiments. Upon the evidence of these earlier results I made a public statement in 1907 which has been often quoted. I cannot, however, justly base any claims for any sort of priority upon it, as my experimental evidence was not given on that occasion. It was indeed not till four years later that I published any experimental data. In explanation of this delay I would ask you to consider the circumstances of the time. The early experiments of Lunin and others had been forgotten by most; the calorimetric studies held the field and tentative suggestions concerning their inadequacy were, I found, received with hesitation among my physiological acquaintances. It seemed that a somewhat rigid proof of the facts would be necessary before publication was desirable. Thus came the great temptation to endeavour to isolate the active substance or substances before publication, and I can claim that throughout the year 1909 I was engaged upon such attempts, though without success. At this time I was using what is now the classic subject for vitamin studies, namely the rat. As I was concerned with the maintenance of growth in the animal, the tests applied to successive products of a fractionation took much longer than those which could be used in studying the cure of polyneuritis in birds by what we have learned to call vitamin B, so the work occupied much time. I may perhaps be allowed to mention what was for me a somewhat unfortunate happening in the beginning of 1910, as it is instructive. A commercial firm had prepared for me a special extract of a very large quantity of yeast made on lines that I had found effective on a small scale. With this I intended to repeat some fractionations which had appeared promising. I thought, however, upon trial that the whole product was inactive and it was thrown away. The real explanation, however, was that instead of using butter, as in earlier experiments, I at this moment determined to use lard, and my supply of this, as I learned to understand much later, was doubtless deficient in vitamins A and D; I was now giving my animals in the main the B-group alone. If I had then had the acumen to suspect that any of the substances I was seeking might be associated with fat I should have progressed faster. Later in 1910, if I may intrude so personal a matter, I suffered a severe breakdown in health, and could do nothing further during the year. On my return to work I felt that the evidence I had by then accumulated would be greatly strengthened by a study of the energy consumption of rats, on the one hand when failing on diets free from the accessory factors (as I had then come to call them), and, on the other hand, when, as the result of the addition of minute quantities of milk, they were growing vigorously. These experiments took a long time, but they showed conclusively, as at that time it seemed necessary to show, that the failure in the former group of animals was not due primarily, or at the outset of the feeding, to any deficiency in the total uptake of food.

My 1912 paper is sometimes unfairly quoted as though its bearing applied only to the influence of minimal quantities of milk upon nutrition. It will be found, however, that it emphasizes on general lines the indispensable nature of food constituents which were then receiving no serious consideration as physiological necessities.

In a personal endeavour to estimate the influence of my publications in 1912 upon the opinion of the time, and their relative importance in the initiation of that great activity in kindred studies which shortly followed their appearance, I have found it necessary to consider at the same time, and particularly, the work and writings of Casimir Funk. It is sure that, until the period 1911–1912, the earlier suggestions in the literature pointing to the existence of vitamins lay buried. There is no evidence, I think, that they were affecting the orientation of any authoritative teaching concerning the phenomena of normal nutrition either at the time in question or indeed, in any effective sense, before.

A few years ago in the American journal *Science*, Funk published a short article in which, after giving me some credit for prophetic vision, he protests against my being called the « discoverer of vitamins ». In this protest he was justified; I have certainly never made any personal claim to be their « discoverer », and all the past circumstances of which I have reminded you, have deprived perhaps every individual worker of that clear title. Funk, however, further remarks in the article mentioned that my chief paper appeared too late to affect the situation to any appreciable degree; a remark which I believe to be entirely unjust.

F. Röhmann, an experienced worker on nutritional problems, and much concerned with the chemical side of them, but one who never fully believed in the claims made for vitamins, wrote in 1916 after discussing the earlier literature « Als der geistige Vater der Vitaminlehre ist wohl Gowland Hopkins zu betrachten, während die Bezeichnung Vitamine von Casimir Funk

herstammt. » Such a statement without extension does of course far less than
justice to Funk's influence, which in many ways was important. Funk himself, however, made no experiments which bore upon the physiological
functions of vitamins until long after my paper had appeared.

Funk's first entry into kindred fields was in a paper published in December
1911, describing his earliest efforts to isolate the curative substance from rice
polishings. He continued this effort, and further papers appeared in 1912,
when Suzuki and others were also describing their endeavours to isolate the
substance. Funk's attempt was extremely praiseworthy, and his publications
doubtless awakened new interest in Eijkman's original discovery. He did
not, however, succeed in isolating any substance which has since been accepted as being in fact any actual vitamin of the B-group, and the papers in
question contain no suggestion bearing on the general physiological importance of vitamins.

In June 1912, however, Funk published a paper dealing with the Etiology
of Deficiency Diseases, which was a valuable summary of the existing knowledge concerning such conditions, and here he showed more clearly than had
been shown before, how definite a group is constituted by these diseases. He
also emphasized in an interesting way and in advance of general opinion, the
view that pellagra would prove to be one of them. This review contains,
however, no account of personal work other than the attempts already mentioned to isolate the anti-beriberi factor. A short final section purports to
relate the available knowledge concerning deficiency diseases to the facts of
normal metabolism and nutrition. I feel justified in saying that this section
is written in a manner which is essentially disingenuous. The author says, for
instance, « I suppose that the substance facilitating growth found in milk is
similar to, if not identical with, the *vitamins* described by me. » But Funk had
then « described » no vitamins. He had in this review – the merits of which I
have already emphasized – on theoretical grounds and with reliance mainly
on the work of others – only made the doubtless very significant suggestion
that each deficiency disease might depend upon the absence of its own specific factor. He admits knowledge of the existence of my experimental results;
knowledge obtained – according to a footnote – through a « private communication ». In the article in *Science* to which I have earlier referred, Funk
states on the other hand that he had in 1912 no knowledge of my work.
I am entitled to say that early in October 1911, I gave a very full account of my results at a meeting of the English Biochemical Club, and I
certainly obtained the impression during the succeeding months that they

had become well-known to English physiologists and biochemists generally.*

In closing these references to Casimir Funk's writings and to his earlier work I would like to make clear my belief that he has not received too much, but too little, credit for his vitamin work as a whole. I venture to think, however, that he was in no sense my predecessor in the physiological field. I may say that till now I have had no intention of commenting on his remarks in the *Science* article of 1925. Only the peculiar circumstances of my present position, which seem incompatible with his view, have led me to discuss it.

Very soon after my chief paper appeared the study of vitamins was, as you know, developed with great energy and success in the United States. We owe especially to Osborne and Mendel, and to McCollum and his co-workers, the all important work which continued during the earlier years of the war. The proof on the part of the American authors of the distinction between what were then known as the «water soluble» and «fat soluble» vitamins was the first clear evidence of diversity among the vitamin factors required for growth. This discovery made all later studies more discriminating, and the pioneer work of the authors mentioned was of the greatest importance. So prominent indeed was the American work at this time, and so large a proportion did it form of the total output from 1912 to near the end of the war that, if I wished to claim that my own publications exerted any real and effective influence in starting a new movement in the study of dietetics, I should have to convince myself that they helped to direct the thoughts of the Harvard and Baltimore investigators. Anyone reading with care the succession of papers describing their experimental studies before and after the appearance of my own publications in 1912, will, I think, become convinced that such directive influence was indeed exerted. This circumstance and much correspondence received at the time from European colleagues made me feel then that my papers had served the purpose I had wished for it, namely, to direct thought concerning normal nutrition into a channel which, if not new, had been long and strangely neglected. For a time indeed

---

* On February 28th and again on March 8th, 1911, the London *Daily Mail* gave great publicity to certain statements of my own which, though distorted for journalistic purposes, made essentially clear the conception of vitamins as based upon my personal experiments. These articles were quoted in the continental and, with especial freedom, in the American press. At the time I much regretted this unsought publicity. My main paper, the publication of which was much delayed in the press, appeared in the month following Funk's review.

I thought that channel to be even new. I was at least a pioneer whose efforts were not wasted, and I am always now content to recall an opinion expressed by the late Franz Hofmeister, the most just, if also the most generous of critics. Hofmeister in 1918, after an exhaustive study of the whole literature, speaks of me as the first to realize the full significance of the facts. If that be true, and if, as well may be, that has been the view of the Nobel Commissioners who have thought me worthy of so great a reward, I can happily enjoy my good fortune.

Biography

Frederick Gowland Hopkins was born on June 20, 1861, at Eastbourne, England. His father, a bookseller in Bishopsgate Street, London, was much interested in science, but he died when Gowland was an infant. For the next ten years Gowland lived with his mother at Eastbourne, showing as a child literary rather than scientific tastes, although, when his mother gave him a microscope, he studied life on the seashore. But he read much and wrote rhymes, and in later life speculated as to whether he might not have become, if he had been encouraged to do so, a classical scholar or a naturalist. Later in his life, however, his literary ability added much to all his scientific papers and addresses.

In 1871 his mother went to live at Enfield, and Hopkins went to the City of London School. He was a bright schoolboy in several subjects, and was given a first-class in chemistry in 1874. Later, as a result of an examination at the College of Preceptors he was given a prize for science, and, at the early age of 17, when he finally left school, he published a paper in *The Entomologist* on the bombardier beetle.

After working for six months as an insurance clerk, Hopkins was articled to a consulting chemist, and subsequently, after taking a course in chemistry at the Royal School of Mines, South Kensington, London, he went to University College, London, where he took the Associateship Examination of the Institute of Chemistry, and did so well that Sir Thomas Stevenson, Home Office Analyst and expert on poisoning, engaged him as his assistant. He was then 22 years old, and he took part in several important legal cases. He then decided to take his London B.Sc. degree and graduated in the shortest possible time. In 1888, when he was 28, he went as a medical student to Guy's Hospital, London, and was immediately given the Sir William Gull Studentship there. He was awarded, during this period, a Gold Medal for Chemistry, and Honours in Materia Medica.

In 1894, when he was 32 years old, he graduated in medicine and taught for four years physiology and toxicology at Guy's Hospital. For two years he was in charge of the Chemical Department of the Clinical Research Association. In 1896 he published, with H. W. Brook, work on the halogen derivatives of proteins, and in 1898, work with S.N. Pinkus on the crystallization of blood albumins. In 1898, when attending a meeting of the Physiological Society at Cambridge, he was invited by Sir Michael Foster to move to Cambridge to develop there the chemical aspects of physiology. Biochemistry was not, at that time, recognized as a separate branch of science and Hopkins accepted the appointment. Given a lectureship at a salary of £200 a year, he added to his income by supervising undergraduates and giving tutorials, doing also for a few years, after the death of Sir Thomas Stevenson, part-time work for the Home Office. Later he was appointed Fellow and Tutor at Emmanuel College, Cambridge.

In 1902 he was given a readership in biochemistry, and in 1910 he became a Fellow of Trinity College, and an Honorary Fellow of Emmanuel College. In 1914 he was elected to the Chair of Biochemistry at Cambridge University. During all this time he had to be content with, at first, one small room in the Department of Physiology, and later with accommodation in the Balfour Laboratory; but in 1925 he was able to move his Department into the new Sir William Dunn Institute of Biochemistry which had been built to accommodate it.

Among his outstanding contributions to science was his discovery of a method for isolating tryptophan and for identifying its structure.

Subsequently he did the work which was to gain him in 1929, together with Christiaan Eijkman, who had demonstrated the association between beriberi and the consumption of decorticated rice, the Nobel Prize.

Later, Hopkins worked with Walter Fletcher on the metabolic changes occurring in muscular contractions and rigor mortis. Hopkins supplied exact methods of analysis, and devised a new colour reaction for lactic acid, and the pioneer work then done laid the foundations for the work of the Nobel Laureates, A.V. Hill and Otto Meyerhof, and also for that of many other later workers.

In 1921 he isolated a substance which he named glutathione, which is, he showed, widely distributed in the cells of plants and animals that are rapidly multiplying. Later he proved it to be the tripeptide of glutamic acid, glycine and cystein. He also discovered xanthine oxidase, a specific enzyme widely distributed in tissues and milk, which catalyzes the oxidation of the purine bases xanthine and hypoxanthine to uric acid. Hopkins thus returned to the uric acids of his earliest work, a method of determining uric acid in urine, which he first published in 1891.

Hopkins was knighted in 1925 and received the Order of Merit in 1935.

He was awarded the Royal Medal of the Royal Society of London in 1918, and its Copley Medal in 1926. From 1930 until 1935 he was President of the Royal Society, and found little time for research. During this period, however, he exerted great influence on his contemporaries.

In 1898 Hopkins married Jessie Anne Stevens. They had two daughters, one of whom Jacquetta Hawkes, is married to J. B. Priestley, the author.

Hopkins died in 1947, at the age of 86.

Physiology or Medicine 1930

KARL LANDSTEINER

«for his discovery of human blood groups»

Physiology or Medicine 1930

Presentation Speech by Professor G. Hedrén, Chairman of the Nobel Committee for Physiology or Medicine of the Royal Caroline Institute

Your Majesty, Your Royal Highnesses, Ladies and Gentlemen.

Thirty years ago, in 1900, in the course of his serological studies Landsteiner observed that when, under normal physiological conditions, blood serum of a human was added to normal blood of another human the red corpuscles in some cases coalesced into larger or smaller clusters. This observation of Landsteiner was the starting-point of his discovery of the human blood groups. In the following year, i.e. 1901, Landsteiner published his discovery that in man, blood types could be classified into three groups according to their different agglutinating properties. These agglutinating properties were identified more closely by two specific blood-cell structures, which can occur either singly or simultaneously in the same individual. A year later von Decastello and Sturli showed that there was yet another blood group. The number of blood groups in man is therefore four.

Landsteiner's discovery of the blood groups was immediately confirmed but it was a long time before anyone began to realize the great importance of the discovery. The first incentive to pay greater attention to this discovery was provided by von Dungern and Hirszfeld when in 1910 they published their investigations into the hereditary transmission of blood groups. Thereafter the blood groups became the subject of exhaustive studies, on a scale increasing year by year, in more or less all civilized countries. In order to avoid, in the publication of research on this subject, detailed descriptions – which would otherwise be necessary – of the four blood groups and their appropriate cell structures, certain short designations for the blood groups and corresponding specific cell structures have been introduced. Thus, one of the two specific cell structures, characterizing the agglutinating properties of human blood is designated by the letter A and another by B, and accordingly we speak of «blood group A» and «blood group B». These two cell structures can also occur simultaneously in the same individual, and this structure as well as the corresponding blood group is described as AB. The fourth blood-cell structure and the corresponding blood group is known as O, which is intended to indicate that people belonging to this group lack the specific blood characteristics typical of each of the other blood groups. Landsteiner had shown that under normal physiological conditions the blood serum will not agglutinate the erythrocytes of the same individual or those of other individuals with the same structure. Thus, the blood serum of people whose erythrocytes have group structure A will not agglutinate erythrocytes of this structure but it will agglutinate those of group structure B, and where the erythrocytes have group structure B the corresponding serum does not agglutinate these erythrocytes but it does agglutinate those with group structure A. Blood serum of persons whose erythrocytes have structures A as well as B, i.e. who have structure AB, does not agglutinate erythrocytes having structures A, B, or AB. Blood serum of persons belonging to blood group O agglutinates erythrocytes of persons belonging to any of the groups A, B, or AB, but erythrocytes of persons belonging to blood group O are not agglutinated by normal human blood serum. These facts constitute the actual basic principles of Landsteiner's discovery of the blood groups of mankind.

When the scientific importance of the discovery of the blood groups had been recognized – thanks to the investigations by von Dungern and Hirszfeld – research in connection with the blood groups was directed first to studies of the hereditary transmission of blood groups and also of the relative occurrence of the individual blood group in different countries and among different peoples and races. The group characteristics are handed down in accordance with Mendel's laws. The characteristics of blood groups A, B, and AB are dominant, and opposing these dominant characteristics are the recessive ones which characterize blood group O. An individual cannot belong to blood group A, B, or AB, unless the specific characteristics of these groups are present in the parents, whereas the recessive characteristics of blood group O can occur if the parents belong to any one of the four groups. If both parents belong to group O, then the children never have the characteristics of A, B, or AB. The children must then likewise belong to blood group O. If one of the parents belongs to group A and the other to group B, then the child may belong to group A or B or it may possess both characteristics and therefore belong to group AB. If one of the parents belongs to group AB and the other to group O, then in accordance with Mendel's law of segregation the AB characteristic can be segregated and the components can occur as separate characteristics in the children. If a child has the A-group structure (either A or AB), then the A-group characteristic must be present in at least one of the parents, i.e. one of them must belong to group A or AB. If the child belongs to group AB, then one of the parents must belong to group A and the other to group B, or one of the parents must belong to group AB and the other to group A or B, or else both parents must belong to group AB. Application of the discovery of blood groups in questions relating to the establishing of paternity is based on these principles governing the hereditary transmission of blood groups.

The four blood groups have been demonstrated in the populations of all countries where tests have been made. These cover the greater part of the world. It is clearly a constant physiological characteristic of man that every individual belongs to a particular blood group. However, the percentage distribution of the four blood groups varies within different populations and races. In the population of Europe, for instance, a larger proportion of individuals belongs to group A than in other parts of the Old World, and in the northern and western parts of Europe a larger proportion of individuals belongs to blood group A than in the southern and eastern parts. The varying frequency of the individual blood groups in different races points to essential constitutional differences. Here Landsteiner's discovery opened up new fields for research on the determination of the racial purity of a people. Blood group determinations have shown that if an alien race is present within a population this race retains its specific blood group characteristics, even if it has lived away from its main and original homeland for centuries. In the field of genetics the discovery of the blood groups has also proved to be of importance from the point of view of methodology in the study of the hereditary transmission of other characteristics. Landsteiner's discovery of the blood groups also prompted research on the question–important for the study of constitution – whether other body cells in addition to erythrocytes, and in particular the germinal cells, can be differentiated according to specific groups.

However, the discovery of the blood groups has also brought with it important scientific advances in the purely practical field – first and foremost in connection with blood-transfusion therapy, identification of blood, and establishing of paternity.

The transfer of blood from one person to another for therapeutic purposes began to be practised on a considerable scale during the 17th century. It was found, however, that such blood transfusion involved serious risks and not infrequently resulted in the death of the patient. Therapeutic application of the blood transfusion had therefore been almost entirely given up by the time of Landsteiner's discovery. As a result of the discovery of the blood groups it was now possible, at least in the majority of cases, to explain the cause of the dangers linked with this therapeutic measure as previous experience had shown, and at the same time to avoid them. A person from whom blood is taken must in fact belong to the same blood group as the patient. Thanks to Landsteiner's discovery of the blood groups, blood transfusions have come back into use and have saved a great many lives.

Already at the time of publishing his discovery of the blood groups in 1901, Landsteiner pointed out that the blood-group reaction could be used for investigating the origin of a blood sample, for instance of a blood stain. However, it is not possible to prove by determining the blood group that a blood sample comes from a particular individual, but it is possible to prove that it is not from a particular individual. If, for instance, the blood of a blood stain is from an individual belonging to blood group A, then it cannot be from an individual who is found to belong to group B, but a blood-group determination will not tell us from which person of blood group A the blood came.

The establishing of paternity for legal purposes has in all ages presented the legislator with insurmountable difficulties owing to the fact that paternity cannot be proved objectively. In this sphere, therefore, the legislator has had to content himself with possibilities or, at best, greater or lesser probabilities. In view of this situation with regard to proof in cases of disputed paternity it is only natural that the possibility of using the determination of blood group in such cases should have aroused general interest, from both the theoretical and the practical point of view. The use of blood-group determination in paternity actions also constitutes a significant advance in this field, even though the proof is of a negative character. A blood-group determination can, in fact, never establish paternity, but can exclude the possibility of it. However, a blood-group determination does not give results suitable for use as evidence under all circumstances. If the child in question belongs to blood group O, then a determination of the group gives no proof, because the recessive blood group in the child provides no basis for any conclusions regarding the parents, who in this case can belong to any one of the four blood groups. Only in those cases where the child belongs to a dominant blood group, i.e. A, B, or AB, and the specific blood structure of the group is not present in the mother, are the results of any value. A group structure which is present in the child but absent in the mother must have been inherited by the child from its father. If the man who is claimed to be the father belongs to a blood group different from that of the child in question, then the child cannot have inherited its blood-group characteristic from this man, and the possibility of his paternity must therefore be ruled out.

Landsteiner's discovery of the blood groups – as will be clear from what has been said – has opened up new avenues for research in several branches of science and has brought with it important advances in the purely practical field. However, it is only recently that the scientific importance of Landsteiner's discovery has been fully realized. In view of all the circumstances outlined above, the Staff of Professors of the Caroline Institute has decided to award the Nobel Prize for Physiology or Medicine, 1930, to Professor Karl Landsteiner for his discovery of the human blood groups.

Professor Karl Landsteiner.   Proffering you its felicitations to the discovery of the human blood groups, which discovery has been of such great importance for many branches of medical science, the Royal Caroline Medical Institute now invites you to receive from the hands of His Majesty the King the Nobel Prize in Physiology or Medicine.

KARL LANDSTEINER

On individual differences in human blood

Nobel Lecture, December 11, 1930

Owing to the difficulty of dealing with substances of high molecular weight we are still a long way from having determined the chemical characteristics and the constitution of proteins, which are regarded as the principal constituents of living organisms. Thus it was not the usual chemical methods but the use of serological reagents which led to an important general result in protein chemistry, namely to the knowledge that the proteins in individual animal and plant species differ and are characteristic of each species. The diversity is increased still further by the fact that the different organs contain special proteins, and it therefore appears that in living organisms specific building materials are necessary for each particular form and function, whereas man-made machines, performing a wide variety of operations, can be produced from a limited number of substances.

The problem raised by the discovery of biochemical specificity peculiar to a species – the subject of the investigations which we are about to discuss – was to establish whether the differentiation extends beyond the species and whether the individuals within a species show similar though smaller differences. Since no observations whatever had been made in this direction, I selected the simplest experimental arrangements available and the material which offered the best prospects. Accordingly, my experiment consisted of causing the blood serum and erythrocytes of different human subjects to react with one another.

The result was only to some extent as expected. With many samples there was no perceptible alteration, in other words the result was exactly the same as if the blood cells had been mixed with their own serum, but frequently a phenomenon known as agglutination – in which the serum causes the cells of the alien individual to group into clusters – occurred.

The surprising thing was that agglutination, when it occurred at all, was just as pronounced as the already familiar reactions which take place during the interaction between serum and cells of different animal species, whereas in the other cases there seemed to be no difference between the bloods of different persons. First of all, therefore, it was necessary to consider whether the physiological differences discovered between individuals were in fact those which were being sought and whether the phenomena, although observed in the case of blood of healthy persons, might not be due to endured illnesses. It soon became clear, however, that the reactions follow a pattern, which is valid for the blood of all humans, and that the peculiarities discovered are just as characteristic of the individual as are the serological features peculiar to an animal species. Basically, in fact, there are four different types of human blood, the so-called blood groups. The number of the groups follows from the fact that the erythrocytes evidently contain substances (iso-agglutinogens) with two different structures, of which both may be absent, or one or both present, in the erythrocytes of a person. This alone would still not explain the reactions; the active substances of the sera, the iso-agglutinins, must also be present in a specific distribution. This is actually the case, since every serum contains those agglutinins which react with the agglutinogens not present in the cells – a remarkable phenomenon, the cause of which is not yet known for certain. This results in certain relationships between the blood groups, which make them very easy to determine and which are shown in the following scheme. The groups are named according to the agglutinogens contained in the cells. (The sign $+$ in the table indicates agglutination.)

Serum of group	Agglutinins in serum	Erythrocytes of group			
		O	A	B	AB
O	$\alpha\beta$	–	+	+	+
A	β	–	–	+	+
B	α	–	+	–	+
AB	–	–	–	–	–

The question now arises whether iso-agglutination by normal serum is confined to human blood or whether it also occurs in animals. In fact such reactions are found but are distinct in only a small number of species and are hardly ever as regular as in man. Only the highest anthropoid apes – whose blood corpuscles, though scarcely their proteins, differ from those of man – have blood group characteristics, which, in so far as we have yet been able to establish, correspond completely to those of man.

It can be assumed that a comparative examination of a large number of animal species will help to explain how the groups are formed – a phenomenon which is not fully understood. One noteworthy result of the examination of animal blood has already been obtained. Very soon after the first observations on iso-agglutination had been made, Ehrlich and Morgenroth described experiments in which, by means of blood-solvent antibodies (iso-lysins), they demonstrated differences in the blood of goats which arose when the animals were injected with blood of other individuals of the same species. In this case, however, no typical blood groups but, instead, numerous apparently random differences were found – a result which, except possibly for the intensity of the reactions, is roughly what one might have expected. Similar investigations, especially those conducted by Todd on cattle and chickens (Landsteiner and Miller; Todd) indicated almost complete individual specificity.

The apparent contradiction between the observations on man and those on animals has recently been resolved. There were already some pointers in this direction, and I – working in conjunction with Levine – obtained significant results by using special immune sera which had been produced by injecting human blood into rabbits; these results led to the discovery of three new agglutinable factors present in all four groups. Thus, when the breakdown of groups A and AB each into two subgroups (v. Dungern and Hirszfeld; Guthrie et al.) – which had recently been subjected to a thorough study at my laboratory and by Thomsen – was taken into account, it was found there were at least 36 different types of human blood. In addition it was shown that weak iso-reactions (Unger, Guthrie et al.; Jones and Glynn; Landsteiner and Levine), which do not follow the group rule and which vary in their specificity, are more common than had previously been assumed – irregular reactions, which can indeed easily be distinguished from the typical ones and which in no way affect the validity of the rule of the four blood groups. These findings justify the assertion that very numerous individual blood differences exist in man, too, and that there are certainly other differences which could not yet be detected. Whether each individual blood really has a character of its own, or how often there is complete correspondence, we cannot yet say.

For the time being, at least, these facts have no importance with regard to the therapeutic application of the blood groups, which will be discussed later, and yet they probably have a close bearing on an important field of surgery, namely the grafting of tissues.

It had long been known that grafts, for instance of skin, were much more successful when the material to be grafted came from the same individual, and the results were similar where transplantable tumours were transferred to different strains of an animal species – as first described by Jensen. The experience of surgeons was confirmed by investigations on animals, among which the important series of experiments conducted by L. Loeb merit particular mention. Loeb's experiments consisted in the grafting of different tissues taken from an animal's own body, from related and unrelated animals of the same stock, and from members of different varieties and species. The success of the grafts was generally speaking in reciprocal relation to the degree of affinity, and in the light of the observations as a whole it was possible to conclude that the tissues of separate individuals must possess special biochemical characteristics.

The agreement between results obtained by two independent methods is so striking that the immediate assumption is that the differences which give rise, on the one hand, to the individual differences detectable by serum reactions and, on the other, to the individual-specific behaviour of grafted tissues are substantially of the same type. The reason for this assumption is that the group features can be demonstrated in the organ cells as well as in the blood. However, experiments based on this assumption – tissue grafts with blood groups taken into account – gave no clear result. But this is understandable, since the blood groups account for only some of the actually existing serological differences, and even apparently slight deviations can affect the permanent healing of tissues. This removes the doubt raised by these experiments, and the most probable assumption is that the two series of phenomena – the serological difference between individuals and the graft specificity – are basically related and rest on chemical differences of a similar kind. Consequently, the possibility of using the serum reactions for the important work of graft therapy can certainly not be ruled out, but existing knowledge justifies no more than a hope.

To the question of the chemical nature of the individual-specific substances – which I shall now examine – the answer is entirely of a negative character but it is nevertheless not without interest. The praecipitin reactions – mentioned at the beginning of this lecture – which revealed the species difference between proteins gave rise to the view that the substrates of all serological reactions were proteins or substances closely related to them. At first this view was shaken by the study of blood antigens. The solubility of specific substances in organic solvents and in particular the investigation into the heterogenetic sheep's blood antigen (which had been discovered by the Swedish pathologist Forssman in sheep's blood and organs of different animals), from which a substance specifically binding but not acting directly as an antigen can be separated by extraction with alcohol, led me to the view that the constituents of many cell antigens are not protein-like substances and only as a result of uniting with proteins become antigens, which are appropriately called « complex antigens ». This theory was strongly supported by the fact that I was able to restore the antigen action of the specific substance by mixing with protein-containing solutions.

Analogous results were obtained from a study of certain specific substances present in bacteria (Zinsser). Whereas in the case of bacteria the chemical nature of the specifically binding substances (haptens) has been determined for certain – as being colloidal polysaccharides (Avery and Heidelberger) – a conclusive result in the case of the animal cell antigens has not yet been obtained. It can nevertheless be stated that the biochemical individuality of the animal species rests on the existence of two different classes of species-specific substances (Landsteiner and Van der Scheer; Bordet and Renaux), which show basic differences in the nature of their occurrence.

With regard to the actual subject of this lecture the fact is that group-specific substances too can be extracted from the blood cells by means of alcohol, and in this state normally give rise to the formation of antibodies only in a mixture with antigenic proteins. It can therefore be concluded that the haptens vary within a species, whereas analogous serological differences between proteins are, admittedly, suspected but cannot convincingly be proved. Another peculiarity is the fact that haptens related in their reactions frequently occur in animal species which are very far apart in the zoological system. Thus, the iso-agglutinogen A is related serologically to Forssman's antigen, which is contained in sheep's blood, and therefore immune sera react both with sheep's blood and with human blood of groups A and AB, but not with blood of group O or B (Schiff and Adelsberger). Still more noteworthy is the presence of similar structures in bacteria. This emerges from the fact that lysins from sheep's blood and apparently also agglutinins for blood of group A are present in many antibacterial sera, e.g. immune sera for paratyphus bacilli, and that a dysentery serum (recently described by Eisler), which agglutinates human blood, contains antibodies which influences to a higher degree the one of the two subgroups of group A which is the less susceptible to iso-agglutinin.

According to the result of investigations on artificial complex antigens the genesis of immune iso-antibodies indicating individual differences is probably due to the fact that as a result of combination with other substances proteins peculiar to the species are enabled to induce the formation of antibodies. If, conversely, haptens identical or closely related with those of the animal are injected in association with foreign proteins, it appears that normally no antibodies arise. An example is provided by Witebsky's experiments, which showed that group-specific immune sera form, following injection of blood of group A, only in rabbits whose organs do not contain substances resembling agglutinogen A. However, experiments conducted by Sachs and Klopstock on the occurrence of Wassermann's reaction in rabbits following injection with foreign serum of mixed alcohol extracts of rabbit organs showed that the rule does not apply universally.

Whereas in this case the antibodies react only with organ extracts, O. Fischer succeeded – in experiments in which foreign serum of mixed extracts of rabbit's blood was injected – in producing auto-antibodies in rabbits which acted upon the intact blood cells but had a haemolytic action only after prior cooling, like the haemolysins which I, together with Donath, found to be the cause of the dissolution of blood in paroxysmal cold haemoglobinuria. This result and the difference between immune sera produced from extracts of erythrocytes of group O and group B, on the one hand, and from intact cells, on the other, indicate that the nature of the combining of the substances contained in the cells also has an influence on the antigen properties.

Following these brief observations on individual differences in blood and individual characteristics of the cell antigens I must now discuss the applications of the group reactions.

The relative frequency of the individual blood groups in various races has been dealt with in a well-nigh endless number of communications since L. and H. Hirschfeld made the noteworthy observation that characteristic differences in this connection are found in different races. Their most important finding was that group A is more frequent than B in northern Europeans, whereas the position is reversed in several Asiatic races. Another striking example is that of the American Indians who, when racially pure, belong almost exclusively to group O (Coca; Snyder), from which it is concluded that in the few cases where groups A and B do occur this is due to mixing of races.

I am not qualified to discuss the results of anthropological investigations on blood groups and the conclusions drawn from them, and in any case various authors differ in their opinions regarding the general principles on which interpretation should be based and regarding individual problems. Nevertheless, the majority view seems to be that the behaviour of the blood groups in conjunction with other anthropological features allows us to draw conclusions regarding the relationship and origin of human races and is of some importance to anthropological research.

One practical application of the group characteristics which immediately suggested itself was the distinguishing between human blood stains for forensic purposes. By means of the praecipitin reactions (Kraus; Bordet; Uhlenhuth) it is not difficult to determine whether a blood stain is of human or animal origin, but forensic medicine knew no way of distinguishing between blood stains from different persons. Since the iso-agglutinins and the corresponding agglutinogens will also keep for a considerable time in a dried condition, the problem can in certain cases be solved, in particular when the bloods in question, e.g. that of the accused and that of the victim, belong to different groups. Reasons for using this method do not of course occur very often and in your country in particular the occasions for using it are few and far between, but nevertheless the test–according to a report by Lattes, who was the first to use it in forensic cases–has proved useful in a number of cases and has been the basis of court verdicts and of the acquittal of accused persons.

To a far greater extent the group reactions have been used in forensic medicine for the purpose of establishing paternity. The possibility of arriving at decisions in such cases rests on the studies of the hereditary transmission of the blood groups; the principal factural results in this field we owe to the work of von Dungern and Hirszfeld. As a result of their research it became established that both agglutinogens A and B are dominant hereditary characteristics and that transmission of these characteristics follows Mendel's laws. The importance of this lies in the fact that in man there is scarcely any other unequivocally identifiable physiological characteristic with such simple hereditary behaviour. The genetic theory that there are two independent pairs of genes, formulated by the above-named authors, had to be abandoned following a statistical investigation by Bernstein. Provided that a population is sufficiently mixed the frequency of the inherited characteristics can be calculated on the basis of a certain genetic hypothesis. Bernstein made this calculation and found that the observed figures were constantly different from the figures calculated on the basis of the theory put forward by von Dungern and Hirszfeld. Complete agreement, on the other hand, was found when the calculation was based on the hypothesis that there are three allelomorphic genes localized at one position in the chromosome. The assumption also leads to certain consequences regarding the children of AB parents, and these consequences – except for a few very isolated cases, which, however, may be explained in accordance with Bernstein's theory–have likewise been proved by experience, as extensive investigations by Thomsen, Schiff, Snyder, Furuhata, and Wiener have shown, and therefore the new theory is now almost universally accepted.

As far as the forensic application is concerned the law of dominance of A and B is decisive. Thus, paternity can be excluded in all those cases where a child belongs to group A or B and where these characteristics are absent in the mother as well as in the alleged father. This test is used fairly frequently in a number of countries – especially in Germany and Austria, though also in Scandinavia. In a survey which appeared last year Schiff reported on some 5,000 forensic investigations in which paternity was excluded, in more than 8% of cases, whilst a calculation of cases in which exclusion would have been possible gives a proportion of approximately 15 to 100. In favour of the method, it can be mentioned that it has also been instrumental in inducing some fathers to recognize their illegitimate children.

It will perhaps be of interest to show how a further development of the paternity diagnosis might be possible. In the light of preliminary results (Landsteiner and Levine) on the transmission of two of the above-mentioned blood characteristics, which are detectable with immune sera and are known by the letters M and N, the most probable assumption is that their presence is due to a pair of genes of which neither is dominant with respect to the other, so that when both are present a mixed type occurs. The existence of three phaenotypes M+N−, M−N+, and M+N+ is then explained by the fact that the third corresponds to the heterozygous form whilst the first and second correspond to the homozygous forms. Accordingly the heterozygous form can be recognized directly as a special phaenotype. The consequences of this hypothesis can be seen from the following schematic representation:

Marriages	Offspring to be expected		
	$M + N +$	$M + N -$	$M - N +$
M + N + × M + N +	50	25	25
M + N + × M − N +	50	0	50
M + N + × M + N −	50	50	0
M + N − × M − N +	100	0	0
M + N − × M + N −	0	100	0
M − N + × M − N +	0	0	100

Our own observations showed some exceptions to these rules, and this prevented us from finally accepting the hypothesis. It is possible, however, that these deviations may have been due to illegitimacy or to imperfections in the experimental method, which is not so simple as determination of the group, and in fact Schiff has found complete agreement with expectation in recently communicated observations on heredity and population statistics. New, unpublished results by Wiener are almost as good.

If this hypothesis should further prove correct the possibility of excluding paternity would be approximately doubled, i.e. a judgment would be possible in roughly one case in three. Even on the basis of data already available, however, assertions can be made with a considerable degree of probability. Use of the subgroups of groups A and B may make possible a further advance (Landsteiner and Levine; Thomsen), if future experience confirms the suspected laws.

More important to practical medicine than the subject with which we have just been dealing is the use of the blood-group reaction in transfusions. It would take too long to go into the details of the interesting history of the transfusion, which goes back for centuries, namely to the time of Harvey's discovery of the circulation of the blood. The possibility of carrying out transfusions had already been debated before this, but the first successful transfusions, prompted by Harvey's great discovery, were performed by Lower on dogs, in 1666 in England, and the next year the first transfusions of animal blood to humans were carried out by Denys in France, and by Lower and King in England. Further efforts were then directed to the inventing of special appliances and led to the experience that there is no need to transfer the blood from vessel to vessel but that even defibrinated blood can be used (Bischoff, 1835). The first transfusion with human blood was probably carried out by Blundell during the first half of the 19th century.

How differently the prospects were assessed can be illustrated by two remarks, which I quote from Snyder. In a history of the Royal Society, Sprat (1607) says: « Hence arose many new experiments, and chiefly that of transfusing blood – that will probably end in extraordinary success.» Again in a *History of the Royal Society*, by Thompson (1812), we find the passage: « The expected advantages resulting from this practice have long been known to be visionary.» The aim of introducing the method into regular medical practice was not achieved, despite great efforts and lively discussions of the question, and the idea was finally abandoned and that because the operation, though often very useful, sometimes resulted in serious symptoms and even in the death of the patient.

With regard to the injection of animal blood the reason for the disasters followed from the observations of Landois, who as far back as 1875 discovered the phenomena of agglutination and haemolysis, which frequently occur when blood is brought into contact with serum from an alien species. However, it remained a mystery why the introduction of human blood into the circulation may also be dangerous, since it was regarded as obvious that serum or plasma was an inert medium as far as cells of the same species were concerned, a conviction which may have been strengthened by the fact that such sera were used in histological examinations.

The simple solution to the problem was provided by the discovery of individual differences in blood, and of the blood groups. Animal experiments and more particularly clinical experiments with cases where mistakes were made in group determinations have confirmed this connection and leave no doubt that the transfusion of agglutinable human blood is normally accompanied by harmful consequences. However, the pathogenesis of transfusion shock has not yet been fully explained.

The first blood transfusion in which the agglutinin reaction was taken into account was carried out by Ottenberg, but it was only during the emergencies of the Great War that the method of transfusion with serological selection of donor was widely adopted – a method which has since remained the normal practice.

It would be out of place here to go into such details as the sources of error in group determination, their control by direct comparison of the blood of the recipient with that of the donor, and the precaution of beginning the transfusion by injecting small quantities of blood. It should only be mentioned that it is not absolutely necessary to use blood of the same group, but that in stead of this, blood of group O for instance (see Ottenberg), the cells of which are not affected by the serum of the recipient, can also be used. In this case, however, care must be taken to exclude donors whose serum has a high agglutinin content, as this can be dangerous, especially to severely anaemic and weakened patients. Use of blood from so-called « universal donors » belonging to group O or of non-agglutinable blood of any alien group can be of great value in emergency cases and for recipients belonging to the rare blood groups.

The most obvious indication for blood transfusion is acute or chronic anaemia, e.g. as a result of wounds or lung haemorrhages, in obstetrical cases and in those of gastric and duodenal ulcer. The effect, which in cases of haemorrhage often means the saving of a patient's life, is of course primarily due to the replacement of blood, an important factor here being that the transferred erythrocytes retain for several weeks their functional capacity in the circulation. Other important effects are haemostasis due to increased coagulability and presumably also stimulation of blood regeneration in the bone marrow, as has been concluded from changes in the histological blood picture. However, transfusion therapy which used to be widely used for pernicious anaemia has now become almost superfluous as the result of the discovery of liver therapy.

Another wide field of application is shock following severe injury and operations, and it is assumed that in these cases the introduction of blood has a better effect than injection of isotonic solutions, such as the acacia (gum arabic)-containing solution of common salt recommended by Bayliss during the war. In accordance with this indication transfusions are given, often with great success, after major operations – not only for the purpose of replacing blood but also to serve as a stimulant. American surgeons also recommend the treatment before major operations where the patient is in a weakened condition.

Good results have also been obtained with haemophilia, thrombopenic purpura and to some extent with agranulocytosis, carbon-monoxide poisoning and burns, whereas with a series of other diseases, e.g. septicaemia, for which transfusion therapy has been tried, the results have been doubtful.

Some figures which I quoted in a report to the Microbiology Congress in Paris provide information on the frequency with which transfusions are given and the relative safety which has been achieved with this method– though it must be remembered that this success is partly due to the great advances in surgical methods. There is a slight variation in the statistics, as some authors in contrast to others still have isolated failures to report. Since these differences are probably connected with the transfusion technique I think I am justified in basing my judgment on the favourable reports, provided that they cover a large number of cases.

The number of transfusions given is surprisingly large, and it may well be that use of this technique has been taken too far. According to statistics which were kindly made available to me by Dr. Corwin of the Academy of Medicine, some 10,000 transfusions were given in New York during 1929. In a recently published communication by Tiber from the Bellevue Hospital in New York, 1,467 transfusions carried out during the three-and-a-half years up to July 1929 are reported. Among these there were two deaths, one due to incorrect blood-group determination, the other – also probably avoidable – involving a group-A baby in poor condition which was given a transfusion of blood from a so-called « universal donor » of group O. Three fatalities which occurred among the 1,036 cases quoted in a report by Pemberton at the Mayo Clinic were caused by errors in blood-group determination. At Kiel, as reported to me by Dr. Beck, 2,300 transfusions were carried out over a period of about five years without one fatal accident. Mild aftereffects, such as shivering and pyrexia, were felt by 2–3 % of the patients. One noteworthy case reported by Beck was that of a patient suffering from pernicious anaemia to whom he gave 87 transfusions within three-and-a-half years, without any serious symptoms.

Good though these results may be, isolated serious and even fatal accidents – which may not be due to technical errors – as well as frequent slight disturbances are still reported as we have already mentioned. It is unlikely that the differences in blood indicated by atypical iso-agglutinins were an important factor in these cases, and if this is so they could easily have been avoided. It has not been established for certain whether, as has been assumed, intense pseudo-agglutination has an injurious effect through the serum of the recipient. Some of the disturbances observed were probably due to allergy to nutritive substances present in the injected blood, whilst others were due to the action of antibodies which formed as a result of earlier transfusions. Another problem which has not yet been investigated sufficiently is whether differences exist between individual proteins, and, if so, whether these may cause antibodies to form.

All in all, the results of blood transfusions are already highly satisfactory, and we have reason to hope that a thorough study of cases with undesirable aftereffects will help us to assess the significance of the suspected causes and perhaps reveal unknown causes, and thus finally virtually eliminate the slight risks which transfusion still involves.

Apart from the solution of this practical problem, the subject with which we have been dealing can also be developed by a study of the biological problem of individual serological differences in general, and in particular by the further improvement of techniques for the finer individual differentiation of human blood as well as by a continuation of the genetic analysis of serological blood differences in humans and animals. As a result of work already done, at least two of the human chromosome pairs – apart from the sexual chromosome – can be regarded as characterized by a specific feature (see also F. Bernstein, in Z. Induktive Abstammungs-Vererbungslehre, 57 (1931) 113).

Biography

Karl Landsteiner was born in Vienna on June 14, 1868. His father, Leopold Landsteiner, a doctor of law, was a well-known journalist and newspaper publisher, who died when Karl was six years old. Karl was brought up by his mother, Fanny Hess, to whom he was so devoted that a death mask of her hung on his wall until he died. After leaving school, Landsteiner studied medicine at the Univerisity of Vienna, graduating in 1891. Even while he was a student he had begun to do biochemical research and in 1891 he published a paper on the influence of diet on the composition of blood ash. To gain further knowledge of chemistry he spent the next five years in the laboratories of Hantzsch at Zurich, Emil Fischer at Würzburg, and E. Bamberger at Munich.

Returning to Vienna, Landsteiner resumed his medical studies at the Vienna General Hospital. In 1896 he became an assistant under Max von Gruber in the Hygiene Institute at Vienna. Even at this time he was interested in the mechanisms of immunity and in the nature of antibodies. From 1898 till 1908 he held the post of assistant in the University Department of Pathological Anatomy in Vienna, the Head of which was Professor A. Weichselbaum, who had discovered the bacterial cause of meningitis, and with Fraenckel had discovered the pneumococcus. Here Landsteiner worked on morbid physiology rather than on morbid anatomy. In this he was encouraged by Weichselbaum, in spite of the criticism of others in this Institute. In 1908 Weichselbaum secured his appointment as Prosector in the Wilhelminaspital in Vienna, where he remained until 1919. In 1911 he became Professor of Pathological Anatomy in the University of Vienna, but without the corresponding salary.

Up to the year 1919, after twenty years of work on pathological anatomy, Landsteiner with a number of collaborators had published many papers on his findings in morbid anatomy and on immunology. He discovered new facts about the immunology of syphilis, added to the knowledge of the Wassermann reaction, and discovered the immunological factors which he named haptens (it then became clear that the active substances in the extracts of normal organs used in this reaction were, in fact, haptens). He made fundamental contributions to our knowledge of paroxysmal haemoglobinuria.

He also showed that the cause of poliomyelitis could be transmitted to monkeys by injecting into them material prepared by grinding up the spinal cords of children who had died from this disease, and, lacking in Vienna monkeys for further experiments, he went to the Pasteur Institute in Paris, where monkeys were available. His work there, together with that independently done by Flexner and Lewis, laid the foundations of our knowledge of the cause and immunology of poliomyelitis.

Landsteiner made numerous contributions to both pathological anatomy, histology and immunology, all of which showed, not only his meticulous care in observation and description, but also his biological understanding. But his name will no doubt always be honoured for his discovery in 1901 of, and outstanding work on, the blood groups, for which he was given the Nobel Prize for Physiology or Medicine in 1930.

In 1875 Landois had reported that, when man is given transfusions of the blood of other animals, these foreign blood corpuscles are clumped and broken up in the blood vessels of man with the liberation of haemoglobin. In 1901–1903 Landsteiner pointed out that a similar reaction may occur when the blood of one human individual is transfused, not with the blood of another animal, but with that of another human being, and that this might be the cause of shock, jaundice, and haemoglobinuria that had followed some earlier attempts at blood transfusions.

His suggestions, however, received little attention until, in 1909, he classified the bloods of human beings into the now well-known A, B, AB, and O groups and showed that transfusions between individuals of groups A or B do not result in the destruction of new blood cells and that this catastrophe occurs only when a person is transfused with the blood of a person belonging to a different group. Earlier, in 1901–1903, Landsteiner had suggested that, because the characteristics which determine the blood groups are inherited, the blood groups may be used to decide instances of doubtful paternity. Much of the subsequent work that Landsteiner and his pupils did on blood groups and the immunological uses they made of them was done, not in Vienna, but in New York. For in 1919 conditions in Vienna were such that laboratory work was very difficult and, seeing no future for Austria, Landsteiner obtained the appointment of Prosector to a small Roman Catholic Hospital at The Hague. Here he published, from 1919–1922, twelve papers on new haptens that he had discovered, on conjugates with proteins which were capable of inducing anaphylaxis and on related problems, and also on the serological specificity of the haemoglobins of different species of animals. His work in Holland came to an end when he was offered a post in the Rockefeller Institute for Medical Research in New York and he moved there together with his family. It was here that he did, in collaboration with Levine and Wiener, the further work on the blood groups which greatly extended the number of these groups, and here in collaboration with Wiener studied bleeding in the new-born, leading to the discovery of the Rh-factor in blood, which relates the human blood to the blood of the rhesus monkey.

To the end of his life, Landsteiner continued to investigate blood groups and the chemistry of antigens, antibodies and other immunological factors that occur in the blood. It was one of his great merits that he introduced chemistry into the service of serology.

Rigorously exacting in the demands he made upon himself, Landsteiner possessed untiring energy. Throughout his life he was always making observations in many fields other than those in which his main work was done (he was, for instance, responsible for having introduced dark-field illumination in the study of spirochaetes). By nature somewhat pessimistic, he preferred to live away from people.

Landsteiner married Helen Wlasto in 1916. Dr. E. Landsteiner is a son by this marriage.

In 1939 he became Emeritus Professor at the Rockefeller Institute, but continued to work as energetically as before, keeping eagerly in touch with the progress of science. It is characteristic of him that he died pipette in hand. On June 24, 1943, he had a heart attack in his laboratory and died two days later in the hospital of the Institute in which he had done such distinguished work.

Physiology or Medicine 1931

OTTO HEINRICH WARBURG

«for his discovery of the nature and mode of action of the respiratory enzyme»

Physiology or Medicine 1931

Presentation Speech by Professor E. Hammarsten, member of the Nobel Committee for Physiology or Medicine of the Royal Caroline Institute

Your Majesty, Your Royal Highnesses, Ladies and Gentlemen.

The discovery for which the Nobel Prize for Physiology or Medicine is to be awarded today concerns intracellular combustion: that fundamental vital process by which substances directly supplied to cells or stored in them are broken down into simpler components while using up oxygen. It is by this process that the energy required for other vital processes is made available to the cells in a form capable of immediate utilization.

Many famous names and many discoveries have been associated with research on this vital process, while, before natural philosophical thought was limited by the demands of accurate measurement, it was a fertile field for speculation. The life work of many savants finds a place in the volume of which Otto Warburg has written – for the time being – the last pages. The first were written by John Mayow in 1670, then less than 30 years of age, whose observations on the power of saltpetre to set fire to organic substances led him to the view that certain igneo-aerial particles existed in saltpetre, in the air, and also in organic substances. He inferred that the significance and function of respiration was to bring these particles into the body, and so make combustion therein possible. It is clear that Mayow's igneo-aerial particles correspond with oxygen, which had not yet been discovered. Some thirty years later the ill-famed phlogiston theory of combustion was born, and spread like an epidemic throughout the scientific world, causing the seeking for truth to be diverted from its proper course that had been opened by Mayow's discovery, which had, if one may use a somewhat dubious expression, been made before its time and had received little attention. Comprehension of the mechanism of combustion was thus, quite foolishly as it might seem, delayed for more than a century. Return to the proper path had to await the discovery by Lavoisier of the real nature of the process in connection with the final discovery and isolation of oxygen in the hands of Priestley and Scheele. Otto Warburg's work has met with a kinder fate.

As combustion of foodstuffs outside the body in the presence of atmospheric oxygen occurs only at high temperatures, it must be assumed that during combustion in living cells, something happens that alters the rather inert air-oxygen, or the foodstuff, or perhaps both so that they can react with each other. Fully conscious of the insuperable difficulties of explaining at present the innermost mechanism by which this inertness was overcome, Warburg decided to investigate the nature of the mysterious substance that acts as the *primus motor* in intracellular combustion. Nature often seems to use methods that appear to be indirect and less « natural » than those we should have devised, and such was the case here. It was not possible to isolate the active substance, the catalyst, or respiratory ferment as Warburg called it, by ordinary chemical methods, because it forms less than about a millionth of the weight of the cells to which it is firmly bound, while it is easily destroyed by procedures which might be used for liberating it. So, just as in modern atomic research, indirect methods had to be used.

It had been known, since the days of Davy and Berzelius, that many metals possess the power of initiating or accelerating various reactions, including combustion. Starting from the possibility that had indeed been envisaged earlier, Warburg assumed that intracellular combustion might also be regarded as being due to catalysis by metals, i.e. that it might be initiated by some metallic compound. Definite proof that he was on the track of this well-hidden secret of Nature was obtained by the use of exact measurements of combustion in living cells or, as Warburg calls it, cell respiration. The quantitatively measured variations in the process of combustion under different conditions threw light on the nature of the respiratory ferment. Its tendency to enter into compounds with substances which combine with iron showed that it is itself an iron compound, and that its effects are due to iron. The correspondence between the effects of light on cellular combustion inhibited by carbon monoxide and on carbon-monoxide compounds of certain pigments closely related to blood pigments led, with the aid of a detailed mathematical analysis to the conclusion that the respiratory ferment is a red pigment containing iron, and that it is closely related to our own blood pigment. This was the first demonstration of an effective catalyst, a ferment, in the living organism, and this identification is the more important because it throws light on a process of general significance in the maintenance of life.

Professor Warburg.   From the start, your research has been focussed on problems of central importance. Your bold ideas, but above all, your keen intelligence and rare perfection in the art of exact measurement have won for you exceptional successes, and for the science of biology some of its most valuable material.

I take the liberty of mentioning those two of your discoveries, which seem to be of the greatest value.

The medical world expects great things from your experiments on cancer and other tumours, experiments which seem already to be sufficiently far advanced to be able to furnish an explanation for at least one cause of the destructive and unlimited growth of these tumours.

Your discovery about the nature and effect of the ferment of respiration, which the Caroline Institute is rewarding this year with Alfred Nobel's Prize for Physiology or Medicine, has added a link of brilliant achievement to the chain that binds for all time, John Mayow (England), Antoine Laurent Lavoisier (France), and Otto Warburg (Germany). On behalf of the Caroline Institute I invite you to accept the prize from the hands of our King.

OTTO WARBURG

The oxygen-transferring ferment of respiration

Nobel Lecture, December 10, 1931

References to the transfer of oxygen by iron in the older literature generally apply to the iron in the blood pigment which carries molecular oxygen from the lungs to other parts of the body. The erroneous view that the iron of blood pigment causes combustion of foodstuffs, and thus transfers the oxygen not only spatially but also by chemical catalysis, is the meaning of Liebig's theory of respiration. « The blood corpuscles », said Liebig, « contain an iron compound, no other constituent of living material contains iron. »

It has only recently been recognized that iron is present in all cells, that it is vitally important, and that it is the oxidation catalyst of cellular respiration. Catalytic oxidation in living substances rests upon change of valency in an iron compound which is the respiratory oxygen-transferring ferment.

The concentration of the ferment iron in living substance is very small, being in the region of 1 g to 10 million g of cellular substance. The effects of iron are very great, and it follows that oxidation and reduction of the ferment iron must occur extremely rapidly. In fact, almost every molecule of oxygen that comes into contact with an atom of ferment iron reacts with it. In this way, ferment iron fulfils its function in almost perfect manner. The space required for a given amount of reaction is reduced to a physically possible minimum, and the only limit set to the separation of reaction from non-reaction in the microstructure of living material is the spatial arrangement of the molecules. This is the physiological meaning of the great reactivity of the cell ferments or of the fact that the *concentration* of ferments in the cellular substance is very small.

Mechanism of cell respiration

Of the two processes concerned in the *oxidation* and the *reduction* of ferment iron, the former is not in any way problematical. Complex-bound bivalent iron in compounds reacts, *in vitro* as well as in the cell, with molecular oxygen. The primary respiratory reaction can be imitated in the test-tube with pure substances of known composition, but it is *not* yet possible to reduce *in vitro* trivalent iron with the cell fuel: it is always necessary to add a substance of unknown composition, a ferment, that activates the combustible material for the attack of the iron. It must, therefore, be concluded that activation of the combustible substance in the breathing cell precedes the attack of the ferment iron; this corresponds with «hydrogen activation» as postulated in the theory of Wieland and Thunberg. According to the results of a joint research with W. Christian, this is a cleavage comparable with those known as fermentation.

It is possible that the interplay of *splitting* ferment and *oxygen-transferring* ferment does not fully explain the mechanism of cellular respiration; that the iron that reacts with the molecular oxygen does not directly oxidize the activated combustible substances, but that it exerts its effects indirectly through still other iron compounds – the three non-autoxidizable cell haemins of MacMunn, which occur in living cells according to the spectroscopic observations of MacMunn and Keilin, and which are reduced in the cell under exclusion of oxygen. It is still not possible to answer the question whether the MacMunn haemins form part of the normal respiratory cycle, i.e., whether respiration is not a simple iron catalysis but a four-fold one. The available spectroscopic observations are also consistent with the view that the MacMunn haemins in the cell are only reduced when the concentration of activated combustible substance is physiologically above normal.

This will suffice to indicate that oxygen transfer by the iron of the oxygen-transferring ferment is not the whole story of respiration. Respiration requires not only oxygen-transferring ferment and combustible substance, but oxygen-transferring ferment and the living cell.

Inhibition technique

The usual methods of analytical chemistry have not been employed for investigation of the chemical constitution of the oxygen-transferring ferment, because it was felt that the infinitesimally small concentration of the ferment and its sensitiveness made such procedures unprofitable. Instead, the «inhibition technique» was used, that is, substances were looked for which have a specific and reversible inhibitory effect on the oxygen-transferring ferment, i.e., which inhibit oxidation of the living substance. It is obvious that any substance that inactivates the ferment must react chemically with the ferment, so that inferences as to the chemical nature of the ferment can be drawn from knowledge of the character of the inhibitory substances and of the conditions under which it reacts. That is to say, the chemical reactions of the ferment are investigated, using the inhibition of some catalytic action as the indicator, instead of the usual colour reactions or production of precipitates. Obviously, in such conditions the ferment *concentration* may have any value as long as the intensity of its *action* is sufficiently great. And it is an advantage that the ferment can then be investigated under the most natural conditions in the intact and breathing cell.

I will mention here two substances that can specifically and reversibly inhibit respiration in living substance: hydrocyanic acid (for historical reasons) and carbon monoxide (for other reasons).

Inhibition of cellular respiration by prussic acid was discovered some 50 years ago by Claude Bernard, and has interested both chemists and biologists ever since. It takes place as the result of a reaction between the prussic acid and the oxygen-transferring ferment iron, that is, with the ferment iron in *trivalent* form. If trivalent iron is designated by Fe, the reaction underlying the effect of prussic acid can be written as follows:

$$FeOH + HCN \rightleftarrows FeCN + H_2O \qquad (1)$$

The oxidizing OH-group of the trivalent ferment-iron is replaced by the non-oxidizing CN-group, thus bringing transfer of oxygen to a standstill. Prussic acid inhibits *reduction* of the ferment iron.

Inhibition of respiration by carbon monoxide was discovered only a few years ago. If the initial reaction in respiration is

$$Fe + O_2 \rightarrow FeO_2 \qquad (2)$$

then, in the presence of carbon monoxide, the competing reaction

$$Fe + CO \rightleftarrows FeCO \qquad (3)$$

will also occur and, varying with the pressures of the carbon monoxide and of the oxygen, more or less of the ferment iron will be removed from the catalytic process on account of fixation of carbon monoxide to the ferment iron. Unlike prussic acid, therefore, carbon monoxide affects the bivalent iron of the ferment. Carbon monoxide inhibits *oxidation* of the ferment iron.

Thus inhibition of respiration by carbon monoxide, unlike that by prussic acid, depends upon the partial pressure of oxygen.

The toxic action of prussic acid in the human subject is based on its inhibitory action on cellular respiration. The toxic effect of carbon monoxide on man has nothing to do with inhibition of cellular respiration by carbon monoxide but is based on the reaction of carbon monoxide with blood iron. For, the effect of carbon monoxide on blood iron occurs at pressures of carbon monoxide far from the level at which cellular respiration would be inhibited.

The effect of light on inhibition of cellular respiration by carbon monoxide

Not only cellular respiration but also simpler iron catalyses are reversibly inhibited by carbon monoxide and prussic acid. If one compares such iron catalyses and their inhibitions with cellular respiration and its inhibitions, then it appears that the catalyst of cellular respiration behaves like an iron compound in which the iron is bound to nitrogen. But it would never have been possible to reach a definite conclusion if Nature had not endowed the iron compounds of carbon monoxide with the remarkable property of becoming dissociated – with splitting off from the carbon monoxide – under the action of light.

If carbon monoxide is added to the oxygen in which living cells breathe, respiration ceases, as has already been mentioned, but if exposure to ultraviolet or visible light is administered, respiration recurs. By alternate illumination and darkness it is possible to cause respiration and cessation of respiration in living, breathing cells in mixtures of carbon monoxide and oxygen. In the dark, the iron of the oxygen-transferring ferment becomes bound to carbon monoxide, whereas in the light the carbon monoxide is split off from the iron which is, thus, liberated for oxygen transfer. This fact was discovered in 1926 in collaboration with Fritz Kubowitz.

Photochemical dissociation of iron carbonyl compounds was discovered in 1891 by Mond and Langer, by exposing iron pentacarbonyl. This reaction is specific for carbonyl compounds of *iron*, most of which appear to dissociate in the presence of light, e.g., carbon-monoxide haemoglobin (John Haldane, 1897), carbon-monoxide haemochromogen (Anson and Mirsky, 1925), carbon-monoxide pyridine haemochromogen (H. A. Krebs, 1928), and carbon-monoxide ferrocystein (W. Cremer, 1929). When the photochemical dissociation of iron carbonyl compounds is measured quantitatively (we followed hereby Emil Warburg's photochemical experiments), by using monochromatic light and comparing the amount of light energy absorbed with the amount of carbon monoxide set free, it is found that Einstein's law of photochemical equivalence is very exactly fulfilled. The number of FeCO-groups set free is equal to the number of light quanta absorbed, and this is independent of the wavelength employed. For example, the equation for the light reaction of carbon-monoxide pyridine haemochromogen is

$$FeCO + 1\ h\nu \rightarrow Fe + CO$$

Photochemical dissociation of iron carbonyl compounds can be used to determine the absorption spectrum of a catalytic oxygen-transferring iron compound. One combines the catalyst in the dark with carbon monoxide, and so abolishes the oxygen-transferring power of the iron. If then this is exposed to monochromatic light of various wavelengths and of measured quantum intensity, and the effect of light W measured–the increase in the rate of catalysis – it is found that the effects of the light are proportional to the quanta absorbed.

The arrangement becomes very simple if the catalyst is present, as is usually the case, in infinitesimally low concentration in the exposed system. Then the thickness of the layers related to the amount of absorption of light can be considered to be infinitely thin, the number of quanta *absorbed* is proportional to the number of quanta *supplied by irradiation*, and the ratio of the absorption coefficients (β) of light is:

$$\frac{\beta_1}{\beta_2} = \frac{W_1}{W_2} \cdot \frac{i_2}{i_1} \tag{4}$$

Here, the effects of light W, i.e., the rate of increase of catalysis, and the incident quantum intensities i (both easily determinable figures) are on the right, while β, on the left, is the ratio (that is to be determined) of the coefficient of light β, so that the relative absorption spectrum of the catalyst, the position of the absorption bands and the intensity ratio of the bands can be estimated.

In collaboration with Erwin Negelein, this principle was employed to measure the relative absorption spectrum of the oxygen-transferring respiratory ferment. The respiration of living cells was inhibited by carbon monoxide which was mixed with the oxygen. We then irradiated with monochromatic light of various wavelengths and of measured quantum intensity, and the increase of respiration measured together with the relative absorption spectrum – according to Eq.(4). For only practically colourless cells are suitable for this type of experiment. A prerequisite for Eq.(4) is a layer infinitely thin with regard to light absorption. Thus, for instance, red blood corpuscles and green vegetable cells are not suitable.

Method for determination of the absolute absorption spectrum

The usefulness of the method is not yet attained with the determination of the relative absorption spectrum, rather can it be so elaborated as to supply the absolute absorption coefficient of the ferment.

Imagine living cells whose respiration is inhibited by carbon monoxide. If these are irradiated, respiration does not increase suddenly from the dark- to the light-value, but there is a definite, although very short, interval until the combination of carbon monoxide with the ferment is broken down

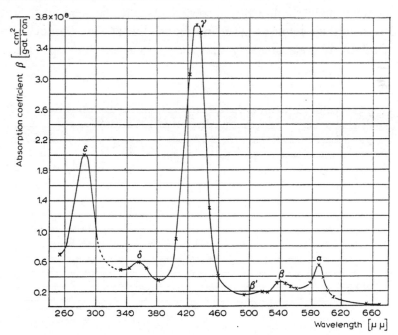

Fig. 1. Carbon-monoxide compound of the oxygen-transferring respiratory ferment.

by the light. Even without calculation, it is obvious that the rate of increase in the effect of light must be related to the depth of colour of the ferment. If the ferment absorbs strongly, the carbon-monoxide compound will be rapidly broken down, and *vice versa*.

The time of increase of the action of light can be measured. The time taken for a given intensity of light to cause dissociation of approximately half the carbon-monoxide compound of the ferment can be measured and, from this time, and from the effective intensity of light, the absolute absorption coefficient of the ferment for every wavelength can be calculated.

The absorption capacity of the ferment, measured in accordance with this principle, was found to be of the same order as the power of light absorption of our strongest pigments. If one imagines a ferment solution of molar concentration, a layer of 2×10^{-6} cm thickness would weaken the blue mercury line 436 $\mu\mu$ by half. The fact that the ferment in spite of this cannot be seen in the cells is due to its low concentration.

Absorption spectrum of the ferment

We have determined the absorption coefficients of the ferment for the region between the ultraviolet line 254 $\mu\mu$ and the red line 660 $\mu\mu$. Monochromatic light of relatively great intensity – 1/100 to 1/10 gramcalories per minute –

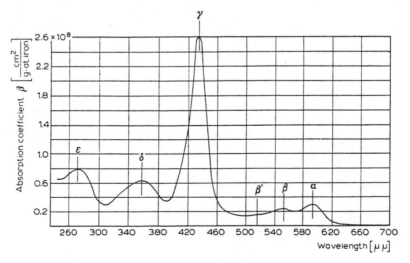

Fig. 2. Carbon-monoxide spirographis haemoglobin.

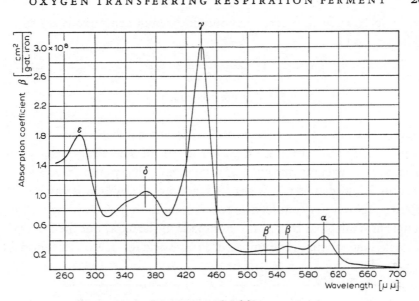

Fig. 3. Carbon-monoxide chlorocruorin.

was required for this purpose. In the first experiments, in collaboration with Negelein, 16 wavelengths were available. F. Kubowitz and E. Haas have isolated a further 15 wavelengths of sufficient intensity and purity, so that it is now possible to determine 31 points of the ferment spectrum. Our light source was a mercury-vapour lamp, and a specially designed discharger with specially high output (Dr. Hans Boas), flame carbons (from the Siemens-Plania Works*), and finally the new Pirani lamps of the Osram Study Group*. Monochromators and colour filters were used to isolate the lines from these sources of light. Table 1 shows the wavelengths isolated up to the present time, together with the absolute absorption coefficients of the carbon-monoxide compound of the ferment.

If the absorption coefficient is entered as a function of the wavelength, the absorption spectrum of the carbon-monoxide compound of the ferment is obtained, as shown in Fig. 1. The principal absorption-band or γ-band lies in the blue, while to the right of this, lie the long-wave subsidiary bands α and β in the green and yellow, and, to the left of the principal band, lie the ultraviolet subsidiary bands δ and ε. This is the spectrum of a haemin compound, according to the position of the bands, the intensity state of the bands, and the absolute magnitude of the absorption coefficients. The ab-

* We thank Dr. Patzelt (of the Siemens-Plania Works) and Dr. Krefft (of the Osram Study Group) for their valuable advice.

sorption spectra of other carbon monoxidehaemin compounds are shown in Figs. 2, 3, and 4.

Model of haemin catalysis

It appeared essential to have a control to ascertain whether haemin as an oxidation catalyst of carbon monoxide and prussic acid really behaves like the ferment. If cystein is dissolved in water containing pyridine, and a trace of haemin is added, and this is shaken with air, the cystein is catalytically oxidized by the oxygen-transferring power of the haemin. According to Krebs, the catalysis is inhibited by carbon monoxide in the dark, but the inhibition ceases when the mixture is illuminated. Prussic acid too acts on this model as

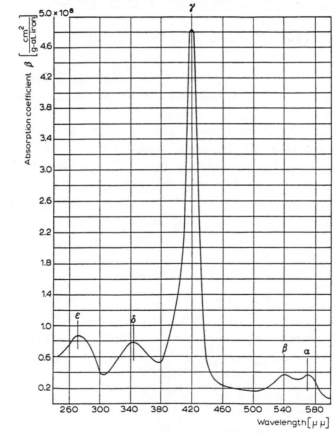

Fig. 4. Carbon-monoxide haemoglobin.

on cellular respiration, inasmuch as it combines with the trivalent haemin and inhibits its *reduction*. Just as in life, inhibition by carbon monoxide is *dependent* on the oxygen pressure, while inhibition by prussic acid is *independent* of the oxygen pressure.

Table 1.

Wave-length ($\mu\mu$)	Light source	Absolute absorption coefficient of the carbon-monoxide compound of the ferment ($cm^2/gramatom\ iron$)
253	Zinc spark	0.70×10^8
283	Magnesium spark	2.00×10^8
309	Flame carbon (aluminium salt)	
313	Mercury-vapour lamp	0.55×10^8
326	Flame carbon (copper salt)	
333	Zinc spark	0.51×10^8
344	Cadmium spark	0.50×10^8
356	Flame carbon (thallium salt)	0.59×10^8
366	Mercury-vapour lamp	0.51×10^8
383	Flame carbon (magnesium salt)	0.35×10^8
405	Mercury-vapour lamp	0.90×10^8
422	Flame carbon (strontium salt)	3.05×10^8
430	Flame carbon (calcium salt)	3.70×10^8
436	Mercury-vapour lamp	3.60×10^8
448	Magnesium spark	1.30×10^8
460	Flame carbon (lithium salt)	0.40×10^8
494	Flame carbon (magnesium salt)	0.15×10^8
517	Flame carbon (magnesium salt)	0.19×10^8
524	Flame carbon (strontium salt)	0.18×10^8
535	Thallium-vapour lamp	0.30×10^8
546	Mercury-vapour lamp	0.30×10^8
553	Flame carbon (magnesium salt)	0.26×10^8
560	Flame carbon (calcium salt)	0.23×10^8
578	Mercury-vapour lamp	0.30×10^8
589	Sodium-vapour lamp	0.54×10^8
596	Flame carbon (strontium salt)	0.38×10^8
603	Flame carbon (calcium salt)	0.20×10^8
610	Flame carbon (lithium salt)	0.12×10^8
652	Flame carbon (strontium salt)	0.02×10^8
670	Flame carbon (lithium salt)	0.005×10^8

(Measurements according to F. Kubowitz and E. Haas.)

In conjunction with Negelein, this model was also used to test the ferment experiments quantitatively. Haemin catalysis in the model was inhibited by carbon monoxide in the dark. Then monochromatic light of known quantum intensity was used to irradiate it, and the absorption spectrum of the catalyst calculated from the effect of the light which was known from direct measurements on the pure substance. The calculation gave the absorption spectrum of the haemin that had been added as a catalyst, and so the method was verified as a technique for the determination of the ferment spectrum, both the calculation and the measurement method.

Ferment bands

It is our intention to make use of the ferment bands in order to determine the chemical constitution of ferment-haemin. A few remarks on haemin and its bands must serve as an introduction.

The absolute height of the bands varies within definite limits – even for one and the same haemin. Variations depend upon the salt concentration, the solvent, etc. If the height of one band decreases, its breadth generally increases, whereby the area within the limits of the band appears to remain constant. The absolute height of the bands is only significant inasmuch as the dimension must harmonize.

The ultraviolet bands of haemin are indicated in free haemin, but are only fully shown when the haemins are bound to protein. The ultraviolet bands shown to the left of the main bands in Figs. 1 to 4 are related to the protein components of the haemin compound and are not of interest here since it is the constitution of the haemin components that is significant.

The principal absorption band in the blue and the subsidiary band of the longest wavelength (the α-band) are suitable for the chemical classification of the compounds of haemin. Later it will be necessary to use the second long-wave band (the β-band) also, but its position in the ferment has not yet been determined with sufficient exactness.

The positions of the principal band and α-band of the ferment are:

	Principal band	α-band
Carbon-monoxide compound of ferment	433 $\mu\mu$	590 $\mu\mu$

These will be referred to as the « ferment bands » because the ferment was the first for which they were determined.

Classification of the haemins

Haemins are the complex iron compounds of the porphyrins, in which two valencies of the iron are bound to nitrogen. The porphyrins, of which Hans Fischer determined the chemical structure, are tetrapyrrol compounds in which the four pyrrol nuclei are held together by four interposed methine groups in the α-position.

Green, red, and mixed shades of haemins are known. If magnesium is replaced by iron in chlorophyll, green haemins are obtained. Their colour is due to a strong band in the red which is already recognized in chlorophyll. The ferment does *not* absorb in the red and cannot, therefore, be a green haemin.

Red haemins are the usual haemins in blood pigment and in its related substances, such as mesohaemin and deuterohaemin. Coprohaemin is also a red haemin which is an iron compound of the coproporphyrin that H. Fischer recognized in the body. Other red haemins are pyrrohaemin, phyllohaemin and rhodohaemin, whose porphyrins Willstätter has prepared by complete reductive breakdown of chlorophyll. The positions of the principal absorption band and of the α-band of the carbon-monoxide compound of red haemins are:

	Principal band	α-band
Carbon-monoxide compound of red haemins	420 $\mu\mu$ and shorter wavelengths	570 $\mu\mu$ and shorter wavelengths

The ferment bands are at least 13 to 20 $\mu\mu$ nearer the red than the haemin bands. It follows that the ferment is *not* a red haemin.

Between the green and red haemins are those of mixed colour which are so called because, in solution, very slight changes in the thickness of a layer make them appear green or red. The corresponding porphyrins – which their discoverer Hans Fischer called pheoporphyrins – are formed when chlorophyll is carefully reduced with hydriodic acid. Phylloerythrin also, is a pheoporphyrin that is formed by reduction of chlorophyll in the intestinal canal of ruminants and which Löbisch and Fischler have isolated from ox bile. The pheoporphyrins are closely related to blood pigment but, as H. Fischer showed, pheoporphyrin a is simply mesoporphyrin in which the *one* propionic acid has been oxidized so that ring closure with the porphyrin nucleus is made possible. Pheoporphyrin a is a reduction product of chlorophyll a or an oxidation product of blood pigment, and connects together, in an amazingly simple manner, the principal pigments of the organic world the blood pigment and the leaf pigment.

The bands of Fischer's pheoporphyrins are shifted towards the bands of the blood pigment to the red, i.e., in the direction of the ferment bands but not to such an extent as to make them identical with them. Chlorophyll *b* has, in general, bands of longer wavelength than chlorophyll *a*, and for this reason, W. Christian and I applied Fischer's reduction method to it. In this way we obtained pheohaemin *b*, which, when linked with protein, corresponds with the ferment in respect to the position of the principal band. The bands of the carbon-monoxide compound of pheohaemoglobin *b* are:

	Principal band	*α-band*
Carbon-monoxide compound of pheohaemoglobin *b*	435 $\mu\mu$	598 $\mu\mu$

While the principal band of pheohaemoglobin corresponds with the ferment bands within the permitted limits, the α-band shifts so far beyond them because it lies too near the red. It is, nevertheless, interesting that when chlorophyll *b* is reduced, one obtains a pheoporphyrin of which the haemin of all the pheohaemins that have been demonstrated up to the present time is the most like the ferment.

Still nearer the ferment in its spectrum, is a haemin occurring in Nature. This is spirographis haemin, which has been isolated from chlorocruorin, the blood pigment of the bristle-worm *Spirographis*, in collaboration with Negelein and Haas. The bands of spirographis haemin, coupled to globin, are:

	Principal band	*α-band*
Carbon-monoxide compound of spirographis haemoglobin	434 $\mu\mu$	594 $\mu\mu$

Constitution of spirographis haemin

It follows from what has already been said that the chemical structure of spirographis haemin is important, but because of the difficulty of obtaining sufficient quantities of crystallized, analytically pure haemin, experiments on it are still incomplete. Working with Negelein, it has so far been found that spirographis haemin and also crystallized spirographis porphyrin (that has also been analysed) contain two carboxyl groups and five oxygen atoms, and therefore a surplus oxygen atom. With hydroxyl amine, this gives an oxime and is then a typical ketone oxygen. Spirographis haemin differs from the red haemins by the surplus or ketone oxygen-atom, and is classified as pheohaemin. Like Fischer's pheohaemins, spirographis haemin is intermediate between chlorophyll and blood pigment in respect of the degree of oxidation of the side-chains.

Formation and disappearance of the ferment bands

The two haemins with a spectrum most like that of the ferment – pheohaemin b and spirographis haemin – possess a remarkable property. If they are dissolved in dilute sodium-hydroxide solution, in the form of ferrous compounds, the absorption bands slowly wander towards the blue, near the bands of blood haemin. In this way, mixed-colour haemins have been converted into red haemins. On acidification, the change reverts: the « blood bands » disappear and the ferment bands appear. This experiment shows that oxidation of the side-chains does not suffice to give rise to the ferment bands, but some process of the type of anhydride formation must also occur.

This reaction, which is the chemical basis of the development of the ferment bands, will not be further discussed here: only the principle on which our work is based will be indicated. Physics brings the ferment bands into existence but organic chemistry is necessary for the identification or creation of these bands. As Anson and Mirsky have said, the procedure is similar to the spectroscopic analysis of the stars. Indeed, the ferment substance – though being so near to us – is, like the substance of the stars, inaccessible for us.

The common origin of haemoglobin and chlorophyll

If oxygen is passed through an aqueous solution of spirographis haemin, at ordinary temperature and under certain conditions, the haemin is oxidized. The previously mixed colour of the solution becomes green, and a band resembling that of chlorophyll appears in the red at 650 $\mu\mu$. On the other hand, if hydrogen is passed through a solution of spirographis haemin at 37°, in the presence of palladium, the spirographis haemin undergoes reduction in the side-chain and a haemin resembling that of blood is formed. This is a genuine red haemin* which does not become mixed-coloured when acidified.

The unique intermediate status of the ferment-like haemins demonstrated by these simple experiments suggests the suspicion that blood pigment and leaf pigments have both arisen from the ferment – blood pigments by reduction, and leaf pigment by oxidation. For evidently, the ferment existed earlier than haemoglobin and chlorophyll.

The investigations on the oxygen-transferring ferment have been supported from the start by the Notgemeinschaft der deutschen Wissenschaft and the Rockefeller Foundation, without whose help they could not have been carried out. I have to thank both organizations here.

---

* According to its spectrum and the hydrochloric-acid number of its porphyrin, this haemin closely resembles mesohaemin, but has a free methine group in the β-position. For this reason spirographis haemin (C_{32}) contains two C atoms less than blood haemin (C_{34}). (Experiments in collaboration with E. Negelein.)

Biography

Otto Heinrich Warburg was born on October 8, 1883, in Freiburg, Baden. His father, the physicist Emil Warburg, was President of the Physikalische Reichsanstalt, Wirklicher Geheimer Oberregierungsrat. Otto studied chemistry under the great Emil Fischer, and gained the degree, Doctor of Chemistry (Berlin), in 1906. He then studied under von Krehl and obtained the degree, Doctor of Medicine (Heidelberg), in 1911. He served in the Prussian Horse Guards during World War I. In 1918 he was appointed Professor at the Kaiser Wilhelm Institute for Biology, Berlin-Dahlem. Since 1931 he is Director of the Kaiser Wilhelm Institute for Cell Physiology, there, a donation of the Rockefeller Foundation to the Kaiser Wilhelm Gesellschaft, founded the previous year.

Warburg's early researches with Fischer were in the polypeptide field. At Heidelberg he worked on the process of oxidation. His special interest in the investigation of vital processes by physical and chemical methods led to attempts to relate these processes to phenomena of the inorganic world. His methods involved detailed studies on the assimilation of carbon dioxide in plants, the metabolism of tumors, and the chemical constituent of the oxygen-transferring respiratory ferment. Warburg was never a teacher, and he has always been grateful for his opportunities to devote his whole time to scientific research. His later researches at the Kaiser Wilhelm Institute have led to the discovery that the flavins and the nicotinamide were the active groups of the hydrogen-transferring enzymes. This, together with the iron-oxygenase discovered earlier, has given a complete account of the oxidations and reductions in the living world. For his discovery of the nature and mode of action of the respiratory enzyme, the Nobel Prize has been awarded to him in 1931. This discovery has opened up new ways in the fields of cellular metabolism and cellular respiration. He has shown, among other things, that cancerous cells can live and develop, even in the absence of oxygen.

In addition to many publications of a minor nature, Warburg is the author of *Stoffwechsel der Tumoren* (1926), *Katalytische Wirkungen der lebendigen Substanz* (1928), *Schwermetalle als Wirkungsgruppen von Fermenten* (1946), *Was-*

serstoffübertragende Fermente (1948), *Mechanism of Photosynthesis* (1951), *Entstehung der Krebszellen* (1955), and *Weiterentwicklung der zellphysiologischen Methoden* (1962). In the last years he added to the problems of his Institute: chemotherapeutics of cancer, and the mechanism of X-ray's action. In photosynthesis he discovered with Dean Burk the 1-quantum reaction that splits the CO_2, activated by the respiration.

Otto Warburg is a Foreign Member of the Royal Society, London (1934) and a member of the Academies of Berlin, Halle, Copenhagen, Rome, and India. He has gained l'Ordre pour le Mérite, the Great Cross, and the Star and Shoulder Ribbon of the Bundesrepublik. In 1965 he was made doctor honoris causa at Oxford University.

He is unmarried and has always been interested in equine sport as a pastime.

Physiology or Medicine 1932

Sir CHARLES SCOTT SHERRINGTON

EDGAR DOUGLAS ADRIAN

«for their discoveries regarding the functions of neurons»

Physiology or Medicine 1932

Presentation Speech by Professor G. Liljestrand, member of the Staff of Professors of the Royal Caroline Institute

Your Majesty, Your Royal Highnesses, Ladies and Gentlemen.

Within the domain of physiology and medicine probably few spheres will be calculated to attract to themselves attention to the same extent as the nervous system, that distributor of rapid messages between the various parts of the body and, beyond that, the material foundation of mental life. An oft-used picture likens the nervous system to a telephone or telegraph system in the body, where the nerves are the cables, while the brain and spinal cord may be regarded as immense stations with numberless coupling possibilities. To obtain a clearer insight into this complicated machinery, its construction, and its own proper features, has been associated with great difficulties. Comprehensive investigations, especially by the two scientists Golgi and Cajal – both of them in their time rewarded by the Nobel Prize – have shown, however, that it is mainly built up of a very large number of characteristic elements or units, which have been given the name «neurones». Each of these consists of a cell, where certain parts are transformed for their special tasks into long processes or runners. Some of these – sometimes a metre or more in length – form part of just the lines which are united in the nerve cables, while others run within the spinal cord and the brain. The afferent or sensory neurones take messages from the surface of the body or internal organs to the stations, the efferent or motor neurones convey orders from these to the muscles and glands. In the stations special neurones can be coupled in between these two kinds.

Of fundamental importance for our knowledge of the workings of the nervous system was the discovery that an external influence, a so-called stimulus, can, without the cooperation of the will, call forth a definite response, such as the contraction of certain muscles. A well-known example is presented by the involuntary blinking at a loud and unexpected noise. The external influence is, so to speak, thrown back or reflected, from which the phenomenon received the name «reflex». For every one of our movements, even under the influence of the will, for numerous processes in the interior of the body, and in all probability also for mental life itself in its various forms, the reflexes play a highly important role. As a rule they are provoked by cooperation between groups of afferent, internuncial, and efferent neurones.

Sir Charles Sherrington has made extraordinary contributions to our knowledge of the reflex phenomena. In exact experiments employing quantitative methods he has investigated numerous reflexes, and also single neurones, with the object of establishing general laws for the origin and cooperation of the reflexes in the organism.

While a muscle which has been at rest becomes quite relaxed immediately after death, this is not the case with the living organ in a healthy person, where the rest is only apparent. Thus, even during sleep, but still more under the influence of a considerable load, e.g. standing, the muscle exhibits a varying degree of persistent but weak tension. This is due to reflexes released, as Sherrington has shown, principally in such a way that every stretching of the muscle affects special formations situated in its interior – a sort of reception apparatus or sense organ – from which signals are sent to the spinal cord, after which a degree of tension suitable to the conditions is mobilized in the muscle. From this the latter acquires a certain plasticity, it gives the body and its constituent parts the necessary stability and is ever ready.

When a reflex movement is provoked, a number of muscles usually contract in varying degrees. But, furthermore, Sherrington has found that this activity is as a rule accompanied by relaxation or inhibition of muscles whose effect is in the opposite sense. In bending, for example, the tension in the extensors is decreased and vice versa. As, in addition, every separate muscle receives a large number of nerve fibres, it is a complicated problem which, even in that apparently simple case, has to be dealt with in the station. Simultaneously, or in rapid succession, thousands of messages are received and deciphered, and the consequent coupling-in carried out in such a way that the movement is precise and appropriate. In more compound movements, such as walking or running, the various reflexes sometimes mesh like the cogs in a precision instrument, an extraordinarily complicated interplay thereby becoming necessary. To Sherrington belongs principally the credit of having solved the problem of how this is accomplished.

It has appeared from his investigations that a discharge from a motor neurone to the muscular fibres occurs when a condition of sufficient stimulus or tension is developed in the neurone, as a result of the impulses which have flowed in from various quarters. But here, as so often, different kinds of influences can make themselves felt in conflicting senses, nay, even one and the same external influence can result in conflicting effects on different neurones or even, under varying conditions, on the same neurone. In this connection what is of the greatest importance is the condition of the station itself, such as the degree of fatigue, or, in other cases, of specially increased susceptibility. The neurone has the capacity of to a certain extent gathering up and summing these different, simultaneous or rapidly succeeding impulses; the inhibiting and stimulating forces can then wholly or partly counterbalance each other, and the result will be decided by which of them obtains the upper hand for the time being. Both are equally necessary for the normal course of the reflexes and they must cooperate intimately. In many cases they obtain the mastery in turn, as in the case of rhythmic reflexes.

I must content myself with this short indication of Sherrington's considerable contributions. His discoveries have ushered in a new epoch in the physiology of the nervous system. On the firm foundation he has laid, many have already built further – among them should be mentioned particularly Magnus's and de Kleyn's brilliant work on the posture of the body, how it is assumed and maintained. But Sherrington's work has already partly passed through the ordeal of fire which lies in its application to pathological conditions; it has shown itself to be of great importance for the understanding of certain disturbances within the nervous system, and, certainly, matters here are still in their infancy.

While Sherrington has devoted his attention particularly to the reflexes as a whole, and very specially investigated how the coupling-in at the stations takes place under the influence of various factors, his fellow-countryman, Edgar Douglas Adrian, has attempted to illuminate the question of the nature of the processes connected with the lines to and from the stations and also within the receiving apparatuses, i.e. the sense organs. He has availed himself of the fact which has been well-known since the middle of last century, that activity in an organ is usually accompanied by electric changes, in that a region in action becomes negatively charged in relation to one that is at rest.

This was proved as regards the sense organs by our fellow-countryman Fritiof Holmgren in 1866. Such so-called « action currents » appear also in the nerves, where they flow with moderate speed; in the same way as one can listen to a conversation from a telephone wire one ought to be able to obtain a conception of the ingoing and outgoing messages or impulses by diverting the action currents from the nerves. Certainly it is a matter of excessively weak currents, but as the microscope once opened for investigators new fields within the world of form, modern technical progress has afforded unsuspected possibilities for studying the functions. For his purpose Adrian used radio amplifiers, by means of which he could increase the effects thousands of times and yet get them reproduced accurately. When he attempted in this way to divert the currents from a nerve under natural conditions, e.g. the signals which are sent when a muscle is stretched, he obtained irregular effects, difficult of interpretation. The explanation of this had already been given: the impulses in the different nerve fibres do not come simultaneously, they can therefore nullify or amplify each other. The situation may be compared with an attempt to construct the separate conversations by listening to the different wires in a telephone cable simultaneously. It was therefore necessary to try to obtain impulses corresponding to one single conversation or one sending station, and in this, by means of special artifices, Adrian and his collaborators were successful for both afferent and efferent neurones – thereby preparing the way for important discoveries. Adrian and his school were able to show that if the receiving apparatuses in the muscle were stimulated by means of various powerful loads, the size of the impulses was nevertheless unchanged. This was in agreement with the conclusion which had already been arrived at – inter alia through Adrian's own work: the single nerve fibre gives, as the expression is, all or none. The light which falls on the retina of the eye, the slight contact of the skin, or the factors which cause pain in a wound, all exercise their influence, as Adrian has shown, by giving rise to impulses of fundamentally the same kind in the nerve fibres by the mediation of the special sense organs. Of them all, it is also true that a more intensive external influence, such as a stronger flood of light, or a more powerful pressure, calls forth an ever more rapid stream of impulses up to a maximum value determined by the character of the nerve; in addition the stronger stimulus engages ever more single nerve fibres. But the orders issued to the muscles and nerves are also of this character. Thus the signals are the same everywhere, but the receiving stations change and the results with them. The sending stations, also, may be differently constituted; if the external influence which gives rise to the impulses remains unchanged, the rapidity of the impulses gradually diminishes, but the rate varies for different cases. The sense organs have thus a varying power of adapting themselves to their milieu and only respond to changes in it. These circumstances afford important points of contact as to the physiological tasks of the various sense organs and the connection between external influences and our sensations.

Adrian's investigations have given us a highly important insight into the question of the nerve principle and the adaptability of the sense organs. In reality they open new paths within important fields which have only to a slight extent been accessible for research hitherto.

As will appear from the above, Sherrington's and Adrian's discoveries concerning the function of the neurone deal mainly with different sides of the matter. Together, however, they provide a complete picture of the course of events, which implies a great step forward and gives research a new starting-point of the greatest importance in its perpetual struggle for clearer insight.

Sir Charles Sherrington and Professor Adrian. Twenty-six years ago the Nobel Prize for Physiology or Medicine was given to Golgi and Cajal who laid the foundation for the modern conception of the structure of the nervous system. It is with the function of that system that your work is concerned.

You, Sir Charles, in famous researches, in part already classic, in part still proceeding with outstanding success, have contributed more than anybody else to our knowledge of what you have termed the integrative action of the nervous system. Your numerous discoveries in this field have profoundly influenced our science and will certainly continue to do so in the future.

To you, Professor Adrian, is due the opening up of new lines of research of great importance and promise for neurophysiology. This has been amply demonstrated by your own discoveries concerning the nature of the nervous impulses and the physical basis of sensation.

The Caroline Institute has decided to award this year's Nobel Prize for Physiology or Medicine to you jointly for your discoveries regarding the function of the neurone.

On behalf of the Institute I offer you its hearty congratulations on your proud achievements, so worthy of the great English school of physiology. With these words I have the honour of asking you to accept the prize from the hands of His Majesty the King.

CHARLES S. SHERRINGTON

Inhibition as a coordinative factor

Nobel Lecture, December 12, 1932

That a muscle on irritation of its nerve contracts had already long been familiar to physiology when the 19th century found a nerve which when irritated prevented its muscle from contracting. This observation seemed for a time too strange to be believed. Its truth did not gain acceptance for ten years; but at last in 1848 the Webers accepted the fact at its face value and proclaimed the vagus nerve to be inhibitory of the heart muscle. Two hundred years earlier Descartes, in writing the *De Homine*, had assumed that muscle was supplied with nerves which caused muscular relaxation. An analogous suggestion was put forward by Charles Bell in 1819. The inhibition suggested was in each case «peripheral». «Peripheral» inhibition, despite its inherent probability, was however to prove void of the fact for skeletal muscle. As just said, it did in fact prove true for the heart; it was found somewhat later to hold good likewise for visceral muscle; and, somewhat later still, was found for the constrictor muscles of the blood vessels. Peripheral inhibition became thus by the sixties and seventies of the 19th century a recognized fact, save for the one important exception of the skeletal muscles.

The first experimental indication of inhibition as a process working *within* the nervous system itself appeared in 1863. Setschenov then noted in the frog that the local reflexes of the limb are depressed by stimulation of the exposed midbrain. Later (1881), somewhat similarly, stimulation of the foot (dog) was found to restrain movements of the foot excited from the brain (Bubnoff and Heidenhain). Matters had, broadly put, reached and remained at that stage, when in the century's last decade experimental examination of mammalian reflexes detected (1892) examples of inhibition of surprising potency and machine-like regularity, readily obtainable from the mammalian spinal cord in its action on the extensors of the hind limb; the inhibitory relaxation of the extension was linked with concomitant reflex contraction of their antagonistic muscles, the flexors. This «reciprocal innervation» was quickly found to be of wide occurrence in reflex actions operating the skeletal musculature. Its openness to examination in preparations with « tonic » background (decerebrate rigidity) made it a welcome and immediate opportunity for the more precise study of inhibition as a central nervous process.

The seat of this inhibition was soon shown to be central, e.g. for spinal reflexes, in the grey matter of the spinal cord. The resulting relaxation of the muscle was found to be both in range and nicety as amenable to grading as is reflex contraction itself. In other words the inhibitory process was found capable of no less delicate quantitative adjustment than is the excitatory process. In « reciprocal innervation » the two effects, excitation and inhibition, ran broadly *pari passu*; a weak stimulus evoked weak inhibitory relaxation along with weak excitatory contraction in the antagonist muscle; a strong stimulus evoked greater and quicker relaxation accompanying greater and speedier contraction of the antagonist. No evidence was forthcoming that the centripetal nervous impulses which on their central arrival give rise to inhibition differ in nature from nerve impulses giving rise centrally to « excitation », or indeed differ from the impulses travelling nerve fibres elsewhere. An « inhibitory » afferent nerve emerged simply as an afferent nerve whose impulses at certain central loci cause, directly or indirectly, inhibition, while at other central loci the same nerve, probably even the same nerve fibre can produce excitation. There was no satisfactory evidence that an afferent nerve fibre whose end-effect is inhibitory ever for its end-effect at that same locus evokes excitation or indeed any other effect than inhibition. That is to say its inhibitory influence never changes to an excitatory influence, or vice versa. Fixity of central effect, inhibitory or excitatory respectively, has to be accepted for the individual afferent fibre acting in a specified direction, i.e. on a specified individual effector unit. That does not of course exclude the contingency that an inhibitory influence on a given unit may under some circumstances be unable to produce effective inhibition there owing to its being too weak to overcome concurrent excitation.

I will not dwell upon the features of reciprocal innervation; they are well known. I would only remark that owing to the wide occurrence of reciprocal innervation it was not unnatural to suppose at first that the entire scope of reflex inhibition lay within the ambit of the taxis of antagonistic muscles and antagonistic movements. Further study of central nervous action, however, finds central inhibition too extensive and ubiquitous to make it likely that it is confined solely to the taxis of antagonistic muscles.

In instance let us take a reflex especially facile and regular to type, the well-known spinal flexion-reflex of the leg, evoked by stimulation of any afferent nerve of the leg itself. Its experimental stimulus may be reduced to a single induction shock evoking a single volley of centripetal impulses in the bared afferent nerve. The reflex effect, observed in an isolated flexor muscle, e.g. of the ankle, is apart from exceptional circumstances, a single contraction wave indicating discharge of a single volley of motor impulses from the spinal centre. This « twitch-reflex », recorded isometrically by the myograph, exhibits a tension proportional to the number of motor units engaged, in other words to the size of the single centrifugal impulse volley. The contraction of each motor unit is on the all-or-nothing principle. The maximal contraction-tension for the reflex twitch will be reached only when all of the motor units composing the muscle are activated. The contraction-tension developed by the reflex being proportional to the number of motor units engaged, an average contraction-tension value for the individual motor unit can be found. The contraction developed by the reflex twitch is less the weaker the induction shock exciting the afferent nerve, in other words the fewer the afferent fibres excited, in short, the smaller the size of the centripetal impulse volley. With a given single-shock stimulus the tension developed by the reflex twitch remains closely constant when sampled at not too frequent intervals. In the case of the spinal flexion-reflex therefore, though with many other reflexes it is not so, a standard reflex twitch of desired size (tension) can be obtained at repeated intervals.

The only index available at present for inhibition is its effect on excitation; thus, a standard twitch-reflex, representing a standard-sized volley of centrifugal discharge, can serve as a quantitative test for reflex inhibition. It serves for this with less ambiguity than does a reflex tetanus. In the tetanus the tension developed will depend within limits on the repetitive-frequency of the contraction waves forming the tetanus. Maximal tetanic contraction is reached only when the frequency reaches a rate which, in many reflex tetani, some of the units do not attain. In reflexes the rate of tetanic discharge can differ from unit to unit in one and the same muscle at one and the same time. The rate will differ too at different stages of the same reflex and according as the reflex is weak or strong. Reflex inhibition acting against a reflex tetanic contraction may diminish the contraction in one or other or all of several different ways. In some units it may suppress the motor discharge altogether, in some it may merely slow the motor discharge thus lessening the wave frequency of the contraction and so the tension. The same aggregate diminution of tension may thus be brought about variously and by various combinations of ways, a result too equivocal for analysis. The same gross result might accrue (*a*) from total suppression of activity in some units or (*b*) from mere slackening of discharge in a larger number of units. These difficulties of interpretation are avoided by using as gauge for inhibition a standard reflex twitch. The deficit of contraction-tension then observed shows unequivocally the number of motor units inhibited out of the total activated for the standard. Since the direct maximal motor twitch compared with the standard reflex twitch can reveal the proportion of the whole muscle which the standard reflex twitch activates, we can find further what proportion of the whole muscle is reflexly inhibited. Of course subliminal excitation and subliminal inhibition are not revealed by the test and require other means for detection.

A stable excitatory twitch-reflex as standard allows us to proceed further in our quantitative examination of inhibition. We then find that inhibition can be admixt in our simple-seeming flexion-reflex itself, and indeed usually is so. To detect it we have simply to add to the earlier excitation of the reflex a following one at not too long interval; we then find the response to the second stimulus-volley partly cut down by an inhibition latent in the first.

This is usually evident with intervals between 300–1,200σ. The very shortest interval at which the inhibitory effect occurs is difficult to determine, for the reason that the excitatory effect has a subliminal fringe and the second stimulus repeats the subliminal effect of the first, and the two subliminal effects can sum to liminal. The second response is therefore enlarged by summation of subliminal fringe in some of the responsive motor units. This activation by the second stimulus of some motor units facilitated for it by the first though not activated by the first alone tends of course to obscure the inhibitory inactivation; the shrinkage due to the latter is offset by the increment due to the former. The inhibition is traceable only by the net diminution of the second reflex twitch. How quickly the inhibitory element in the stimulus develops centrally is not fully ascertainable, because the sooner the second reflex follows on the first the more the facilitation from it that it gets. This increment will conceal at least in part the decrement due to inhibition. Similarly the beginning of the inhibition may be concealed from observation by concomitant excitatory facilitation. This uncertainty does not attach to the longer intervals between the two stimuli because the central inhibitory process considerably outlasts the central excitatory facilitation.

The reflex therefore, which at first sight seems a purely excitatory reaction, proves on closer examination to be in fact a commingled excitation and inhibition. Usually clearly demonstrable in the simple spinal condition of the reflex, this complexity of character is yet more evident in the decerebrate condition.

We may hesitate to generalize from this example, because a stimulus applied to a bared afferent nerve is of course «artificial» in as much as it is applied to an anatomical collection of nerve fibres not homogeneous in function; and, we may suppose, not usually excited together. If cutaneous, its fibres will belong to such different species of sense as «touch» and «pain» which often provoke movements of opposite direction and are therefore in their effect on a given muscle opposed in effect. That a strong stimulus to such an afferent nerve, exciting most or all of its fibres, should in regard to a given muscle develop inhibition and excitation concurrently is not surprising.

With weak stimuli the case is somewhat different. Such stimuli excite only a few of the constituent fibres of the afferent nerve, and those of similar calibre, presumably an indication of some functional likeness. Nevertheless, as shown above, the reflex result even then exhibits admixed excitatory and inhibitory influence on one and the same given muscle. And this admixture of excitation and inhibition persists when the stimulus is reduced in strength still further so as to be merely liminal. It still is so when the afferent nerve chosen is homogeneous in the sense that it is a purely muscular afferent, e.g. the afferent from one head of the gastrocnemius muscle. But we must remember that the afferent nerve from an extensor muscle has been shown to contain fibres which exert opposite reflex influences upon their own muscle, some exciting and some inhibiting that muscle's contraction. This brings with it the question whether admixture of exciting and inhibiting influence in the reflex effect obtains when instead of stimulation of a bared nerve some more «natural» stimulation is employed.

For this the reflex evoked by passive flexion of the knee in the decerebrate preparation has been taken. The single-joint extensor (vasto-crureus) of each knee is isolated; and nothing but that muscle pair thus retained is still innervated in the whole of the two limbs. The preparation thus obtained is a tonic preparation; one of the two muscles is then stretched by passively flexing a knee. This passive flexion excites in the extensor muscle which it stretches a reflex relaxation, i.e. the lengthening reaction; this relaxation at one knee is accompanied in the opposite fellow vasto-crureus by a reflex contraction enhancing the existing «tonic» contraction. The reflex contraction thus provoked is characteristically deliberate and smooth in performance and passes without overshoot into a maintained extension posture. Let however the manœuvre be then repeated with the one difference of condition, that the muscle contralateral to that which is passively stretched has been deafferented. In the deafferented muscle contraction is still obtained, and more easily than before, but the deafferented condition of the muscle alters the course of its contraction in two respects. The course is no longer deliberate. The contraction is an abrupt rush, with overshoot of the succeeding postural contraction, and this latter is hardly maintained at all. The severance of the afferent nerve has removed a reflex self-restraint from the contracting muscle. Normally the proprioceptives of the contracting muscle put a brake on the speed of the contracting muscle (autogenous inhibition). The explosive rush and momentum of these deafferented extensor reflexes recall the ataxy of *tabes*. They recall also the abruptness and overshooting of the « willed » movements of a deafferented limb. In both cases a normal self-braking has been lost along with the deprivation of the muscle of its own proprioceptive afferents. These latter mediate both a self-braking and a self-exciting (autogenous excitation) reflex action of the muscle. Thus here again there is admixture of reflex inhibition and excitation, and in this case the admixture obtains in response to a « natural » stimulation. Here therefore the admixture of central inhibition with central excitation is a normal feature of a natural reflex.

This makes it clear that for the study of normal nervous coordinations we require to know how central inhibition and excitation interact. As said above, the centripetal impulses which evoke inhibition do not differ in nature from those which evoke excitation. Inhibition like excitation can be induced in a « resting » centre. The only test we have for the inhibition is excitation. Existence of an excited state is not a prerequisite for the production of inhibition; inhibition can exist apart from excitation no less than, when called forth against an excitation already in progress, it can suppress or moderate it. The centripetal volley which excites a « centre » finds, if preceded by an inhibitory volley, the centre so treated is already irresponsive or partly so.

A first question is, are there degrees of « central inhibitory state »; and are they, like central excitatory state, capable of summation. This can be examined in several ways. Thus: against the central inhibition caused by a given single volley of inhibitory impulses a standard single volley of excitatory impulses can be launched at an appropriate interval. The relatively long duration of the central inhibitory state allows a second inhibitory volley to be interpolated between the original inhibitory volley and the standard excitatory volley. The standard excitation is found to be then diminished (as shown by the twitch-contraction which it evokes) more than it is if subjected to either one inhibitory volley only. This holds even when the second inhibitory volley, launched from the same cathode as the first, is arranged to be clearly smaller than the first. Since the distribution of the effect of the smaller impulse volley (launched from the same cathode as the larger) among the motoneurones of the centre must lie completely included within that of the first, the added inhibition due to the second volley indicates that the combined influence of the two volleys prevents activation of some motoneurones which neither inhibitory volley acting alone was able to prevent from being activated. Evidently therefore central inhibition sums; consequently it is capable of subliminal existence. Also, successive subliminal degrees of inhibition can by temporal overlap sum to supraliminal degree. In these ways central inhibition presents analogy with its converse « central excitations»; both exhibit various degrees of intensity in respect to the individual motoneurone.

Summation of inhibition is well exhibited when a given twitch-reflex is evoked at various times during and after a tetanic inhibition. The cutting down of the reflex twitch is progressively greater, as within limits, the inhibitory tetanus proceeds. After cessation of the tetanus the inhibitory state, similarly tested, passes off gradually, more quickly at first than later.

The relatively long persistence of the central inhibitory state induced by a single centripetal impulse volley allows examination of the effect on it of two successive excitation volleys as compared with one of the two alone. An excitatory volley is interpolated between the inhibitory volley and a subsequent standard excitatory volley. The interpolated excitatory volley is found to lessen the inhibitory effect upon the final excitatory volley. The interpolated excitation volley neutralizes some of the inhibition which otherwise would have counteracted the final test excitation. Just as central inhibitory state (c.i.s.) counteracts central excitatory state (c.e.s.) so c.e.s. neutralizes c.i.s. The mutual inactivation is quantitative. There occurs at the individual neurone an algebraic summation of the values of the two opposed influences.

It is still early to venture any definite view of the intimate nature of « central inhibition ». It is commonly held that nerve excitation consists essentially in the local depolarization of a polarized membrane on the surface of the neurone. As to « central excitation », it is difficult to suppose such depolarization of the cell surface can be graded any more than can that of the fibre.

But its antecedent step (facilitation) might be graded, e.g. subliminal. Local depolarization having occurred the difference of potential thus arisen gives a current which disrupts the adjacent polarization membrane, and so the « excitation » travels. As to inhibition the suggestion is made that it consists in the temporary stabilization of the surface membrane which excitation would break down. As tested against a standard excitation the inhibitory stabilization is found to present various degrees of stability. The inhibitory stabilization of the membrane might be pictured as a heightening of the « resting » polarization, somewhat on the lines of an electrotonus. Unlike the excitation-depolarization it would not travel; and, in fact, the inhibitory state does not travel.

The quantitative character of the interaction between opposed inhibition and excitation is experimentally demonstrable. Thus: a given inhibitory tetanus exerted on a certain set of motoneurones fails to prevent their excitation in response to strong stimulation of a given afferent nerve; but when the stimulation of the excitatory afferent is weaker the given standard inhibitory tetanus does prevent the response of the motor neurones to the excitatory stimulation. With the weaker stimulation of the afferent nerve there are fewer of its fibres acting, and therefore fewer converge for central effect on some of the units. On these the standard c.i.s. has therefore less c.e.s. to counteract.

Many features characteristic of reflex myographic records of various type become interpretable in light of the stimulus volley from a single afferent nerve trunk, even small, evoking an admixture of inhibition and excitation, with consequent central conflict and interaction between them. Features which find facile explanation in this way are the following. (A) The flexion-reflex (spinal) commonly has a *d'emblée* opening; that is, a steep initial contraction passes abruptly into a plateau, giving an approximately rectangular beginning to the myogram. Here the initial reflex excitation is closely followed by an ensuing reflex inhibition commingled with and partially counteracting the concurrent excitation. (B) Allied to this and of analogous explanation is the so-called « fountain »-form of flexion-reflex. After the first uprush of contraction a component of reflex inhibition grows relatively more potent and the contraction-tension drops low before continuing-level. Between these extreme forms there are intermediates. The key to the production of them all is admixture of central excitation with central inhibition; the excitation is prepotent earlier, and later suffers from encroaching inhibition.

(C) Again, the typical opening of the crossed extensor reflex (decerebrate)

« recruits ». A variably long latent period precedes a contraction which climbs slowly, taking perhaps seconds to reach its plateau. Here, struggling with excitation, inhibition has the upper hand at first. The action currents of the muscle marking the serial stimuli to the afferent nerve are not choked by secondary waves of after-discharge. The concurrent inhibition cuts them out. The inhibition is traceable partly to the proprioceptive reflex mechanism attached to the contracting muscle itself; the progress of the reflex contraction is partly freed from inhibition by deafferenting the muscle, but still not wholly freed. A residuum of inhibition in the reflex is traceable to the crossed afferent nerve employed. This again illustrates the ubiquitous commingling of inhibition and excitation in the spinal and decerebrate reflexes evoked by direct stimulation of afferent nerves.

An instance of combination of excitation and inhibition for coordinative effect is the rhythmic reflex of stepping. In the « spinal » cat and dog there occurs « stepping » of the hind limbs; it starts when the « spinal » hind limbs, lifted from the ground, hang freely, the animal being supported vertically from the shoulders. The extensor phase in one limb occurs with the flexor phase in the other. This « stepping » can also be evoked by a stigmatic electrode carrying a mild tetanic current to a point in the cross-face of the cut spinal cord. The « stepping » then opens with flexion in the ipsilateral hind limb accompanied by extension in the contralateral. To reproduce this stepping movement by appropriately timed repetitions of tetanization of, for instance, a flexion producing afferent of one limb or an extension-producing afferent of the other never succeeds even remotely in exciting the rhythmic stepping. In the true rhythmic movement itself, which has been examined particularly by Graham Brown, the contraction in each phase develops smoothly to a climax and then as smoothly declines, waxing and waning much as does the activity of the diaphragm in normal inspiration. But although this rhythmically intermittent tetanus affecting alternately the flexors and extensors of the limb and giving the reflex step cannot be copied reflexly by employing excitation alone, it can be easily and faithfully reproduced and with perfect alternation of phase and with its characteristic asymmetrical bilaterality, by employing a stimulation in which reflex excitation and reflex inhibition are admixt in approximately balanced intensity. The result is then a rhythmic sea-saw about a neutral point. The effect on the individual motor unit appears then to run its course thus: if we start to trace the cycle with the moment when c.e. and c.i. are so equal as to cancel out, the state of the motoneurone is a zero state, for which the term « rest », although often applied to it, is perhaps better avoided. With supervention of preponderance of c.e. over c.i. the motor neurone's discharge commences and under progressive increase of that preponderance the frequency of discharge increases in the individual motor neurone, and more motor neurones are «recruited» for action until in due course the preponderance of c.e. begins to fail and c.i. in its turn asserts itself more. The recruitment and frequency of discharge begin to wane, and then reach their lowest, and may cease, and an interval of zero state or quiescense may ensue. The quiescence may be inhibitory or merely lack of excitation. Which of these it were could be directly determined only by testing the threshold of excitation. However brought about, it is synchronous with the excitation-phase in the antagonistic muscle and with the excitation-phase in the symmetrical fellow muscles of the opposite limb. Since reciprocal innervation has been observed to obtain between these muscles, the phase of lapse of excitation is probably one of fuller active inhibition. The rhythm induced by stimulation of the «stepping»-point in the cut face of the lateral column of the cord would seem to act therefore by evoking concurrently excitation and inhibition, and so playing them off one against the other as to induce alternate dominance of each. Intensifying the mild current applied to the point quickens the tempo of the rhythm, i.e. of the alternation.

Another class of events revealing inhibition as a factor wide and decisive in the working of the central nervous system is presented by the «release» phenomenon of Hughlings Jackson. The depression of activity called «shock» supervenes on injury of a distant but related part; conversely there supervenes often an over-action due likewise to injury or destruction of some distant but related part. «Shock» is traceable to loss of excitatory influence, which, though perhaps commonly subliminal in itself, lowers the threshold for other excitation. The over-action conversely is traceable to loss of inhibitory influence, perhaps subliminal in itself and yet helping concurrent influences of like direction to maintain a normal restraint, the normal height of threshold against excitation. Where the relation between one group of muscles and another, e.g. between flexors and extensors, is reciprocal, the effect of removal (by trauma or disease) of some influence exerted by another part of the nervous system is commonly two-fold in direction. There is «shock», i.e. depression of excitability in one field of the double mechanism and «release», i.e. exaltation of excitability, in another. Thus spinal transection, cutting off the hind-limb spinal reflexes from prespinal centres inflicts «shocks» on the extensor half-centre and produces «release» of the flexor half-centre. In this case the direction both of the «shock» and of the «release» runs aborally; but it can run the other way, as in the influence that the hind-limb centres have on the fore-limb. Which way it runs, of course, depends simply on the relative anatomical situation of the influencing and the influenced centres.

The role of inhibition in the working of the central nervous system has proved to be more and more extensive and more and more fundamental as experiment has advanced in examining it. Reflex inhibition can no longer be regarded merely as a factor specially developed for dealing with the antagonism of opponent muscles acting at various hinge-joints. Its role as a coordinative factor comprises that, and goes beyond that. In the working of the central nervous machinery inhibition seems as ubiquitous and as frequent as is excitation itself. The whole quantitative grading of the operations of the spinal cord and brain appears to rest upon mutual interaction between the two central processes «excitation» and «inhibition», the one no less important than the other. For example, no operation can be more important as a basis of coordination for a motor act than adjustment of the quantity of contraction, e.g. of the number of motor units employed and the intensity of their individual tetanic activity. This now appears as the outcome of nice co-adjustment of excitation and inhibition upon each of all the individual units which cooperate in the act.

In reflexes, even under simple spinal or decerebrate conditions, interplay between excitation and inhibition is commonly induced even by the simplest stimulus. It need not surprise us therefore that variability of reflex result is met by the experimenter. Indeed, that it troubles him by being partly beyond his control, need not surprise him in view of the multiplicity and complicity of the sources of the inhibition and of the excitation. This variability seems underestimated by those who regard reflex action as too rigid to provide a prototype for cerebral behaviour. It is in virtue of their containing inhibition and excitation admixt that, in accord with central conditions prevailing for the time being, a limb-reflex provoked by a given stimulus in the decerebrate preparation can on one occasion be opposite in direction to what it is on another, e.g. extension instead of flexion («reversal»). Excitation and inhibition are both present from the very stimulus outset and are pitted against one another. The central circumstances may favour one at one time, the other at another. Again, if the quantity of contraction needed normally for a given act be reached by algebraic summation of central excitation and inhibition, it can obviously be attained by variously compounded quantities of those two. Hence when disease or injury has caused a deficit of excitation, a readjustment of concurrent inhibition offers a means of arriving once more at the normal quantity required. The admixture of inhibition and excitation as a mechanism for coordination thus provides a means of understanding the remarkable « compensations » which restore in course of time, and even quickly, the muscular competence for execution of an act which has been damaged by central nervous lesions. More than one way for doing the same thing is provided by the natural constitution of the nervous system. This luxury of means of compassing a given combination seems to offer the means of restitution of an act after its impairment or loss in one of its several forms.

Biography

Charles Scott Sherrington was born on November 27, 1857, at Islington, London. He was the son of James Norton Sherrington, of Caister, Great Yarmouth, who died when Sherrington was a young child. Sherrington's mother later married Dr. Caleb Rose of Ipswich, a good classical scholar and a noted archaeologist, whose interest in the English artists of the Norwich School no doubt gave Sherrington the interest in art that he retained throughout his life.

In 1876 Sherrington began medical studies at St. Thomas's Hospital and in 1878 passed the primary examination of the Royal College of Surgeons, and a year later the primary examination for the Fellowship of that College. After a short stay at Edinburgh he went, in 1879, to Cambridge as a non-collegiate student studying physiology under Michael Foster, and in 1880 entered Gonville and Caius College there.

In 1881 he attended a medical congress in London at which Sir Michael Foster discussed the work of Sir Charles Bell and others on the experimental study of the functions of nerves that was then being done in England and elsewhere in Europe. At this congress controversy arose about the effects of excisions of parts of the cortex of the brains of dogs and monkeys done by Ferrier and Goltz of Strasbourg. Subsequently, Sherrington worked on this problem in Cambridge with Langley, and with him published, in 1884, a paper on it. In this manner Sherrington was introduced to the neurological work to which he afterwards devoted his life.

In 1883 Sherrington became Demonstrator of Anatomy at Cambridge under Professor Sir George Humphrey, and during the winter session of 1883–1884 at St. Thomas's Hospital he demonstrated histology.

The years 1884 and 1885 were eventful ones for Sherrington, for during the winter of 1884–1885 he worked with Goltz at Strasbourg, in 1884 he obtained his M.R.C.S., and in 1885 a First Class in the Natural Sciences Tripos at Cambridge with distinction. During this year he published a paper of his own on the subject of Goltz's dogs. In 1885 he also took his M.B. degree at Cambridge and in 1886 his L.R.C.P.

In 1885 Sherrington went, as a member of a Committee of the Association for Research in Medicine, to Spain to study an outbreak of cholera, and in 1886 he visited the Venice district also to investigate the same disease, the material then obtained being examined in Berlin under the supervision of Virchow, who later sent Sherrington to Robert Koch for a six weeks' course in technique. Sherrington stayed with Koch to do research in bacteriology for a year, and in 1887 he was appointed Lecturer in Systematic Physiology at St. Thomas's Hospital, London, and also was elected a Fellow of Gonville and Caius College, Cambridge. In 1891 he was appointed in succession to Sir Victor Horsley, Professor and Superintendent of the Brown Institute for Advanced Physiological and Pathological Research in London. In 1895 he became Professor of Physiology at the University of Liverpool.

During his earlier years in Cambridge, Sherrington, influenced by W. H. Gaskell and by the Spanish neurologist, Ramón y Cajal, whom he had met during his visit to Spain, took up the study of the spinal cord. By 1891 his mind had turned to the problems of spinal reflexes, which were being much discussed at that time, and Sherrington published several papers on this subject and, during 1892–1894, others on the efferent nerve supply of muscles. Later, from 1893–1897, he studied the distribution of the segmented skin fields, and made the important discovery that about one-third of the nerve fibres in a nerve supplying a muscle are efferent, the remainder being motor.

At Liverpool he returned to his earlier study of the problem of the innervation of antagonistic muscles and showed that reflex inhibition played an important part in this. In addition to this, however, he was studying the connection between the brain and the spinal cord by way of the pyramidal tract, and he was at this time visited by the American surgeon Harvey Cushing, then a young man, who stayed with him for eight months.

In 1906 he published his well-known book: *The Integrative Action of the Nervous System*, being his Silliman Lectures held at Yale University the previous year, and in 1913 he was invited to become Waynfleet Professor of Physiology at Oxford, a post for which he had unsuccessfully applied in 1895, and here he remained until his retirement in 1936. Here he wrote, and published in 1919, his classic book entitled *Mammalian Physiology: a Course of Practical Exercises*, and here he regularly taught the students for whom this book was written.

In physique Sherrington was a well-built, but not very tall man with a strong constitution which enabled him to carry out prolonged researches.

During the First World War, as Chairman of the Industrial Fatigue Board, he worked for a time in a shell factory at Birmingham, and the daily shift of 13 hours, with a Sunday shift of 9 hours, did not, at the age of 57, tire him. From his early years he was short-sighted, but he often worked without spectacles.

The predominant notes of his character as a man were his humility and friendliness and the generosity with which he gave to others his advice and valuable time. An interesting feature of him is that he published, in 1925, a book of verse entitled *The Assaying of Brabantius and other Verse*, which caused one reviewer to hope that « Miss Sherrington » would publish more verse. He was also sensitive to the music of prose, and this and the poet in him, but also the biologist and philosopher, were evident in his Rede Lecture at Cambridge in 1933 on *The Brain and its Mechanism*, in which he denied our scientific right to join mental with physiological experience.

The philosopher in him ultimately found expression in his great book, *Man on his Nature*, which was the published title of the Gifford Lectures for 1937–1938, which Sherrington gave. As is well known, this book, published in 1940, centres round the life and views of the 16th century French physician Jean Fernel and round Sherrington's own views. In 1946 Sherrington published another volume entitled *The Endeavour of Jean Fernel*.

Sherrington was elected a Fellow of the Royal Society of London in 1893, where he gave the Croonian Lecture in 1897, and was awarded the Royal Medal in 1905 and the Copley Medal in 1927. In 1922 the Knight Grand Cross of the Order of the British Empire and in 1924 the Order of Merit were conferred upon him. He held honorary doctorates of the Universities of Oxford, London, Sheffield, Birmingham, Manchester, Liverpool, Wales, Edinburgh, Glasgow, Paris, Strasbourg, Louvain, Uppsala, Lyons, Budapest, Athens, Brussels, Berne, Toronto, Montreal, and Harvard.

As a boy and a young man Sherrington was a notable athlete both at Queen Elizabeth's School, Ipswich, where he went in 1871, and later at Gonville and Caius College, Cambridge, for which College he rowed and played rugby football; he was also a pioneer of winter sports at Grindelwald.

In 1892 Sherrington married Ethel Mary, daughter of John Ely Wright, of Preston Manor, Suffolk. After some years of frail health, during which, however, he remained mentally very alert, he died suddenly of heart failure at Eastbourne in 1952.

Edgar D. Adrian

The activity of the nerve fibres

Nobel Lecture, December 12, 1932

The sense organs respond to certain changes in their environment by sending messages or signals to the central nervous system. The signals travel rapidly over the long threads of protoplasm which form the sensory nerve fibres, and fresh signals are sent out by the motor fibres to arouse contraction in the appropriate muscles. What kind of signals are these, and how are they elaborated in the same organs and nerve cells? The first part of this question would have been answered correctly by most physiologists many years ago, but now it can be answered in much greater detail. It can be answered because of a recent improvement in electrical technique. The nerves do their work economically, without visible change and with the smallest expenditure of energy. The signals which they transmit can only be detected as changes of electrical potential, and these changes are very small and of very brief duration. It is little wonder therefore that progress in this branch of physiology has always been governed by the progress of physical technique and that the advent of the triode valve amplifier has opened up new lines in this, as in so many other fields of research.

I shall deal mainly with some of the results which have followed from this new technique, but the present state of our knowledge will be made clearer by a brief survey of the position as it was twenty years ago when I was a student in the Cambridge laboratories.

In the closing years of the last century the improvement of the capillary electrometer had marked a new phase. It was already known that some kind of rapid wave, called the nerve impulse, could be set up in the nerve by an electric stimulus, and there was good reason to suppose that the signals normally transmitted were made up of similar impulses. The disturbance due to an electric stimulus travelled at much the same rate as the natural signals, and it would produce similar effects on the muscles or on the central nervous system. It could be detected in the nerve by the change of potential which accompanied it; in fact Bernstein had already elaborated the «membrane hypothesis» which regards the impulse as a wave of surface disintegration spreading by reason of the electric disturbance which it creates. With the

development of the capillary electrometer it became possible to make direct
and accurate records of this electric disturbance. Before long the work of
Gotch and Burch, Garten, Samojloff, and finally of Keith Lucas had given a
detailed knowledge of its time relations and of its connection with the impulse. It was made clear that the wave of activity is invariably accompanied
by a change of potential, that the activity at any point lasts only for a few
thousandths of a second, and that it is followed by a refractory state which
must pass away before another wave of activity can occur. The existence of
a refractory period in the heart muscle had been recognized long before and
its discovery in the nerve was of fundamental importance. It showed that the
nerve fibre, when stimulated electrically, could only work in a succession of
jerks separated by periods of enforced rest, and this was true both for the
waves of potential change and for the underlying impulse which produced
them.

In the same period came Gotch's observation that the potential wave in a
nerve had an equal duration whether it was set up by a strong or a weak stimulus. As it seemed unlikely that a feeble and an intense disturbance would
last for the same time, Gotch suggested that in each nerve fibre the disturbance was always of the same intensity, and that a strong stimulus set up a
larger potential wave merely because it brought more fibres into activity.
This agreed with the fact that the rate of conduction and the length of the
refractory period were also uninfluenced by the strength of the stimulus. It
seemed, therefore, that each pulse of activity in a nerve fibre must be of constant intensity, involving the entire resources of the fibre whatever the
strength of the stimulus which set it in motion. The fibre was a unit giving
always its maximal response, behaving like the heart muscle in this respect as
well as in that of its refractory state. Conclusive proof was lacking, but
Gotch's work made it likely that the same all-or-nothing behaviour might
be found in skeletal muscle fibres. Keith Lucas recorded the contraction of a
band of muscle containing only a few fibres and found that with an increasing stimulus the contraction increased in sudden steps. The number of steps
was never greater than the number of fibres in the preparation. It was clear,
then, that skeletal muscle fibres followed the all-or-nothing rule.

I have mentioned this work of Keith Lucas (confirmed later by Pratt)
because it was the first direct evidence of the ungraded character of the wave
of activity in excitable tissues other than the heart. It was also the first successful attempt to record the behaviour of the units in muscle and nerve
instead of inferring the behaviour of the units from that of the whole ag-

gregate. A few years later I had the great good fortune to work with him, to appreciate his technical skill and his penetrating thought. I cannot let this occasion pass by without recording how much I owe to his inspiration. In my own work I have tried to follow the lines which Keith Lucas would have developed if he had lived, and I am happy to think that in honouring me with the Nobel Prize you have honoured the master as well as the pupil.

After Keith Lucas's work on muscle, attempts were made to secure more evidence as to the all-or-nothing reaction of the nerve fibre. Verworn and his school showed that the strength of the stimulus made no difference to the ability of the impulse to pass through a narcotized area, and Lucas and I made use of the same method. Its value seemed to lie in its offering a means of measuring the impulse in terms of its ability to travel, but Kato has since pointed out the fallacies which arose from supposing that the impulse became progressively smaller as it passed through the affected region.

More direct evidence was lacking, but at the end of this period we had good reason to believe that the nerve impulse was a brief wave of activity depending in no way on the intensity of the stimulus which set it up. We did not know for certain that the nervous signalling in the intact animal was carried out by means of such impulses, but it seemed highly probable – so much that we could elaborate hypotheses to explain the working of the central nervous system in terms of the interference and reinforcement of trains of impulses.

It was at this point that the need arose for a more sensitive electrical technique. When a nerve trunk is stimulated by an electric shock every fibre is thrown into action simultaneously and the total potential change in the whole nerve is large enough to be recorded directly. But in more normal circumstances the nerve fibres work as independent conducting units, and simultaneous activity in many fibres is a rare event. Potential changes could be detected when there was reason to believe that signals were passing, but to analyse these changes was a far more difficult problem. Granting that they were caused by the passage of impulses of the familiar type, there was little or nothing to show how the impulses were spaced. Records of the electric changes in contracting muscle seemed to one school to imply a very high frequency of discharge in each nerve fibre. Others believed that the frequency was lower, but neither side could find convincing evidence. To show clearly what kind of signals passed from the sense organs to the brain and from the brain to the muscles it would have been necessary to record the electrical events in the individual nerve fibres. The potentials to be dealt with are of the order of a few microvolts lasting for a few thousandths of a second. They were quite beyond the reach of the instruments available at the time, and other lines of evidence had to be followed. These were indirect and, in fact, most of them led nowhere.

The revolution in technique has come about not from any increase in the sensitivity of galvanometers and electrometers but from the use of the thermionic valve to amplify potential changes. The recording instruments used nowadays are actually far less sensitive than their predecessors. Since the energy available is almost unlimited, any system can be chosen which will react rapidly enough and the limiting factor has become not the period of the instrument but that of the amplifying circuits. There is a lower limit to the sensitivity of a valve, but fortunately a change as small as one or two microvolts is within the range of useful amplification. Many workers have contributed to the introduction of this technique into physiology, notably Forbes of Harvard, Gasser of St. Louis, who was the first to use very high amplification, and Matthews of Cambridge who devised the convenient moving-iron oscillograph which is now in common use; to all these my own work is deeply indebted.

Seven years ago it became clear to me that a combination of the capillary electrometer with an amplifier would permit the recording of far smaller potential changes than had been dealt with previously, and might enable us to work on the units of the nerve trunk instead of on the aggregate. A preliminary survey confirmed this, for it showed that the normal activity of sensory and motor fibres was always accompanied by potential changes of the familiar type. The problem was then to limit the activity to only one or two nerve fibres. In this I was happy to have the cooperation of Dr. Zotterman of the Caroline Institute. We found that the sterno-cutaneous muscle of the frog could be divided progressively until it contained only one sense organ; this could be stimulated by stretching the muscle, and we could record the succession of impulses which passed up the single sensory nerve fibre.

A variety of methods now exists for studying in this way the activity of individual sensory and motor nerve fibres. Many records have been made of the signals which they transmit in the normal working of the organism and in every case the signals are found to be extremely simple. They consist of nerve impulses repeated more or less rapidly, impulses which differ in no way from those already studied by the classical methods of electro-physiology. This may have seemed no more than a proof of what was already obvious, but our records showed another point which was more illuminating. To illustrate this we may take the discharge produced by stretching a muscle spindle. A record of the potential changes in the nerve shows a succession of brief diphasic waves, each due to the passage of a single impulse along the nerve fibre. The waves are of constant size and duration, but they begin at a frequency of about 10 a second, and as the extension increases, their frequency rises to 50 a second or more. The frequency depends on the extent and on the rapidity of the stretch; it depends, that is to say, on the intensity of excitation in the sense organ, and in this way the impulse message can signal far more than the mere fact that excitation has occurred.

In all the sense organs which give a prolonged discharge under constant stimulation the message in the nerve fibre is composed of a rhythmic series of impulses of varying frequency. Hartline, for instance, has shown that the discharge from one of the light-sensitive receptor organs in the eye of Limulus is a fairly close copy of that from a frog's muscle spindle. With some kinds of sense organ there is a rapid adaptation to the stimulus, and the nervous discharge is too brief to show a definite rhythm, though it consists as before of repeated impulses of unvarying size.

The nerve fibre is clearly a signalling mechanism of limited scope. It can only transmit a succession of brief explosive waves, and the message can only be varied by changes in the frequency and in the total number of these waves. Moreover, the frequency depends on the rate of development of the stimulus, as well as on its intensity; also the briefer the discharge the less opportunity will there be for signalling by change of frequency. But this limitation is really a small matter, for in the body the nervous units do not act in isolation as they do in our experiments. A sensory stimulus will usually affect a number of receptor organs, and its result will depend on the composite message in many nerve fibres. A good example of this is to be found in the discharge which passes up the nerve from the carotid sinus at each heart beat. Bronk and Stella have shown that as the blood pressure rises, the impulses in each nerve fibre increase in frequency and more and more fibres come into action. Since rapid potential changes can be made audible as sound waves, a gramophone record will illustrate this, and you will be able to hear the two kinds of gradation, the changes in frequency in each unit and in the number of units in action.

The sense organs which are most easily investigated in this way are those which react to mechanical deformation – tactile endings, muscle spindles and the like. They are supplied by the larger nerve fibres in which the potential change can be readily detected. But there are many sensory nerve fibres which are exceedingly small. The recent work of Erlanger and Gasser and of Ranson has made it highly probable that some of these fibres are concerned with pain, and this alone makes it essential to learn more about their normal activities. For such problems our present methods are still scarcely adequate, for in the smallest fibres the potential changes are probably too small to appear above the random fluctuations due to the operation of the thermionic valve. But we may hope that this failure will be remedied before long.

There is another field of sensory physiology which seemed at first to offer special difficulties but is now more promising. This is the field of the special sense organs. With Mrs. Matthews I investigated the activity of the vertebrate optic nerve, but although the usual impulse messages could be recorded they gave very little information about the working of the receptor organs in the retina. The reason is that the retina is a complex nervous structure. The messages in the optic nerve fibres have been elaborated by the interaction of many nerve cells, even though the stimulus is restricted so as to fall on a very small number of rods and cones. We learnt something of the processes which take place in groups of nerve cells with synaptic connections, but little about the action of light as a sensory stimulus. Fortunately this difficulty has been overcome by Hartline, who finds that in the eye of Limulus there is no evidence of such interaction and no reason to expect it on grounds of structure. And since his work is showing what takes place in the receptors themselves, the complexities of the vertebrate retina become less formidable.

The messages in the vertebrate optic nerve have come not from receptor organs but from nerve cells. They are comparable, therefore, with the messages which are sent from the motor nerve cells to the muscles. The grading and coordination of muscular activity is a subject which has been so greatly illuminated by my friend Sir Charles Sherrington that I mention my own work as a very small supplement to his. It has dealt as before with the signals which are sent by the individual nerve fibres, and its results emphasize the close correspondence between the sensory and motor activities of the nervous system. The messages which pass down the motor fibres to the muscles have, of course, the same limitations as the sensory messages, and again we find that the effect is graded by changes in the frequency of the impulse discharge and in the number of units in action. In a contraction of gradually increasing force the nerve fibres transmit a succession of impulses beginning at a very low frequency (5 to 10 a second) and rising to 40 or 50 a second at the height of the contraction; and as the frequency rises in one nerve fibre, another will start at a low frequency and then more and more, until it becomes impossible to distinguish the individual rhythms. The force of contraction varies with the impulse frequency, because in a muscle fibre each impulse produces a mechanical effect of relatively long duration and the successive effects of a series of impulses can be summed to give a greater contraction. Thus the result of the intermittent message in each nerve fibre is a much less intermittent contraction in a group of muscle fibres, and in the whole muscle there are so many of these fibre groups working independently that the contraction rises and subsides smoothly.

On the whole it appears that the frequency of the impulses varies over a more restricted range in the motor than in the sensory discharge, but the two are so closely alike that the mechanism of the sense organ and of the motor nerve cell must have much in common. They have, of course, the common factor of a nerve fibre which can only respond in one way, but the likeness goes beyond this. Also the particular frequencies which commonly occur are lower than they would be if determined solely by the characteristics of the nerve fibre. In quiet breathing, for instance, at each expansion of the lungs the sense organs of the vagus send up a train of impulses rising to a frequency of about 20 a second at the height of inspiration, and simultaneously the movement of expansion is being produced by trains of motor impulses rising to much the same frequency and almost indistinguishable from the discharge in the sensory fibres. In fact the motor nerve cells seem to be acting just like a collection of sense organs responding to a rhythmic stretch.

Resemblances of this kind show that there is an underlying unity of response in the various parts of the neurone in spite of their differentiation into axon, dendrites or terminal arborizations. They show, too, that a knowledge of the mechanism of the sensory end organ might lead us very far in our search for the mechanisms of the central nervous system. Here we must enter a more speculative region, but there are certain pointers to guide us. In the nerve fibre, for instance, a rhythmic discharge of impulses may arise from an injured region. Electrically such a region behaves as though it were permanently instead of momentarily active. It is at a negative potential to the rest of the fibre owing to the destruction of the polarized surface membrane, and we have fair grounds for supposing that the rhythmic discharge is a consequence of this depolarization. A closer parallel with the sense organ is afforded by a muscle fibre bathed in a solution of NaCl instead of its usual Ringer's fluid. Sooner or later such a fibre becomes spontaneously active, the activity consisting of a serial discharge of impulses from some point. At an earlier stage, however, the activity can be started, as with a sense organ, by mechanical deformation, and it ceases when the deformation is over. Thus a muscle fibre may discharge impulses in response to stretch or touch almost as though it had been transformed into a muscle spindle or a touch receptor, though naturally it is a far less perfect instrument for translating mechanical stress into an impulse message. Here again there is reason to suppose that discharge of impulses is due to a breakdown in the polarized surface, a breakdown which is repaired as soon as the mechanical stress is removed.

Analogies of this kind suggest that sense organs and nerve cells send out impulses because some part of their surface has become depolarized. There are certain difficulties to be faced before this can be treated as more than a crude working hypothesis, but it is one which has important consequences. If the regions from which the discharge originates remain partly or wholly depolarized as long as they are excited, it should be possible to detect potential changes of relatively long duration in sense organs and in the motor nerve centres. Such changes are well known to occur in the eye, and they have been found in the vertebrate brain stem and in the nerve ganglia of insects. Unfortunately the structures in which they occur are so complex that it is difficult to be sure of their interpretation, but at least they suggest the possibility of obtaining direct records of the activities of the grey matter. To extract much information from such records is likely to be a far harder task than it has been in the case of peripheral nerve. In the latter our chief concern is to find out what is happening in the units, and this turns out to be a fairly simple series of events. Within the central nervous system the events in each unit are not so important. We are more concerned with the interactions of large numbers, and our problem is to find the way in which such interactions can take place*.

*The lecture was illustrated by lantern slides and gramophone records.

Biography

Edgar Douglas Adrian was born on November 30, 1889, in London. He was the second son of Alfred Douglas Adrian, C.B., K.C., legal adviser to the British Local Government Board. Adrian went to school at Westminster School, London, and in 1908 he went to Trinity College, Cambridge, at which College he had won a Scholarship in Science. At Cambridge University he studied physiology and the other subjects of the Natural Sciences Tripos and in 1911 he took his B.A. degree with first classes in five separate subjects.

In 1913 he was elected to a Fellowship of Trinity College on account of his investigation of the « all or none » principle in nerve. He then studied medicine, doing his clinical work at St. Bartholomew's Hospital, London, and taking his medical degree in 1915. After working for a time on clinical neurology, he returned to Cambridge in 1919, to lecture on the nervous system. He was made Fellow of the Royal Society in 1923. In 1925 he began investigating the sense organs by electrical methods.

In 1929 he was elected Foulerton Professor of the Royal Society. In 1937 he succeeded Sir Joseph Barcroft as Professor of Physiology at the University of Cambridge, a post which he held until 1951.

In 1951 Adrian was elected Master of Trinity College, Cambridge, a post which he still, at the time of writing, holds.

When Adrian graduated at Cambridge, the Department of Physiology there included several distinguished research workers. Among them were J.N.Langley (1852–1925), who had succeeded Sir Michael Foster (1836–1907), W.H.Gaskell (1847–1914), Sir Hugh K.Anderson (1865–1928), Sir Walter Morley Fletcher (1873–1933), Sir Joseph Barcroft (1872–1947), Keith Lucas (1879–1916) and Archibald Vivian Hill (b. 1886) who was then beginning his work on heat production in muscle. Sir Frederick Gowland Hopkins (1861–1947) was then doing his pioneer work on the vitamins.

Adrian's first research work was done with Keith Lucas, who was working on the impulses transmitted by motor nerves; he showed that, when a muscle fibre contracts, the passage of the nerve impulse that causes the contractions leaves the motor nerve in a state of diminished excitability. Keith Lucas was, at the time of the First World War, thinking of improving the study of the electrical currents in nerves by amplifying them by means of valves, a method which Adrian was later to employ.

First, however, Adrian went to London to take his medical degree and was, until the end of the First World War occupied with work on military patients suffering from nerve injuries or nervous disorders. Returning to Cambridge in 1919 to take over Keith Lucas's laboratory, he began the work with which his name will always be associated. In order to obtain a more sensitive detection of nerve impulses, he used the cathode ray tube, the capillary electrometer and amplification of the electrical impulses by means of thermionic valves, and was thus able to amplify them 5,000 times. He succeeded in setting up a preparation consisting of a single end organ in a muscle of the frog, together with the single nerve fibre related to it and he found that, when the end organ is stimulated, the nerve fibre showed regular impulses with a variable frequency.

With this apparatus he was able to record the electrical discharges in single nerve fibres which were produced by tension on the muscle, pressure on it, touch, the movement of a hair and pricking with a needle. By 1928 he was able to publish his conclusion that a stimulus of constant intensity applied to the skin, immediately excites the end organ, but that this excitation progressively decreases for as long as the stimulation continues. At the same time sensory impulses of constant intensity pass along the nerve from the end organ. These sensory impulses are at first very frequent, but their frequency gradually decreases and as they decrease the sensation in the brain progressively diminishes. As A.V.Hill (*The Ethical Dilemma of Science*, 1960) has said, Adrian, by thus showing that the afferent effect in a given neurone depends on the pattern in time of the impulses travelling in it, has provided a new quantitative basis of nervous behaviour.

Later Adrian extended his investigations to a study of the electrical impulses caused by stimuli likely to cause pain, he concluded that, as Sir Henry Head had postulated as a result of his clinical studies, the nerve fibres which conduct impulses excited by pain probably do not pass further into the brain than the optic thalamus, but that all other sensory impulses can be distinguished in the sensory area of the cortex of the brain and he showed that the part of the cerebral cortex devoted to any particular kind of end organ is related to the special needs of the animal concerned. Thus in man and the monkey the sensory area of the cerebral cortex devoted to the face and hand is relatively large, and relatively little is given to the trunk of the body. In the pony the area devoted to the nostrils is as large as that devoted to the rest of the body; in the pig almost the whole of the sensory area of the cerebral cortex devoted to the sense of touch is given to fibres from the snout, which the pig uses to explore its environment.

Subsequently, Adrian studied the sense of smell and the electrical activity of the brain and the variations and abnormalities of waves shown in the encephalogram, which Hans Berger, of Jena, had described in 1929. This work opened up new fields of investigation in the study of epilepsy and other lesions of the brain.

For his work about the functions of neurones Adrian was awarded, jointly with Sir Charles Sherrington, the Nobel Prize for 1932.

The results of Adrian's brilliant researches on the electrophysiology of the brain and nervous system were published in numerous scientific papers and in his three books, *The Basis of Sensation* (1927), *The Mechanism of Nervous Action* (1932) and *The Physical Basis of Perception* (1947). With others he wrote *Factors Determining Human Behaviour* (1937).

Adrian had numerous honours bestowed upon him. During 1950–1955 he was President of the Royal Society, and during 1960–1962 of the Royal Society of Medicine. In 1954, he was President of the British Association for the Advancement of Science. He is Chevalier of the French Legion of Honour and a trustee of the Rockefeller Institute. He holds honorary degrees, memberships, and fellowships of numerous universities and other learned bodies. He was knighted Baron of Cambridge in 1955.

A man of tireless energy and continuous industry, Adrian has, throughout his busy life, and as a Member of the Medical Research Council and many other scientific advisory bodies, exerted great influence, not only on his pupils and collaborators, but also on the development of physiological research and the sciences in general.

To the citizens of Cambridge he has long been familiar as a lean, small figure, dominated by the forward thrust of the nose and chin and the set expression of purpose, as he threads his way at high speed on a bicycle through the crowded streets of the city. An expert fencer, he is also an enthusiastic mountaineer, a recreation which he shares with Lady Adrian, who is a Justice of the Peace and does much social work in the City. Among Lord Adrian's other recreations are sailing and his great interest in the arts. A superb after-dinner speaker, all his lectures and speeches have been the result of very careful preparation.

In 1923 Adrian married Hester Agnes Pinsent, daughter of the late Hume Pinsent of Birmingham, England, and a descendant of the philosoper David Hume. They have one son and two daughters.

Physiology or Medicine 1933

THOMAS HUNT MORGAN

«for his discoveries concerning the role played by the chromosome in heredity»

Physiology or Medicine 1933

Presentation Speech by F. Henschen, member of the Staff of Professors of the Royal Caroline Institute

Your Majesty, Your Royal Highnesses, Honourable Audience.

As long as human beings have existed they will have observed children's resemblance to their parents, the resemblance or non-resemblance of brothers and sisters, and the appearance of characteristic qualities in certain families and races. They will also early have asked for an explanation of these circumstances, which has produced a kind of primitive theory of heredity chiefly on a speculative basis. This has been characteristic of the theories of heredity right up to our time, and as long as there existed no scientific analysis of the hereditary conditions, the mechanism of fertilization remained impenetrable mysticism.

Old Greek medicine and science took much interest in these questions. In Hippocrates, the father of the healing art, you can find a theory of heredity that probably can be traced back to primitive ideas. According to Hippocrates, inherited qualities, in some way or other, must have been transmitted to the new individual from different parts of the organisms of the father and the mother. Similar ideas of the transmission of qualities from parents to children are to be found in other Greek scientists, and, modified, also in Aristotle, the greatest biologist of the olden times.

Later on, this so-called transmission theory has been dominating. The only theory of heredity that has perhaps rivalled it, is the so-called preformation theory, an old scholastic idea that can be followed back to Augustine, the father of the Church. This theory maintained that, by the creation of the first woman, all following generations were also preformed in this first mother of ours. In modified form the preformation theory dominated the biology of the eighteenth century. Nevertheless, the transmission theory survived. Its last great representative was Darwin. He also seems to have understood heredity as a transmission of the personal qualities of the parents to the offspring through a kind of extract from the different organs of the body.

This conception, however, that is thus deeply rooted in the biology of past times and that will still be adopted rather generally, is fundamentally false; it has been reserved to the genetic researches of our time to prove this.

Modern hereditary researches are of a recent date, they are not yet seventy years old. Their founder is the Augustine monk Gregor Mendel, Professor at Brünn, who published (1866) his experiments on hybridization among plants, fundamental for this whole science. In the same year, in Kentucky, the man was born, who became Mendel's heir and founder of the school in heredity researches that has been called higher Mendelism, the winner of this year's Nobel Prize in Physiology or Medicine, Thomas Hunt Morgan.

Mendel's observations are of revolutionizing importance. As a matter of fact they completely upset the older theories of heredity, although this was not at all appreciated by his contemporaries. Mendel's discoveries usually are stated in two heredity laws or better rules of heredity. The first of his rules, the cleaving rule, means that if two different hereditary dispositions or hereditary factors (*genes*) for a certain quality – for instance for size – are combined in one generation, they separate in the following generation. If, for instance, a constantly tall race is crossed with a constantly short race, the individuals of next generation become altogether medium-sized, or, if the factor « tall » is dominant, exclusively tall. In the following generation, however, a cleaving takes place, so that once more the size of the individuals becomes variable according to certain numerical proportions, then of four descendants: one tall, two medium-sized, and one short.

The second of Mendel's rules, the rule of free combinations, means that, when new generations arise, the different hereditary factors can form new combinations independent of each other. If, for instance, a tall, red-flowered plant is crossed with a short, white-flowered one, the factors red and white can be inherited independent of the factors large and small. The second generation then, besides tall red-flowered and short white-flowered plants, produces short red-flowered and tall white-flowered ones.

Mendel's immortal merit is his exact registration of the special qualities and consequent following of their appearance from generation to generation. In this way he discovered the relatively simple, recurrent, numerical proportions, which give us the key to a true understanding of the course of heredity. The experimental genetics of our century then has proved that, taken as a whole, these Mendel rules are applicable to all many-celled organisms, to mosses and flowering plants, to insects, mollusks, crabs, amphibia, birds, and mammals.

Mendel's rules, however, met with the same fate as many other great discoveries that have been made before their time. Their significance was not understood, they fell into oblivion, and after pater Mendel had died in 1884, nobody mentioned them any more. Darwin apparently knew nothing about his great contemporary; otherwise he could have made use of Mendel's works for his own researches, and the rediscovery of Mendel's work was made only about 1900.

By that time, however, the qualifications for the application and perfection of Mendel's theories were quite different from those of their first publication. The general biological attitude had changed, and, above all, the knowledge of the cell and the cell nucleus had made excellent progress. The mechanism of fertilization had been discovered by Hertwig in 1875, and in the eighteen-eighties Weismann had asserted the opinion that the nuclei of the sex cells must be the bearers of the hereditary qualities. The indirect or mitotic cell division and the chromosomes – the strange, threadlike, colourable structures that then appear – had been discovered by Schneider in 1873 already. Only several decades later, however, was the meaning of the remarkable cleaving, wandering, and fusion of these chromosomes during the different phases of the cell division and the fertilization understood.

When, at last, Mendel's discoveries came to light, their significance was soon perceived. Behind Mendel's rules there must be some relatively simple, cellular mechanism for the exact distribution of the hereditary factors at the genesis of the new individual. This mechanism was found just in the proportion of chromosomes in the sex cells before and after the fertilization. The opinion that the chromosomes are the real bearers of heredity was first clearly pronounced by Sutton in 1903, and by Boveri in 1904. This opinion was enthusiastically received by the students of the cell. Only by this discovery organic life got the unity, the continuity that human thought demands and that is more real and more provable than the hypothetic common descent of Darwinism.

The further development of the chromosome theory during the first decade of this century may here be skipped. However, the soil was well prepared when, in 1910, the American zoologist Thomas Hunt Morgan began his researches in heredity. These soon led him to the great discoveries regarding the functions of the chromosomes as the bearers of heredity that have now been rewarded with the Nobel Prize for Medicine in 1933.

Morgan's greatness and the explanation of his astonishing success is partly to be found in the fact that, from the beginning, he has understood to join two important methods in hereditary research, the statistic-genetic method adopted by Mendel, and the microscopic method, and that he has always looked for an answer to the question: which microscopic processes in cells and chromosomes result in the phenomenons appearing at the crossings?

Another cause for Morgan's success is no doubt to be found in the ingenious choice of object for his experiments. From the beginning Morgan chose the so-called banana-fly, *Drosophila melanogaster*, which has proved superior to all other genetic objects known so far. This animal can easily be kept alive in laboratories, it can well endure the experiments that must be made. It propagates all the year round without intervals. Thus a new generation can be had about every twelfth day or at least 30 generations a year. The female lays about 1,000 eggs, males and females can easily be distinguished from each other, and the number of chromosomes in this animal is only four. This fortunate choice made it possible to Morgan to overtake other prominent genetical scientists, who had begun earlier but employed plants or less suitable animals as experimental objects.

Finally, few have like Morgan had the power of assembling around them a staff of very prominent pupils and co-operators, who have carried out his ideas with enthusiasm. This explains to a large extent the extraordinarily rapid development of his theories. His pupils Sturtevant, Muller, Bridges, and many others stand beside him with honour and have a substantial share in his success. With perfect justice we speak about the Morgan school, and it is often difficult to distinguish what is Morgan's work and what is that of his associates. But nobody has doubted that Morgan is the ingenious leader.

As Mendelism can be summed up in Mendel's two rules, Morganism, at least to a certain extent, can be expressed in laws or rules. The Morgan school usually speaks of four rules, the combination rule, the rule of the limited number of the combination groups, the crossing-over rule, and the rule of the linear arrangement of the genes in the chromosomes. These rules complete the Mendel rules in an extraordinarily important way. They are all inextricably connected, and form together a close biological unity.

It is true that Morgan's combination rule, according to which certain hereditary dispositions are more or less firmly combined, limits to a large degree Mendel's second rule that, at the formation of new hereditary substances, the genes may be freely combined. It is completed by the rule of the limited number of the combination groups, which has turned out to be corresponding to the number of chromosomes. On the other hand, the combination rule is confined by the strange phenomenon that Morgan calls crossing-over or the exchange of genes, which he imagines as a real exchange of parts between the chromosomes. This crossing-over theory has met with much resistance. During the last few years, however, it has got a firm support through direct microscopic observations. Also the theory of the linear arrangement of the hereditary factors seemed in the beginning a fantastic speculation, and the publication of Morgan's so-called genetic chromosome map, upon which the different hereditary factors are checked in the chromosomes like beads in a necklace, was greeted with justified scepticism. The fact was that Morgan had arrived at these sensational conclusions by statistic analysis of his Drosophila crossings and not by direct examination of the chromosomes, which, besides, is possible only in exceptional cases. But also on this point later researches have acknowledged him to be in the right, and nowadays also other genetic scientists admit that the theory of the localization of the hereditary factors within the chromosomes is not an abstract way of thinking but corresponds to a stereometric reality.

The results of the Morgan school are daring, even fantastic, they are of a greatness that puts most other biological discoveries into the shade. Who could dream some ten years ago that science would be able to penetrate the problems of heredity in that way, and find the mechanism that lies behind the crossing results of plants and animals; that it would be possible to localize in these chromosomes, which are so small that they must be measured by the millesimal millimetre, hundreds of hereditary factors, which we must imagine as corresponding to infinitesimal corpuscular elements. And this localization Morgan had found in a statistic way! A German scientist has appropriately compared this to the astronomical calculation of celestial bodies still unseen but later on found by the tube – but he adds: Morgan's predictions exceed this by far, because they mean something principally new, something that has not been observed before.

Morgan's researches chiefly occupy themselves with the family of Drosophila, and perhaps it may seem strange that his discoveries have been rewarded with the Nobel Prize for Medicine, which is to be bestowed on the man who «has done the greatest service to mankind» and «has made the most important discoveries in the field of physiology or medicine». To this may first be alleged that numerous later examinations of other genetic objects, of lower and higher plants and animals, have given evidence of the fact that, as a principle, Morgan's rules are applicable to all many-celled organisms.

Further, comparative biological research has for a long time shown a far-extending fundamental correspondence between man and other beings. We can therefore consider it as a matter of course that also such an elementary function of the cell as the transmission of hereditary dispositions is similar, that, in other words, Nature uses the same mechanism with man as with other beings to preserve species, and that Mendel's and Morgan's rules thus are applicable also to man.

Human hereditary researches have already made great use of Morgan's investigations. Without them modern human genetics and also human eugenics would be impractical– it may be that eugenics still chiefly remain a future goal. Mendel's and Morgan's discoveries are simply fundamental and decisive for the investigation and understanding of the hereditary diseases of man. And considering the present attitude of medicine and the dominating place of the constitutional researches, the role of the inner, hereditary factors as to health and disease appears in a still clearer light. For the general understanding of maladies, for prophylactic medicine, and for the treatment of diseases, hereditary research thus gains still greater importance.

Mr. Steinhardt. The Caroline Institute regrets very much that Professor Morgan is not able to be here today in person. I beg Your Excellency, as the official representative of the United States of America, to accept the Nobel Prize for Professor Morgan. May I also ask Your Excellency, in forwarding the prize to him, to convey with it the admiring congratulations of our Institute.

THOMAS H. MORGAN

The relation of genetics to physiology and medicine

Nobel Lecture, June 4, 1934

The study of heredity, now called genetics, has undergone such an extraordinary development in the present century, both in theory and in practice, that it is not possible in a short address to review even briefly all of its outstanding achievements. At most I can do no more than take up a few outstanding topics for discussion.

Since the group of men with whom I have worked for twenty years has been interested for the most part in the chromosome mechanism of heredity, I shall first briefly describe the relation between the facts of heredity and the theory of the gene. Then I should like to discuss one of the physiological problems implied in the theory of the gene; and finally, I hope to say a few words about the applications of genetics to medicine.

The modern theory of genetics dates from the opening years of the present century, with the discovery of Mendel's long-lost paper that had been overlooked for thirty-five years. The data obtained by De Vries in Holland, Correns in Germany, and Tschermak in Austria showed that Mendel's laws are not confined to garden peas, but apply to other plants. A year or two later the work of Bateson and Punnett in England, and Cuénot in France, made it evident that the same laws apply to animals.

In 1902 a young student, William Sutton, working in the laboratory of E. B. Wilson, pointed out clearly and completely that the known behavior of the chromosomes at the time of maturation of the germ cells furnishes us with a mechanism that accounts for the kind of separation of the hereditary units postulated in Mendel's theory.

The discovery of a mechanism, that suffices to explain both the first and the second law of Mendel, has had far-reaching consequences for genetic theory, especially in relation to the discovery of additional laws; because, the recognition of a mechanism that can be seen and followed demands that any extension of Mendel's theories must conform to such a recognized mechanism; and also because the apparent exceptions to Mendel's laws, that came to light before long, might, in the absence of a known mechanism, have called forth purely fictitious modifications of Mendel's laws, or even seemed to invalidate their generality. We now know that some of these « exceptions » are due to newly discovered and demonstrable properties of the chromosome mechanism, and others to recognizable irregularities in the machine.

Mendel knew of no processes taking place in the formation of pollen and egg cell that could furnish a basis for his primary assumption that the hereditary elements separate in the germ cells in such a way that each ripe germ cell comes to contain only one of each kind of element: but he justified the validity of this assumption by putting it to a crucial test. His analysis was a wonderful feat of reasoning. He verified his reasoning by the recognized experimental procedure of science.

As a matter of fact it would not have been possible in Mendel's time to give an objective demonstration of the basic mechanism involved in the separation of the hereditary elements in the germ cells. The preparation for this demonstration took all of the thirty-five years between Mendel's paper in 1865, and 1900. It is here that the names of the most prominent European cytologists stand out as the discoverers of the role of the chromosomes in the maturation of the germ cells. It is largely a result of their work that it was possible in 1902 to relate the well-known cytological evidence to Mendel's laws. So much in retrospect.

The most significant additions that have been made to Mendel's two laws may be called linkage and crossing-over. In 1906 Bateson and Punnett reported a two-factor case in sweet peas that did not give the expected ratio for two pairs of characters entering the cross at the same time.

By 1911 two genes had been found in Drosophila that gave sex-linked inheritance. It had earlier been shown that such genes lie in the X-chromosomes. Ratios were found in the second generation that did not conform to Mendel's second law when these two pairs of characters are present, and the suggestion was made that the ratios in such cases could be explained on the basis of interchange between the two X-chromosomes in the female. It was also pointed out that the further apart the genes for such characters happen to lie in the chromosome, the greater the chance for interchange to take place. This would give the approximate location of the genes with respect to other genes. By further extension and clarification of this idea it became possible, as more evidence accumulated, to demonstrate that the genes lie in a single line in each chromosome.

Two years previously (1909) a Belgian investigator, Janssens, had described a phenomenon in the conjugating chromosomes of a salamander,

Batracoseps, which he interpreted to mean that interchanges take place between homologous chromosomes. This he called chiasmatypie–a phenomenon that has occupied the attention of cytologists down to the present day. Janssens' observations were destined shortly to supply an objective support to the demonstration of genetic interchange between linked genes carried in the sex chromosomes of the female Drosophila.

Today we arrange the genes in a chart or map. The numbers attached express the distance of each gene from some arbitrary point taken as zero. These numbers make it possible to foretell how any new character that may appear will be inherited with respect to all other characters, as soon as its crossing-over value with respect to any other two characters is determined. This ability to predict would in itself justify the construction of such maps, even if there were no other facts concerning the location of the genes; but there is today direct evidence in support of the view that the genes lie in a serial order in the chromosomes.

What are the genes?

What is the nature of the elements of heredity that Mendel postulated as purely theoretical units? What are genes? Now that we locate them in the chromosomes are we justified in regarding them as material units; as chemical bodies of a higher order than molecules? Frankly, these are questions with which the working geneticist has not much concern himself, except now and then to speculate as to the nature of the postulated elements. There is no consensus of opinion amongst geneticists as to what the genes are– whether they are real or purely fictitious–because at the level at which the genetic experiments lie, it does not make the slightest difference whether the gene is a hypothetical unit, or whether the gene is a material particle. In either case the unit is associated with a specific chromosome, and can be localized there by purely genetic analysis. Hence, if the gene is a material unit, it is a piece of a chromosome; if it is a fictitious unit, it must be referred to a definite location in a chromosome – the same place as on the other hypothesis. Therefore, it makes no difference in the actual work in genetics which point of view is taken.

Between the characters that are used by the geneticist and the genes that his theory postulates lies the whole field of embryonic development, where the properties implicit in the genes become explicit in the protoplasm of the cells. Here we appear to approach a physiological problem, but one that is new and strange to the classical physiology of the schools.

We ascribe certain general properties to the genes, in part from genetic evidence and in part from microscopical observations. These properties we may next consider.

Since chromosomes divide in such a way that the line of genes is split (each daughter chromosome receiving exactly half of the original line) we can scarcely avoid the inference that the genes divide into exactly equal parts; but just how this takes place is not known. The analogy of cell division creates a presumption that the gene divides in the same way, but we should not forget that the relatively gross process involved in cell division may seem quite inadequate to cover the refined separation of the gene into equal halves. As we do not know of any comparable division phenomena in organic molecules, we must also be careful in ascribing a simple molecular constitution to the gene. On the other hand, the elaborate chains of molecules built up in organic material may give us, some day, a better opportunity to picture the molecular or aggregate structure of the gene and furnish a clue concerning its mode of division.

Since by infinite subdivisions the genes do not diminish in size or alter as to their properties, they must, in some sense, compensate by growing between successive divisions. We might call this property autocatalysis, but, since we do not know how the gene grows, it is somewhat hazardous to assume that its property of growth after division is the same process that the chemist calls autocatalytic. The comparison is at present too vague to be reliable.

The relative stability of the gene is an inference from genetic evidence. For thousands–perhaps many millions–of subdivisions of its material it remains constant. Nevertheless, on rare occasions, it may change. We call this change a *mutation*, following De Vries' terminology. The point to emphasize here is that the mutated gene retains, in the great majority of cases studied, the property of growth and division, and more important still the property of stability. It is, however, not necessary to assume, either for the original genes or for the mutated genes, that they are all equally stable. In fact, there is a good deal of evidence for the view that some genes mutate oftener than others, and in a few cases the phenomenon is not infrequent, both in the germ cells and in somatic tissues. Here the significant fact is that these repetitional changes are in definite and specific directions.

The constancy of position of genes with respect to other genes in linear order in the chromosomes is deducible, both from genetic evidence and from cytological observations. Whether the relative position is no more than a *historical accident*, or whether it is due to some relation between each gene and its neighbors, can not be definitely stated. But the evidence from the dislocation of a fragment of the chromosome, and its reattachment to another one indicates that accident rather than mutual interaction has determined their present location: for, when a piece of one chromosome becomes attached to the end of a chain of genes of another chromosome or when a section of a chromosome becomes inverted, the genes in the new position hold as fast together as they do in the normal chromosome.

There is one point of great interest. So far as we can judge from the action of mutated genes, the kind of effect produced has as a rule no relation to location of the gene in the chromosome. A gene may produce its chief effect on the eye color, while one nearby may affect the wing structure, and a third, in the same region, the fertility of the male or of the female. Moreover, genes in different chromosomes may produce almost identical effects on the same organs. One may say, then, that the position of the genes in the hereditary material is inconsequential in relation to the effects that they produce. This leads to a consideration which is more directly significant for the physiology of development.

In the earlier days of genetics it was customary to speak of unit characters in heredity, because certain contrasted characters, rather clearly defined, furnished the data for the Mendelian ratios. Certain students of genetics inferred that the Mendelian units responsible for the selected character were genes producing only a single effect. This was careless logic. It took a good deal of hammering to get rid of this erroneous idea. As facts accumulated, it became evident that each gene produces not a single effect, but in some cases a multitude of effects on the characters of the individual. It is true that in most genetic work only one of these character effects is selected for study – the one that is most sharply defined and separable from its contrasted character – but in most cases minor differences are also recognizable that are just as much the product of the same gene as is the major effect. In fact, the major difference selected for classification of the contrasted character-pairs may be of small importance for the welfare of the individual, while some of the concomitant effects may be of vital importance for the economy of the individual, affecting its vitality, its length of life, or its fertility. I need not dwell at length on these relations because they are recognized today by all geneticists. It is important, nevertheless, to take cognizance of them, because the whole problem of the physiology of development is involved.

The coming together of the chromosomes at the maturation devision, and their subsequent movement apart to opposite poles of the meiotic figure, insures the regular distribution of one set of chromosomes to each daughter cell and the fulfilment of Mendel's second law. These movements have the appearance of physical events. Cytologists speak of these two phenomena as *attraction* and *repulsion* of the members of individual chromosomes, but we have no knowledge of the kind of physical processes involved. The terms attraction and repulsion are purely descriptive, and mean no more at present than that like chromosomes come together and later separate.

In earlier times, when the constitution of the chromosomes was not known, it was supposed that the chromosomes come together at random in pairs. There was the implication that any two chromosomes may mate. The comparison with conjugation of male and female protozoa, or egg and sperm cell, was obvious, and since in all diploid cells one member of each pair of chromosomes has come from the father and one from the mother, it must have seemed that somehow maleness and femaleness are involved in the conjugation of the chromosomes also. But today we have abundant evidence to prove that this idea is entirely erroneous, since there are cases where both chromosomes that conjugate have come from the female, and even where both have been sister strands of the same chromosome.

Recent genetic analysis shows not only that the conjugating chromosomes

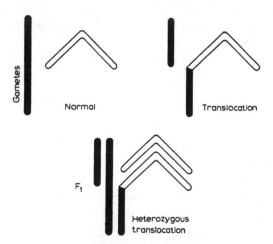

Fig. 1. Diagram to illustrate the case when a piece of one chromosome (*black*) has been translocated to another chromosome (*white*). In the lower part of the figure the method of conjugation of these chromosomes is shown.

are like chromosomes, i.e., chains of the same genes, but also that a very exact process is involved. The genes come together, point for point, unless some physical obstacle prevents. The last few years have furnished some beautiful illustrations showing that it is genes rather than whole chromosomes that come to lie side by side when the chromosomes come together. For example: occasionally a chromosome may have a piece broken off (Fig. 1 above) which becomes attached to another chromosome. A new linkage group is thus established. When conjugation takes place, this piece has no corresponding piece in the sister chromosome. It has been shown (Fig. 1 below) that it then conjugates with that part of the parental chromosomes from which it came.

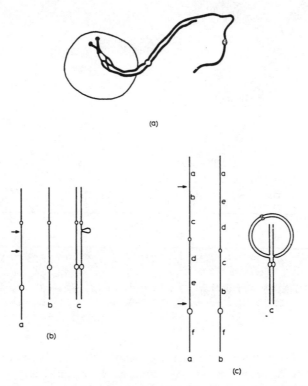

Fig. 2. (*a*) Two conjugating chromosomes of Indian corn (after McClintock). One chromosome has a terminal deficiency. (*b*) Two chromosomes of Indian corn, one having a deficiency near its middle. When these two chromosomes conjugate there is a loop in the longer chromosome opposite the deficiency in the other one. (*c*) Two chromosomes of Indian corn, one having a long inverted region. When they conjugate they come together as shown in the figure to the right (after McClintock), like genes coming together.

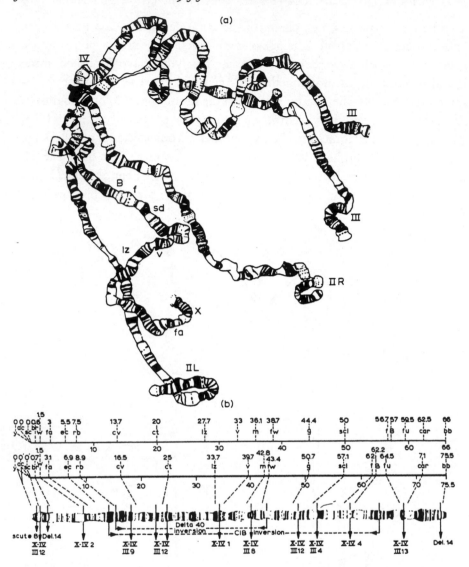

Fig. 3. (*a*) The chromosomes of the salivary gland of the female larva of *Drosophila melanogaster* (after Painter). The two X-chromosomes are fused into a single body. This chromosome is attached at one end to the common chromidial mass at its «attachment end». The 2nd and 3rd chromosomes have the attached point near the middle and are fused with the common chromidial mass at this point, leaving two free ends of each chromosome. Like limbs of each of these free ends are fused, giving four free ends in all. (*b*) The banded salivary X-chromosome of *Drosophila melanogaster* is below, with the genetic map above (after Painter). Oblique broken lines connect the loci of the genetic map with corresponding or homologous loci of the salivary chromosome.

When a chromosome has lost one end, it conjugates with its mate only in part (Fig. 2 a), i.e., where like genes are present. When a chromosome has lost a small region, somewhere along its length, so that it is shorter than the original chromosome, the larger chromosome shows a loop which is opposite the region of deficiency in the shorter chromosome as shown in Fig. 2 b. Thus like genes, or corresponding loci, are enabled to come together through the rest of the chromosome. More remarkable still is the case where the middle region of a chromosome has become turned around (inversion). When such a chromosome is brought together with its normal homologue, as shown in Fig. 2 c, like regions come together by the inverted piece reversing itself, so to speak, so that like genes come together as shown to the right in Fig. 2 c. In this same connection the conjugation of the chromosomes in species of *Oenothera* furnish beautiful examples of the way in which like series of genes find each other, even when halves of different chromosomes have been interchanged.

The very recent work of Heitz, Painter, and Bridges has brought to light some astonishing evidence relating to the constitution of the chromosomes in the salivary glands of *Drosophila melanogaster*.

The nuclei of the cells of the salivary glands of the old larvae are very large and their contained chromosomes (Fig. 3) may be 70 to 150 times as large as those of the ordinary chromosomes in process of division. Heitz has shown that there are regions of some of the chromosomes of the ganglion cells – more especially of the X- and the Y-chromosomes – that stain deeply, and other regions faintly, and that these regions correspond to regions of the genetic map that do not and do contain genes. Painter has made the further important contribution that the series of bands of the salivary chromosomes can be homologized with the genetically known series of genes of the linkage maps (Fig. 3 a, b), and that the empty regions of the X and Y do not have the banded structure. He has further shown that when a part of the linkage map is reversed, the sequence of the bands is also reversed; that when pieces are translocated they can be identified by characteristic bands: and that when pieces of linked genes are lost there is a corresponding loss of bands. Bridges has carried the analysis further by an intensive study of regions of particular chromosomes, and has shown a close agreement between bands and gene location. With improved methods he has identified twice as many bands, thus making a more complete analysis of the relation of bands and gene location. Thus, whether or not the bands are the actual genes, the evidence is clear in showing a remarkable agreement between the location of genes and the location of corresponding bands. The analysis of the banded structure has confirmed the genetic evidence, showing that when certain alterations of the order of the genes takes place, there is a corresponding change in the sequence of the bands which holds for the finest details of the bands.

The number of chromosomes in the salivary nuclei is half that of the full number (as reported by Heitz) which Painter interprets as due to homologous chromosomes conjugating (Fig. 3 a). Moreover, the bands in each of the component halves show an identical sequence which is strikingly evi-

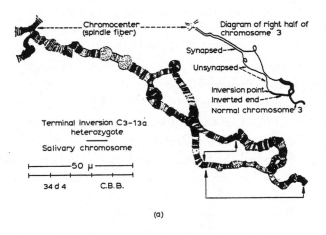

(a)

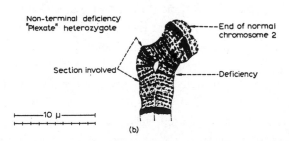

(b)

Fig. 4. (*a*) Salivary gland preparation of the right half of the third chromosome. The two components are united through a part of their length (*above left*). One component had a terminal inversion. This part conjugated with the corresponding normal chromosome by turning back on itself, as shown in the small diagram above (*to the right*). (*b*) Salivary gland preparation showing a part of chromosome 2; one component is «deficient». At the level of the deficiency the other component is bent outward so that above and below these level like bands meet. (After Bridges.)

dent when the halves are not closely apposed. It has been suggested by Bridges and by Koltzoff that homologous chromosomes have not only united, but that they have each divided two or three times, giving in some cases as many as 16 or 32 strands (Fig. 4 a, b). The bands may then be said to be composed each of 16 or 32 genes; or, if this identification of the bands as genes is questioned in so far as the genes are concerned, the bands are multiples of some kind of unit of which the chromosomes are composed.

A few examples may serve to illustrate the way in which the banded chromosomes confirm the genetic conclusions as to occasional changes that have taken place in the serial order of the genes. In Fig. 4 a the right half of chromosome 3 from the salivary gland is represented. In part the two components are fused, in part are separate. In the lower part of the figure a reversed piece of one component is present (terminal inversion). Like bands conjugate with like and, as shown in the smaller diagram above, in Fig. 4 a, this is made possible by the end of one component turning back on itself. In Fig. 4 b is drawn a short region of chromosome 2. One component has a deficiency for certain genes; the opposite normal chromosome forms a bulge in the region of the deficiency, allowing like bands to come together above and below the deficiency level.

The physiological properties of the genes

If, as is generally implied in genetic work (although not often explicitly stated), all of the genes are active all the time; and if the characters of the individual are determined by the genes, then why are not all the cells of the body exactly alike?

The same paradox appears when we turn to the development of the egg into an embryo. The egg appears to be an unspecialized cell, destined to undergo a prescribed and known series of changes leading to the differentiation of organs and tissues. At every division of the egg, the chromosomes split lengthwise into exactly equivalent halves. Every cell comes to contain the same kind of genes. Why then, is it that some cells become muscle cells, some nerve cells, and others remain reproductive cells?

The answer to these questions seemed relatively simple at the end of the last century. The protoplasm of the egg is visibly different at different levels. The fate of the cells in each region is determined, it was said, by the differences in different protoplasmic regions of the egg.

Such a view is consistent with the idea that the genes are all acting; the initial stages of development being the outcome of a reaction between the identical output of the genes and the different regions of the egg. This seemed to give a satisfactory *picture* of development, even if it did not give us a *scientific explanation* of the kind of reactions taking place.

But there is an alternative view that can not be ignored. It is conceivable that different batteries of genes come into action one after the other, as the embryo passes through its stages of development. This sequence might be assumed to be an automatic property of the chain of genes. Such an assumption would, without proof, beg the whole question of embryonic development, and could not be regarded as a satisfactory solution. But it might be that in different regions of the egg there is a reaction between the kind of protoplasm present in those regions and specific genes in the nuclei; certain genes being more affected in one region of the egg, other genes in other regions. Such a view might give also a purely formal hypothesis to account for the differentiation of the cells of the embryo. The initial steps would be given in the regional constitution of the egg.

The first responsive output of the genes would then be supposed to affect the protoplasm of the cells in which they lie. The changed protoplasm would now act reciprocally on the genes, bringing into activity additional or other batteries of genes. If true this would give a pleasing picture of the developmental process. A variation of this view would be to assume that the product of one set of genes is gradually in time overtaken and nullified or changed by the slower development of the output of other genes, as Goldschmidt, for example, has postulated for the sex genes. In the last case the theory is dealing with the development of hybrid embryos whose sex genes are assumed to have different rates of activity.

A third view may also be permissible. Instead of all the genes acting in the same way all the time, or instead of certain kinds of genes coming successively into action, we might postulate that the kind of activity of all the genes is changed in response to the kind of protoplasm in which they lie. This interpretation may seem less forced than the others, and in better accord with the functional activity of the adult organ systems.

We must wait until experiments can be devised that will help us to discriminate between these several possibilities. In fact, geneticists all over the world are today trying to find methods that will help to determine the relation of genes to embryonic and adult characters. The problem (or problems) is being approached both from a study of chemical changes that take place near the final steps in organ formation, especially in the development of pigments, and from a study of the early differentiation of the cell groups of the embryo.

We have come to realize that the problem of development is not as simple as I have so far assumed to be the case, for it depends, not only on independent cell differentiation of individual cells, but also on interactions between cells, both in the early stages of development and on the action of hormones on the adult organ systems. At the end of the last century, when experimental embryology greatly flourished, some of the most thoughtful students of embryology laid emphasis on the importance of the interaction of the parts on each other, in contrast to the theories of Roux and Weismann that attempted to explain development as a progressive series of events that are the outcome of self-differentiating processes, or as we would say today, by the sorting out of genes during the cleavage of the egg. At that time there was almost no experimental evidence as to the nature of the postulated interaction of the cells. The idea was a generalization rather than an experimentally determined conclusion, and, unfortunately, took a metaphysical turn.

Today this has changed, and owing mainly to the extensive experiments of the Spemann school of Germany, and to the brilliant results of Hörstadius of Stockholm, we have positive evidence of the far-reaching importance of interactions between the cells of different regions of the developing egg. This implies that original differences are already present, either in the undivided egg, or in the early formed cells of different regions. From the point of view under consideration, results of this kind are of interest because they bring up once more, in a slightly different form, the problem as to whether the organizer acts first on the protoplasm of the neighboring region with which it comes in contact, and through the protoplasm of the cells on the genes; or whether the influence is more directly on the genes. In either case the problem under discussion remains exactly where it was before. The evidence from the organizer has not as yet helped to solve the more fundamental relation between genes and differentiation, although it certainly marks an important step forward in our understanding of embryonic development.

The physiological action of the genes on the protoplasm, and reciprocally that of the protoplasm on the genes, is a problem of functional physiology in a very profound sense. For it is a problem that involves not only the irreversible changes of embryonic development, but also the recurrent changes in the organ systems of the adult body.

Genetics and medicine

That man inherits his characters in the same way as do other animals there can be no doubt. The medical literature contains hundreds of family pedigrees, in which certain characters, usually malformations, appear more frequently than in the general population. Most of these are structural defects; a few are physiological traits (such as haemophilia); others are psychopathic. Enough is already known to show that they follow genetic principles.

Man is a poor breeder – hence many of these family pedigrees are too meagre to furnish good material for genetic analysis. When an attempt is made to combine pedigrees from different sources in order to insure sufficient data, the question of correct diagnosis sometimes presents serious difficulties, especially in the older materials; but with the very great advances that have been made in medical diagnosis in recent years this difficulty will certainly be less serious in the future.

The most important contribution to medicine that genetics has made is, in my opinion, intellectual. I do not mean to imply that the practical applications are unimportant, and I shall in a moment point out some of the more obvious connections, but the whole subject of human heredity in the past (and even at the present time in uninformed quarters) has been so vague and tainted by myths and superstitions that a scientific understanding of the subject is an achievement of the first order. Owing to genetic knowledge, medicine is today emancipated from the superstition of the inheritance of maternal impressions: it is free from the myth of the transmission of acquired characters, and in time the medical man will absorb the genetic meaning of the role of internal environment in the coming to expression of genetic characters.

The importance of this relation will be seen when it is recalled that the germ plasm or, as we say, the genic composition of man is a very complex mixture – much more so than that of most other animals, because in very recent times there has been a great amalgamation of many different races owing to the extensive migration of the human animal, and also because man's social institutions help to keep alive defective types of many kinds that would be eliminated in wild species through competition. Medicine has been, in fact, largely instrumental in devising means for the preservation of weak types of individuals, and in the near future medical men will, I suggest, often be asked for advice as to how to get rid of this increasing load of defectives. Possibly the doctor may then want to call in his genetic friends for consultation! The point I want to make clear is that the complexity of the genic composition of man makes it somewhat hazardous to apply only the simpler rules of Mendelian inheritance; for, the development of many inherited characters depends both on the presence of modifying factors and on the external environment for their expression.

I have already pointed out that the gene generally produces more than one visible effect on the individual, and that there may be also many invisible effects of the same gene. In cases where a condition of susceptibility to certain diseases is present, it may be that a careful scrutiny will detect some minor visible effects produced by the same gene. As yet our knowledge on this score is inadequate, but it is a promising field for further medical investigation. Even the phenomenon of linkage may some day be helpful in diagnosis. It is true there are known as yet in man no certain cases of linkage, but there can be little doubt that there will in time be discovered hundreds of linkages and some of these, we may anticipate, will tie together visible and invisible hereditary characteristics. I am aware, of course, of the ancient attempts to identify certain gross physical human types – the bilious, the lymphatic, the nervous and the sanguine dispositions, and of more modern attempts to classify human beings into the cerebral, respiratory, digestive and muscular, or, more briefly, into asthenics and pycnics. Some of these are supposed to be more susceptible to certain ailments or diseases than are other types, which in turn have their own constitutional characteristics. These well-intended efforts are, however, so far in advance of our genetic information that the geneticist may be excused if he refuses to discuss them seriously.

In medial practice the physician is often called upon for advice as to the suitability of certain marriages where a hereditary taint is present in the ancestry. He is often called upon to decide as to the risk of transmitting certain abnormalities that have appeared in the first-born child. Here genetics will, I think, be increasingly helpful in making known the risk incurred, and in distinguishing between environmental and hereditary traits.

Again, a knowledge of the laws of transmission of hereditary characters may sometimes give information that may be helpful in the diagnosis of certain diseases in their incipient stages. If, for example, certain stigmata appear, whose diagnosis is uncertain, an examination of the family pedigree of the individual may help materially in judging as to the probability of the diagnosis.

I need scarcely point out those legal questions concerning the paternity of an illegitimate child. In such cases a knowledge of the inheritance of blood groups, about which we now have very exact genetic information, may often furnish the needed information.

Geneticists can now produce by suitable breeding, strains of populations of animals and plants that are free from certain hereditary defects; and they can also produce, by breeding, plant populations that are resistant or immune to certain diseases. In man it is not desirable, in practice, to attempt to do this, except in so far as here and there a hereditary defective may be discouraged from breeding. The same end is accomplished by the discovery and removal of the external causes of the disease (as in the case of yellow fever and malaria) rather than by attempting to breed an immune race. Also, in another way the same purpose is attained in producing immunity by inoculation and by various serum treatments. The claims of a few enthusiasts that the human race can be entirely purified or renovated, at this later date, by proper breeding, have I think been greatly exaggerated. Rather must we look to medical research to discover remedial measures to insure better health and more happiness for mankind.

While it is true, as I have said, some little amelioration can be brought about by discouraging or preventing from propagating well-recognized hereditary defects (as has been done for a long time by confinement of the insane), nevertheless it is, I think, through public hygiene and protective measures of various kinds that we can more successfully cope with some of the evils that human flesh is heir to. Medical science will here take the lead – but I hope that genetics can at times offer a helping hand.

Biography

Thomas Hunt Morgan was born on September 25, 1866, at Lexington, Kentucky, U.S.A. He was the eldest son of Charlton Hunt Morgan.

He was educated at the University of Kentucky, where he took his B.S. degree in 1886, subsequently doing postgraduate work at Johns Hopkins University, where he studied morphology with W. K. Brooks, and physiology with H. Newell Martin.

As a child he had shown an immense interest in natural history and even at the age of ten, he collected birds, birds' eggs, and fossils during his life in the country; and in 1887, the year after his graduation, he spent some time at the seashore laboratory of Alphaeus Hyatt at Annisquam, Mass. During the years 1888–1889, he was engaged in research for the United States Fish Commission at Woods Hole, a laboratory with which he was continuously associated from 1902 onwards, making expeditions to Jamaica and the Bahamas. In 1890 he obtained his Ph.D. degree at Johns Hopkins University. In that same year he was awarded the Adam Bruce Fellowship and visited Europe, working especially at the Marine Zoological Laboratory at Naples which he visited again in 1895 and 1900. At Naples he met Hans Driesch and Curt Herbst. The influence of Driesch with whom he later collaborated, no doubt turned his mind in the direction of experimental embryology.

In 1891 he became Associate Professor of Biology at Bryn Mawr College for Women, where he stayed until 1904, when he became Professor of Experimental Zoology at Columbia University, New York. He remained there until 1928, when he was appointed Professor of Biology and Director of the G. Kerckhoff Laboratories at the California Institute of Technology, at Pasadena. Here he remained until 1945. During his later years he had his private laboratory at Corona del Mar, California.

During Morgan's 24-years period at Columbia University his attention was drawn toward the bearing of cytology on the broader aspects of biological interpretation. His close contact with E. B. Wilson offered exceptional opportunities to come into more direct contact with the kind of work which was being actively carried out in the zoological department, at that time.

Morgan was a many-sided character who was, as a student, critical and independent. His early published work showed him to be critical of Mendelian conceptions of heredity, and in 1905 he challenged the assumption then current that the germ cells are pure and uncrossed and, like Bateson was sceptical of the view that species arise by natural selection. « Nature », he said, « makes new species outright. » In 1909 he began the work on the fruit-fly *Drosophila melanogaster* with which his name will always be associated.

It appears that Drosophila was first bred in quantity by C. W. Woodworth, who was working from 1900–1901, at Harvard University, and Woodworth there suggested to W. E. Castle that Drosophila might be used for genetical work. Castle and his associates used it for their work on the effects of inbreeding, and through them F. E. Lutz became interested in it and the latter introduced it to Morgan, who was looking for less expensive material that could be bred in the very limited space at his command. Shortly after he commenced work with this new material (1909), a number of striking mutants turned up. His subsequent studies on this phenomenon ultimately enabled him to determine the precise behaviour and exact localization of genes.

The importance of Morgan's earlier work with Drosophila was that it demonstrated that the associations known as *coupling* and *repulsion*, discovered by English workers in 1909 and 1910 using the Sweet Pea, are in reality the obverse and reverse of the same phenomenon, which was later called *linkage*. Morgan's first papers dealt with the demonstration of sex linkage of the gene for white eyes in the fly, the male fly being heterogametic. His work also showed that very large progenies of Drosophila could be bred. The flies were, in fact, bred by the million, and all the material thus obtained was carefully analysed. His work also demonstrated the important fact that spontaneous mutations frequently appeared in the cultures of the flies. On the basis of the analysis of the large body of facts thus obtained, Morgan put forward a theory of the *linear arrangement* of the genes in the chromosomes, expanding this theory in his book, *Mechanism of Mendelian Heredity* (1915).

In addition to this genetical work, however, Morgan made contributions of great importance to experimental embryology and to regeneration. So far as embryology is concerned, he refuted by a simple experiment the theory of Roux and Weismann that, when the embryo of the frog is in the two-cell stage, the blastomeres receive unequal contributions from the parent blastoderm, so that a « mosaic » results. Among his other embryological discoveries was the demonstration that gravity is not, as Roux's work had suggested, important in the early development of the egg.

Although so much of his time and effort was given to genetical work, Morgan never lost his interest in experimental embryology and he gave it, during his last years increasing attention.

To the study of regeneration he made several important contributions, an outstanding one being his demonstration that parts of the organism which are not subject to injury, such as the abdominal appendages of the hermit crab, will nevertheless regenerate, so that regeneration is not an adaptation evolved to meet the risks of loss of parts of the body. On this part of his work he wrote his book *Regeneration*.

Apart from the books previously mentioned Morgan wrote: *Heredity and Sex* (1913), *The Physical Basis of Heredity* (1919), *Embryology and Genetics* (1924), *Evolution and Genetics* (1925), *The Theory of the Gene* (1926), *Experimental Embryology* (1927), *The Scientific Basis of Evolution* (2nd. ed., 1935), all of them classics in the literature of genetics.

Morgan was made a Foreign Member of the Royal Society of London in 1919, where he delivered the Croonian Lecture in 1922. In 1924, he was awarded the Darwin Medal, and in 1939 the Copley Medal of the Society.

For his discoveries concerning the role played by the chromosome in heredity, he was awarded the Nobel Prize in 1933.

Among his collaborators at Columbia may be mentioned H. J. Muller, who was awarded the Nobel Prize in 1946 for his production of mutations by means of X-rays.

Morgan married Lilian Vaughan Sampson, in 1904, who had been a student at Bryn Mawr College, and who often assisted him in his research. They had one son and three daughters.

Professor Morgan died in 1945.

Physiology or Medicine 1934

GEORGE HOYT WHIPPLE

GEORGE RICHARDS MINOT

WILLIAM PARRY MURPHY

« for their discoveries concerning liver therapy in cases of anaemia »

Physiology or Medicine 1934

Presentation Speech by Professor I. Holmgren, member of the Staff of Professors of the Royal Caroline Institute

Your Majesty, Your Royal Highnesses, Ladies and Gentlemen.

The Caroline Institute has awarded this year's prize for Physiology or Medicine to three American investigators, viz. Professor George Minot, of the Harvard Medical School (Boston), Dr. William Murphy, of the same College, and Professor George Whipple, of the School of Medicine (Rochester, New York), in recognition of their discoveries respecting liver therapy in anaemias. By anaemias is to be understood diseases in which the patient is anaemic. The medical man, in speaking of anaemia and anaemic, has not in mind the actual quantity of the blood, which is not easily determinable by simple methods, but rather certain shortcomings in the composition of the blood that allow of being readily established in his routine investigations, in the first place a diminution in the number of the red blood corpuscles per unit-volume of blood, and a diminution of the concentration in the blood of the haemoglobin, i.e. of the red pigment that gives the blood its colour, a diminution consequently in the colour-strength of the blood. In a word, the blood has become diluted.

Of the three prize-winners, it was Whipple who first occupied himself with the investigations for which the prize has now been awarded. He began in 1920 to study the influence of food on blood regeneration, the re-building-up of the blood, in cases of anaemia consequent upon loss of blood. For the fact is that anaemia in the sense just mentioned also arises as a consequence of a loss of blood. The diminution in the quantity of the blood is made up for, comparatively quickly, by an influx of water from the tissues, that is to say by a process of dilution, of which the consequence is a reduction in the number of the red blood corpuscles and the haemoglobin per unit of volume – thus a case of anaemia. That being so, Whipple started to study the effect that various food substances might have on the process of the regeneration of the blood. It was known beforehand, it is true, that a plentiful supply of food is an important factor in restoring the blood to a normal consistency, but it was not known that independent of the quantity of the food and of its caloric value, different articles of food played differing parts. The method

Whipple adopted in his experiments was to bleed dogs, that is to say to with-draw from them a certain quantity of their blood, supplying them afterwards with food of various kinds. By that method he discovered that certain kinds of food were considerably superior to others, inasmuch as they gave stim-ulus to a more vigorous reformation of blood, that is to say stimulated the bone marrow – in which the blood corpuscles are produced – to a more vigorous manufacture of red blood corpuscles. It was first and foremost liver, then kidney, then meat, and next after that certain vegetable articles of food too, e.g. apricots, that proved in an especial degree to have a strongly stimulating effect. Whipple's experiments were planned exceedingly well and carried out very accurately, and consequently their results can lay claim to absolute reliability. These investigations and results of Whipple's gave Minot and Murphy the idea, that an experiment could be made to see whether favourable results might not also be obtained in the case of perni-cious anaemia, an anaemia of quite a different type, by making use of foods of the kind that Whipple had found to yield favourable results in his exper-iments regarding anaemia from loss of blood.

Before entering upon a discussion of Minot's and Murphy's investigations in detail, I propose to say a little about pernicious anaemia. As the name tells us, it is a fateful disease, which, previous to the labours of our prize-winners, almost invariably, with only very few exceptions, ended fatally in the course of a few years, or in a still shorter time, a few months. Its cause is not known. It customarily makes its appearance in middle-aged persons, who lose col-our, feel tired, and ultimately consult a doctor, who establishes the fact that their red blood corpuscles have become reduced in number from the normal figure of about five million per mm^3 to considerably lower values, e.g. one million per mm^3, or to still less, eight, seven or six hundred thousand per mm^3, and that the colour-strength of the blood has also diminished, though as a rule not in the same high degree as the number of the red blood cor-puscles. Moreover, on examining the blood microscopically, the investigator finds that the red blood corpuscles in it are very different to normal red blood corpuscles. The latter are all alike in size and in form, whereas in pernicious anaemia there are to be noticed blood corpuscles of a great variety of sizes, some considerably larger than normal and some small ones; and their shapes vary too. They are diseased or immature forms of red blood corpuscles. Hence what the bone marrow has supplied to the blood is not of a completely satisfactory make or consistency. The course of the disease is customarily a cyclical one, periods of specially severe anaemia alternating with periods when the composition of the blood is more normal. The circumstance that the disease itself is subject to variations, showing now improvement, now relapses, renders it of course very much more difficult to determine the actual effect produced by any treatment administered to the patient. Previous to the results of Minot's and Murphy's experiments the principal mode of treatment adopted, and one that was practised all over the world, was the giving of large doses of arsenic, while in serious cases it was also customary sometimes to resort to splenectomy, that is to say to removal of the spleen by an operation, or to blood transfusion, i.e. the transfer to the patient of blood from another person, a method that is still to be recommended at a critical stage in severe cases. Hence it was quite a strange conception, and one lying remote from the customary beat, that came into the minds of Minot and Murphy, when they bethought themselves, that it might possibly be feasible to treat a patient suffering from this disease by administering *food* to him. It was an idea, in fact, that had never been conceived of, up to that time. A consultation of the textbooks with regard to the matter, for instance, will show us that very little attention was paid to the diet. Some items of good advice are to be found given, indeed, respecting the patient's diet in pernicious anaemia, not as an integral part of the treatment though, but rather as an element in the nursing required in general. There was, however, one exception to that universal rule, for in one spot in the world the idea had really arisen and been put into practice. To that I shall return later on.

The first work published by Minot and Murphy in regard to this question dates from 1926. It was a short paper of nineteen pages, entitled: « Treatment of Pernicious Anaemia with a Special Diet ». No mention here of a liver diet, but of a *special* diet, the special diet they had in view being one derived from Whipple's investigations, based on Whipple's diet, consisting of liver, kidney, meat and vegetables, the last two also in large quantities. As reports of observations began to come flooding in, showing that results had really been achieved by the application of the diet, there occurred by degrees a variation in the composition of the dietary in favour of liver, that food substance which, according to the showing of Whipple's investigations, had the strongest stimulating effect upon the erythropoietic, red-corpuscle forming function of the bone marrow. In their later publications we see, moreover, that they employ the designation: « a diet rich in liver ». Hence their diet became more and more preponderatingly a liver diet. There were, it is clear, great difficulties in the way to prevent their arriving at any real discovery in this regard, for the fact is, as we know, that in order to be able to achieve any palpable results from a liver diet, it is requisite for the patient to have liver administered in very considerable quantities every day, in quantities running to three, four, five, or six hundred grams, or even still more, per diem – consequently, in the twenty-four hours, upwards of half a kilogram of liver, either in a raw state or in some cooked form. We can understand what obstacles that presented for the successful issue of the inquiry, since such quantities of liver seemed very outrageous at that juncture, when there were no particularly strong reasons for expecting that the diet would have any important influence on a sufferer from pernicious anaemia. The opinion held respecting pernicious anaemia was, that it was so essentially different in nature to the blood-loss anaemia, that there did not exist, from a therapeutic point of view, any real reason for combining them together at all. As consequently there was nothing that could with any degree of certainty be expected to result from the application of the diet in question, and as the method of treatment demanded such unreasonably large quantities, it is clear that the experimenters must of necessity have been possessed of an extraordinary measure of far-sightedness, an extraordinary degree of energy and an extraordinarily clear grasp of all the circumstances of the case, as they were enabled to succeed in inducing the patients to submit to such a regimen notwithstanding its disagreeableness. If Minot and Murphy had not been imbued with such an irresistible urge to bring matters to a head, so to say, their discovery would never have been achieved. It was found, however, that, on the diet being put contemporaneously to the test at a large number of hospitals in the United States, results were actually obtained from the treatment that were astonishing, showing a more rapid improvement and blood formation, i.e. a more complete restoration to normal conditions, than had been seen to result from any other methods of treatment. It was also by degrees observed that, when once they had recovered a normal state of the blood, the patients remained well in health, which had not been the case with the methods previously applied, for with them, even in instances where very considerable improvement occurred, there had been the periodic relapses, which, as I mentioned before, are a characteristic feature of the disease. No sooner had these results been achieved in America, and been made public, than liver diet began to be tested in all parts of the world as a cure for pernicious anaemia, and from all quarters there came reports of the same experiences as to results. From everywhere confirmation came of Minot's and Murphy's observations being correct.

The success of the treatment, i.e. that by administering liver one could actually secure the disappearance of the symptoms of pernicious anaemia, meant not only a great therapeutic triumph, but also a reversal of the theory up to then held respecting the nature of the disease. The former idea had been, as a fact, that the essential point in the conditions predisposing to pernicious anaemia was the presence in the organism of a poisonous agent, a poison arising in one way or another or originating from one quarter or another – ideas respecting that point were divergent, but in general there was agreement that there must be some poison that interfered with the proper functioning of the bone marrow, entailing as a consequence the production on its part of diseased, imperfect, immature red blood corpuscles. The discovery that the disease could be cured by the sufferer from it being given a diet of liver, led scientists, very naturally, at once to query in their minds whether the theory regarding the disease that had up to then been prevalent could be correct. Reflection upon the matter must as a matter of course lead to the conclusion, that it was not probable that the *presence* of a supposed *poisonous substance* could be the cause of the disease, it being seemingly due rather to the *absence of a substance that was requisite* for a satisfactory production of red blood corpuscles, a substance that must be present in liver, seeing that the patient's condition became normal when liver was supplied as a food. In fact, a new function of the liver had thus been revealed. It is interesting to observe the reaction of medical science to this altered aspect of the matter. There exist certain representatives of quasi-medical or medico-religious bodies, who are in the habit of alleging that medical science is a species of religion or philosophical system, based upon irrefrangible tenets that do not allow of alteration or modification. That may have been so some hundreds of years ago, but in our days it is so no longer. The framework of medical science, as of other natural sciences, is a body of facts and upon them medical theories are based and built up. When a new and important fact has been established, the effect is somewhat like that of a bomb falling to earth: those theories that do not admit of being reconciled with it are exploded, being replaced at once by others that can be brought into better harmony with the newly acquired item of factual knowledge. And that has been the case in the present instance.

With a view to affording a background against which the discoveries of the prize-winners may stand out in stronger relief, I propose to say a few words as to the function of the liver in general. We are all aware that the function of the liver is to secrete bile, to be discharged into the bowel. As is well known, the bile is of great importance in aiding the process of digestion. This function of the liver is termed its external secretion, its products by degrees reaching the surface of the body. Besides that, however, the liver has other functions; it has for instance also a so-termed internal secretion. The first fact respecting an internal secretion that science was able to reveal to mankind bore reference actually to the liver, being the discovery, made in 1855 by the great French physiologist Claude Bernard, of the liver's glycogenic function, i.e. of the important part played by the liver in the metabolism of sugar in the body. On demand the liver supplies the sugar that the body requires for its normal functioning. Claude Bernard gave that process the name of «une sécrétion interne», thereby creating the term: «internal secretion», of which we hear so much at the present day. Since his time the theory that he first enunciated has been built up and amplified very materially. Thus, it is known now, respecting a number of glandular organs, that, in addition to any external secretion that they may be capable of, they also possess an internal secretion, that they consequently manufacture products that are delivered directly into the blood, and which are subsequently conveyed via the blood vessels, to remote parts of the body, where in other organs they give rise to impulses, accomplish effects, that are of the very greatest, indeed often of a vital, importance for the body. To those products, which the internal secretion supplies, the English physiologist Starling has given the name of *hormones*. We are now acquainted with a very considerable number of hormones. Time, however, does not admit of my entering into the subject further; I will only call attention in passing to the latterly so much discussed sexual hormones, those hormones, that is to say, that regulate the sexual functions, as to which in recent years especially Professor Zondek has made such fundamental investigations. Further, I may mention insulin, familiar to us all, a pancreas hormone, which medical men make use of in the treatment of diabetes. When the pancreas itself is incapable of producing this hormone, which is an essential requisite of the body, diabetes establishes itself. If under those circumstances the hormone is supplied to the body by the injection of insulin, derived from the pancreas of some animal, the symptoms of the disease can be kept away. This method of treatment, which consists of making the patient, in whose body some organ is not functioning satisfactorily, consume portions of the said organ, or absorb by injection, preparations of the said organ, is termed organotherapy. Another name is substitutional or replacement therapy, since the method consists in making substitution for, or replacing, the production that is wanting by supplies from outside. That type of treatment is not so absolutely new as many people may think. Thus, we may recall that a French physiologist, Brown-Séquard, as long ago as 1889, carried out investigations, which aroused great astonishment at the time, as to the effect of an injection into the body of testicular juice, got from the male genital glands. He gave himself injections of testicular juice and observed considerable rejuvenating effects both physically and mentally. That constituted the first achievement in the direction in question that science accomplished. Hence, it is Brown-Séquard who laid the foundation of organotherapy, and the rejuvenation treatments that we hear so much about at the present time, had forerunners at an earlier date. In this particular case, as in general in the field of medicine, it is seen that, by delving into the ancient records of the cultural achievements of the past, we may discover the nucleus of many of the methods that attract so much attention in our own times as novel and epoch-making. If we study Ebers, the papyrus of the ancient Egyptians, a venerable documentary record of the world of many thousand years ago, we shall find numerous evidences to show that the ancient Egyptians made use of organotherapy. Owing, however, to their defective knowledge of details they were unable to secure such brilliant results or to carry on their labours so purposefully, as, thanks to our incomparably greater knowledge, we are now in a position to do.

On now examining, in the light of the above facts, the discoveries made by our Nobel Prize winners, we thus find that they have come upon a new, hitherto unknown internal secretory function of the liver, have discovered that the liver yields a substance that is of the utmost importance for the normal work of the bone marrow in forming new blood.

The liver treatment for pernicious anaemia, which from its very start showed itself to be of immense value, was as already said not an easy or agreeable one for the patients to submit to, owing to the large quantities of liver that they had to consume. Even after the composition of the blood had been restored by the liver treatment to a normal state, it was obligatory to make the patients continue to eat large quantities of liver, in order to keep up their recovered health. During the years that have elapsed since then, the technique of the treatment has undergone development: the active agent or stimulating substance in the liver has been successfully extracted, an extract being thereby obtained that contains the active substance in a more concentrated form. Latterly the concentration has been brought down to such small volume, that an injection of less than a gram of the substance in solution proves a sufficient quantity for maintaining the blood in a normal condition for a period of a fortnight. Hence one injection a fortnight should suffice for keeping such patients in a state of health. The fact of so small quantities being found to be all-sufficient leads one to think of the hormones to which I referred above, they being capable of accomplishing very great results with very small quantities, and one puts the query to oneself: Is it a hormone, this substance in the liver that is made use of for curing pernicious anaemia? As to that, it is not as yet quite decided what it would be most correct to call it – a hormone, or a vitamin, or something else, and I do not propose to enter upon a discussion of that here; it may be that there are not essential differences either between the various concepts named.

In the foregoing I mentioned the fact, that the idea of treating pernicious anaemia by *food* had arisen in one quarter previous to Minot's and Murphy's day. That quarter was Stockholm, for the late Dr.Warfvinge, superintendent physician at the Sabbatsberg Hospital, who died early in the present century, had actually conceived the said idea. He was a very distinguished clinician, specially interested in diseases of the blood, and he used to urge with great emphasis that in cases of pernicious anaemia the most essential thing was for the patients to eat *meat*, meat in large quantities and at every meal, that being more important than the administration of medicine or than any other form of treatment. He was very insistent in urging it upon his assistants at the hospital, that they should bring all the influence of their authority to bear on the effort to induce the patients to eat meat, to feed themselves up on it, to force themselves to get down as large quantities of it as they possibly could. What was that but a harbinger of the very treatment, to the discoverers of which there has now been awarded the Nobel Prize? Whipple's investigations have taught us that, next to liver and kidney, meat is the article of our regimen that exercises the greatest influence as a stimulus to the blood regeneration carried on by the bone marrow. With his keen clinical eye, Warfvinge had discerned that there was something of special importance in meat, and consequently he placed in the forefront of the treatment the item that the consumption of meat constituted. It appeared to me that it would possess a certain interest for those who are present here this evening to be reminded of that circumstance.

What is then the significance of this new method of treatment? In the first place there is the fact that, thanks to it, a sufferer from pernicious anaemia can with tolerable certainty be rescued from a premature death. Hence, for the individual, the method is of an exceedingly great importance. Its importance in a wider sense must of course be dependent on the seriousness of the disease in its effect on the community at large, i.e. on the relative frequency of cases of pernicious anaemia. It is in fact quite a common disease, for in the United States, for instance, it is estimated that, previous to 1926, the year in which Minot's and Murphy's methods of treatment were first applied, about six thousand persons died of pernicious anaemia every year. I have calculated – very approximatively – that, since the date when their methods began to be generally applied, some fifteen or twenty thousand persons must have been saved from death in the United States alone. As regards conditions in Stockholm, there have been 450 cases of pernicious anaemia that have been treated at the Serafimer Hospital since I became attached to it, while, according to a recent estimate, in Sweden as a whole there must be something like three thousand persons who are suffering at the present time from pernicious anaemia. Hence the number of the lives to be rescued year by year by the application of the new method of treatment is quite considerable.

When medical progress in regard to the healing of diseases is on the tapis, there is one objection that is not infrequently brought up. People are apt to urge that, although it may be fine and merciful to effect those cures, yet, from the point of view of *the community at large*, it is not a thing to set much store by or that is beneficial, since disease, being a scavengering implement that Nature makes use of, clears off the elements in the population that are of inferior value, and leaves a residue that is of more vigorous stamina. That line of argument would lead us to the conclusion that, from the point of view of race improvement and race development, it is not to be considered any advantage to rescue those who are afflicted with diseases from which they cannot cure themselves. That reasoning is specially prevalent with respect to epidemic diseases. It is supposed by many people that, when such scourges beset a population, the victims that are swept off are the weaker and more delicate individuals, and that consequently, apart from the humanitarian point of view, it is after all advantageous to the community that the disease should run its course. That chain of reasoning, however, is based upon ignorance of how matters stand in reality. For it is not the case that, when an infectious complaint sweeps over a populated area, it is the inferior elements that are snatched away. Those victims that fall a prey are not inferior in any other sense than in regard to their powers of resistance to that particular virus or injurious agent – in other respects they may very well be as physically fit as possible. A person's power of resistance to an infectious disease cannot be estimated from a weighing-up of his general physical powers; neither physical nor mental powers are deciding factors in the case. Any individual may be possessed of a feeble power of resistance to *one* infectious disease and of a strong power of resistance to *another*. Let me take as an example the influenza epidemic of 1918–1919, which most of those present will probably recollect. Now, what was the actual state of things then? Why, it was just among the young, vigorous members of the community that far larger numbers were cut off than was the case among other groups. It would happen for instance that a vigorous, full-blooded, muscular, intelligent, promising young man died of the disease, whereas a feeble specimen, of poor intellect and of no importance to the community, survived. That is how matters stand actually. Moreover, it should not be forgotten either, that those who pass through an infectious disease successfully have nevertheless had their vitality injured by it to a greater or less degree, have consequently forfeited some of their value as members of the community. Hence, everything that medical science can do to ward off and cure disease is of benefit to the community, just from the point of view of its being contributory to the health and vigour of the race. Circumstances are the same respecting pernicious anaemia. The sufferers from the disease are not of inferior value as human beings, except in the particular respect in question. If their inferiority therein can be removed by feeding them with liver, there is no reason why they should be inferior in value in any other respect. Hence liver therapy not only rescues persons suffering from pernicious anaemia from death, it also restores them and renders them capable of living useful and active lives to the benefit of the community.

If, now in conclusion, I sum up the discoveries for which the prize has on this occasion been awarded, they are as follows: We have acquired new knowledge as to the very divergent effects exercised by different food substances in promoting and stimulating the bone-marrow's activity in the regeneration of the blood; we have been made acquainted with a new internal secretory function possessed by the liver, that is of the utmost importance; we have been furnished with a method of treatment for pernicious anaemia, and also for other diseased conditions, that will save the lives of many thousands of persons every year. Hence it must be said that our prize-winners fulfil in an ideal fashion Nobel's criteria for a prize-winner, since he lays it down that the prize shall be bestowed on the person or persons who have conferred the greatest benefit on mankind.

Now I turn to you, Professor Minot, Dr. Murphy, and Professor Whipple. I have tried to give an idea of what you have done, of the importance, of the greatness of what you have done. You have spread a new light over the process of regeneration of the blood, you have discovered a function of the liver, before you unknown to science, you have invented and elaborated a new method for the treatment of anaemia, especially pernicious anaemia, that dreadful disease, which hitherto has killed practically everyone who was afflicted by it. This new method, the liver treatment, has saved already thousands of lives, and will in the future save innumerable human beings from death.

The donator says in his will, that the prize should be given to those, who have conferred the greatest benefit on humanity. The Caroline Institute, in awarding you the Nobel Prize for Physiology or Medicine, has acted then exactly along the line of the intention of the donator. You belong to the very small number of men, of whom can in truth be said, that they have done immense services to mankind.

I ask you now to step down and receive from the hands of the King of Sweden the prize of which you arc so worthy.

GEORGE H. WHIPPLE

Hemoglobin regeneration as influenced by diet and other factors[*]

Nobel Lecture, December 12, 1934

Experiments usually have a past history or a genealogical sequence, and it may be appropriate at this time to review the genealogy of the *liver diet* experiments in anaemia due to loss of blood in dogs. With Dr. Sperry[1] in 1908 we took up a study of the liver injury produced by chloroform anaesthesia, giving particular attention to the regeneration of the liver cells to repair this injury. Icterus is invariably present in dogs with liver injury of this character and this condition was studied further. With Dr. King[2] we studied obstructive jaundice and found that the bile pigments were absorbed from the liver into the blood capillaries direct rather than by way of lymphatics. With Dr. Hooper in 1912 we began a systematic study of bile pigment production in the body as influenced by the Eck fistula[3], and we were finally able to show[4] that hemoglobin could be rapidly changed into bile pigment within the circulation of the head and thorax, the liver being completely excluded; also that hemoglobin could be rapidly changed to bile pigment[5] within the pleural or peritoneal cavities.

After leaving Baltimore to work at the University of California (1914), Dr. Hooper and I[6] took up a careful study of bile pigment metabolism by means of bile fistulas in dogs and investigated the effect of diet upon bile pigment output. As these studies were continued[7] and extended to include bile fistulas combined with splenectomy and the Eck fistula[8], it became apparent that we could not understand completely the story of bile pigment metabolism without more knowledge about the construction of blood hemoglobin in the body. Blood hemoglobin is a most important precursor of bile pigment and it was necessary to understand what factors influenced the building of new hemoglobin in the dog.

For this reason we produced simple anemia in dogs by means of blood withdrawal and in short experiments followed the curve of hemoglobin regeneration back to normal. These experiments with Dr. Hooper[9] were

[*] This paper is designed to summarize the author's contributions but does not pretend to give a review in this field nor to describe the work of others.

begun in 1917 and it was found at once that diet had a significant influence on this type of blood regeneration. Because of our interest in liver function and injury[10] we soon began testing *liver* as one of the diet factors and could readily demonstrate that it had a powerful effect upon hemoglobin regeneration[11] (see Fig. 1). These short anemia experiments were relatively crude and gave at best qualitative values for the various diet factors.

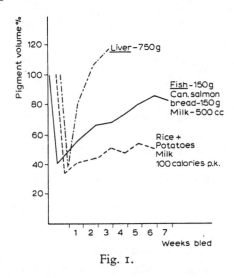

Fig. 1.

After the transfer of the anemia colony of dogs from San Francisco to Rochester, New York (1923), Dr. Frieda Robscheit-Robbins and the writer[12] began to use a different type of anemia. Dogs were bled by aspiration from the jugular vein and gradually reduced from a normal hemoglobin level of 140–150 per cent to about ⅓ normal, or 40–50 per cent, and this anaemia level was maintained a constant for indefinite periods by suitable removal of new-formed hemoglobin. The potency of the diet factor was then accurately measured in terms of the grams hemoglobin removed to preserve the constant anemia level. The stimulus presumably was maximal and uniform, and the reaction of a given dog to a diet factor was shown to be uniform when repeated time after time.

Much effort and time were spent in devising a basal ration adequate for health and maintenance during these long anemia periods lasting throughout the entire life of the dog (5–8 years). This salmon bread (Table 1)[12], moreover, also permits of minimal new hemoglobin regeneration and therefore gives a low base-line hemoglobin output from which to measure the increased output due to liver, kidney, gizzard, or other favorable diet factor.

Table 1. Composition of salmon bread.

Ingredients (g)		Protein (g)	Fat (g)	Carbohydrate (g)
Wheat flour	12,000	1,240	125	8,480
Potato starch	6,000	—	—	5,400
Bran	2,000	300	86	1,080
Sugar	3,000	—	—	3,000
Cod-liver oil	1,000	—	1,000	—
Canned tomatoes	2,000	24	4	80
Canned salmon	2,500	545	302	—
Yeast, compressed	455	55	2	96
Salt mixture*	150	—	—	—
Water	7,500	—	—	—
Total	—	2,164	1,519	18,136

Protein, 10.0 per cent. *Fat*, 6.5 per cent. *Carbohydrate*, 83.4 per cent. *Caloric value*, 4.8 per gram as fed.
* McCollums' and Simmonds' salt mixture, with ferric citrate omitted.

From Table 2 it is obvious that liver[13] again stands out as the most potent diet factor. Kidney[14] is a close second. Gizzard, spleen, and pancreas also rate high as factors which favor abundant new hemoglobin production under these standard anemia conditions. Gradually various diet factors were standardized and this information was placed at the disposal of physicians who were concerned with the therapeutic treatment of human anemias. Iron[15] was found to be the most potent inorganic element.

Pernicious anemia, examined from the point of view of the pathologist[16], was described in 1921 as a disease in which all pigment factors were present in the body in large excess but with a scarcity of stroma-building material or an abnormality of stroma-building cells. This fits quite closely with the modern conception of this interesting disease as developed from the important observations of Castle[17]. When the true factor is isolated I shall be surprised if it does not have to do with the stroma, but it may be related to the globin fabrication.

Hemoglobin utilization in anemia was studied in considerable detail. It was found that the anemic dog can conserve for new hemoglobin production about 100 per cent of injected hemoglobin[18], whether given intravenously or intraperitoneally. Muscle hemoglobin was included in this study and there is a probability that some of the injected muscle hemoglobin is

Table 2. Hemoglobin potency of diet factors (average values).

Diet factor daily intake		Total net hemoglobin average output per 2 wks. (g)	Bread base-line hemoglobin average output per 1 wk. (g)	Hemoglobin output per 2 wks.		Number of expts.
				Maximal (g)	Minimal (g)	
Pig liver	300 g	93	6	124	69	77
Liver extract 55	300 eq.	56	4	72	37	22
Pig kidney	300 g	69	3	92	49	9
Beef heart	300 g	49	5	57	33	7
Apricots dried	100 g	42	4	92	13	31
Iron (Fe)	40 mg	53	6	95	25	43
Iron (Fe)	400 mg	94	7	127	67	6
Salt mixt.-Fe	6 mg	9	3	22	0	16
Salmon bread	400 mg	—	7	19	2	110

also used in this emergency to form new blood hemoglobin[18]. Certain digests of blood hemoglobin, when given intravenously, will be utilized to about 40 per cent to build new hemoglobin in the anemic dog[18]. Foreign hemoglobins (goose and sheep) are also readily utilized[19], when given intravenously to the anemic animal, and we observe nearly 100 per cent conservation. Hemoglobin fed by mouth is poorly digested and we observe only about 10–15 per cent recovery as new-formed hemoglobin in anemia.

Liver fractions and extracts have been studied[20] and the active principles for this type of anemia separated from the active principle of pernicious anemia[21] as contained in the normal liver. The crude secondary anemia fraction[20] contains about 65–75 per cent of the potency of whole liver for new hemoglobin production in this type of experimental anemia and represents only 3 per cent of the whole liver weight.

Anemic dogs produce more new hemoglobin during a fast than during basal-diet periods; this phenomenon has received much study, with the hope that information relating to the internal metabolism of hemoglobin may be acquired. When a standard anemic dog is fed only on sugar plus iron, there will be a large output of new hemoglobin (100 g or more as a result of a two weeks' fast). Obviously this new hemoglobin must be derived from the body protein, and the mechanism of this reaction has been investigated by

Table 3. Hemoglobin construction and decrease in urinary nitrogen due to anemia and iron feeding.

Days on experiment	Fe intake (g)	Total N (mg/wk.)	Urea N + NH₃ — N (mg/wk.)	Urea N + NH₃ — N (%)	Creatinine N (mg/wk.)	Creatine N (mg/wk.)	Creatinine N + creatine N (%)	Uric acid N (mg/wk.)	Undetermined N (mg/wk.)
Non-anemic dog 29–326.									
7	0	19,250	15,990	83.0	1,190	150	7.0	70	1,860
7	2.8	13,630	10,710	78.6	1,020	0	7.5	50	1,850
7	2.8	12,140	9,840	81.0	870	0	7.2	40	1,390
2	0	13,120	10,750	81.9	920	0	7.0	50	1,410
Anemic dog 29–326.									
7	0	25,550	21,320	83.5	1,150	460	6.3	150	2,480
7	2.8	13,830	10,450	75.5	970	50	7.3	120	2,250
5	2.8	11,420	8,180	71.6	770	360	9.9	70	2,040
2	0	11,550	8,230	71.2	770	360	9.8	70	2,120

Total hemoglobin production 112 g, equivalent to 19.0 g of nitrogen in anemic period.

Drs. Daft, Robscheit-Robbins, and Whipple[22]. Nitrogen partition of the urinary nitrogen shows that during such periods there is a conspicuous decrease in the urea-ammonia fraction, which points to a conservation of nitrogenous intermediates, which would otherwise appear as urinary N but under these circumstances are used to build new hemoglobin. The importance of this body reaction is obvious and it is being studied in considerable detail (Table 3).

Human liver material obtained at autopsy has been studied recently[23], and its potency compared with standard animal liver material. If we rate pig liver as 100 per cent (our normal base-line), we may compare any given type of human liver with this control by means of our standardized anemic dogs. In this way it was found that the human liver from young healthy adults gives average values of 160 per cent. Elderly persons with arteriosclerosis and degenerative changes will show values for liver tissue of 117 per cent, as compared with the animal control of 100 per cent. Acute infections of course show swollen livers and this increase in size may account for the « dilution » of the active principle, but the average value for this liver tissue is 117 per cent. Chronic infections give liver values which are practically normal (150 per cent). Cancer invasion of the liver reduces the values of the whole liver in proportion to the replacement by cancer tissue, which by itself appears to be inert. Liver cirrhosis is compatible with normal human values for the liver tissue, 164 per cent of the control animal liver, but when hepatic insufficiency supervenes these values drop markedly (48 per cent, or about ⅓ the human normal). Secondary anemia and leukemia show values somewhat below the human normal (125 per cent), indicating a moderate depletion of these reserve factors within the liver, presumably due to blood loss.

Pernicious and aplastic anemias show a definite heaping up of these potent factors within the liver tissue, which values run above 200 per cent. In aplastic anemia there is no formation of red cells; therefore the hemoglobin-building material piles up in reserve. In pernicious anemia there is a lack of something, so that the marrow cannot produce the needed red cells; therefore the hemoglobin-building material heaps up in the liver store-house (Tables 4 and 5).

Dogs with abnormal conditions are being included within the anemia colony and observations are accumulating to show in what measure splenec-

Table 4. Hemoglobin production factors in abnormal human liver – Pernicious anemia.

Number	Cause of death	Iron content human liver		Liver intake per day		Hemoglobin output per seven days' feeding		
		Fresh tissue (mg %)	Daily intake (mg)	Human (g)	Control (g)	From human (g)	From control (g)	Ratio human to control (%)
A-371	No therapy	162.0	208	129	300	63	35	420
A-1800	No therapy	36.7	92	250	300	97	74	157
A-1045	Sl. therapy	47.3	130	290	300	112	56	208
X-2479	Nephritis	36.5	70	190	105	56	30	104
A-1472	Sl. therapy	17.5	27	158	300	52	50	200
A-425	Sl. therapy	34.8	52	150	300	45	34	265
A-1122	Embolism	24.6	33	130	300	25	30	192
A-1173	No therapy	—	—	160	300	37	46	148
	Average	51.3	87	182	—	—	—	218

Table 5. Hemoglobin production factors in human liver.

Diagnosis	Cases no.	Average	
		Iron content (mg %)	Ratio human to control (%)
Normal	9	12	162
Normal?	11	12	117
Acute infections	11	—	117
Chronic infections	16	12	149
C.P.C. liver	6	—	94
Amyloid—fat liver	10	—	111
Cancer liver	8	15	75
Cirrhosis	20	9	164
Hepatitis—insufficiency	10	10	48
Pernicious anemia	8	51	218
Aplastic anemia	4	70	201
Secondary anemia	10	7	135
Leukemia	14	13	129

tomy, the Eck fistula and the bile fistula may introduce factors having a bearing on the production of new hemoglobin under these standardized conditions. Acute and chronic infection, liver injury, and chronic nephritis are also being observed in the anemia colony. The list of abnormal states is a long one and includes disease conditions developing spontaneously as well as acute conditions of purely experimental nature. New hemoglobin regeneration can be influenced by many of these disease conditions, but it would be premature at this time to attempt evaluation of these effects. It is an interesting field, full of difficulties but also of promise for the future.

Amino acids deserve particular attention in this type of investigation and it should be possible to give certain amino acids intravenously and thereby influence hemoglobin production in anemia. We are proceeding with a systematic investigation of amino acids as diet factors in our standard anemic dogs. It would be premature to make any statement about amino acids at this time, but certain amino acids do exert a definite influence upon hemoglobin regeneration, when added in moderate amounts to the basal ration. Phenylalanine, tyrosin, and proline may be mentioned, but we have as yet no adequate data to establish any definite claim. The literature already contains too many claims for the potency of one or another amino acid in anemia; the experimental data, however, are wholly inadequate.

It is obvious to any student of anemia that a beginning has been made, but our knowledge of pigment metabolism and hemoglobin regeneration is inadequate in every respect. This is a stimulating outlook for the numerous investigators in this field and we may confidently expect much progress in the near future.

1. G. H. Whipple and J. A. Sperry, *Bull. Johns Hopkins Hosp.*, 20 (1909) 278.
2. G. H. Whipple and J. H. King, *J. Exptl. Med.*, 13 (1911) 115.
3. G. H. Whipple and C. W. Hooper, *J. Exptl. Med.*, 17 (1913) 593.
4. G. H. Whipple and C. W. Hooper, *J. Exptl. Med.*, 17 (1913) 612.
5. C. W. Hooper and G. H. Whipple, *J. Exptl. Med.*, 23 (1916) 137.
6. C. W. Hooper and G. H. Whipple, *Am. J. Physiol.*, 40 (1916) 332.
7. G. H. Whipple and C. W. Hooper, *Am. J. Physiol.*, 42 (1917) 256.
8. C. W. Hooper and G. H. Whipple, *Am. J. Physiol.*, 43 (1917) 275.
9. C. W. Hooper and G. H. Whipple, *Am. J. Physiol.*, 45 (1918) 573.
10. N. C. Davis and G. H. Whipple, *Arch. Internal Med.*, 23 (1919) 612.
11. G. H. Whipple, C. W. Hooper, and F. S. Robscheit, *Am. J. Physiol.*, 53 (1920) 151 and 236.
12. G. H. Whipple and F. S. Robscheit-Robbins, *Am. J. Physiol.*, 72 (1925) 395.
13. F. S. Robscheit-Robbins and G. H. Whipple, *Am. J. Physiol.*, 72 (1925) 408.
14. F. S. Robscheit-Robbins and G. H. Whipple, *Am. J. Physiol.*, 79 (1927) 271.
15. G. H. Whipple and F. S. Robscheit-Robbins, *Am. J. Physiol.*, 72 (1925) 419.
16. G. H. Whipple, *Arch. Internal Med.*, 29 (1922) 711.
17. W. B. Castle, *Am. J. Med. Sci.*, 178 (1929) 748.
18. G. H. Whipple and F. S. Robscheit-Robbins, *Am. J. Physiol.*, 83 (1927) 60.
19. G. B. Taylor, E. J. Manwell, F. S. Robscheit-Robbins, and G. H. Whipple, *Am. J. Physiol.*, 92 (1930) 408.
20. G. H. Whipple, F. S. Robscheit-Robbins, and G. B. Walden, *Am. J. Med. Sci.*, 179 (1930) 628.
21. E. J. Cohn, G. R. Minot, G. A. Alles, and W. T. Salter, *J. Biol. Chem.*, 77 (1928) 325.
22. F. S. Daft, F. S. Robscheit-Robbins, and G. H. Whipple, *J. Biol. Chem.*, 103 (1933) 495.
23. G. H. Whipple and F. S. Robscheit-Robbins, *J. Exptl. Med.*, 57 (1933) 637.

Biography

George Hoyt Whipple was born on August 28, 1878, in Ashland, New Hampshire, U.S.A., the son of Dr. Ashley Cooper Whipple and his wife Frances Hoyt. His paternal grandfather and his father, both physicians, were born and bred in New Hampshire.

Whipple was educated at Andover Academy and then went to Yale University, where he took his A.B. degree in 1900. Subsequently he went to Johns Hopkins University, where he took his M.D. degree in 1905.

In 1905 he was appointed Assistant in Pathology at the Johns Hopkins Medical School and, although he spent a year as pathologist to the Ancon Hospital, Panama, he remained at Johns Hopkins University until 1914, being successively Assistant, Instructor, Associate and Associate Professor in Pathology.

In 1914 he was appointed Professor of Research Medicine at the University of California Medical School, and Director of the Hooper Foundation for Medical Research at that University, being Dean of the Medical School during the years 1920 and 1921. In 1921 he was appointed Professor of Pathology and Dean of the School of Medicine and Dentistry at the University of Rochester.

Whipple's main researches were concerned with anaemia and the physiology and pathology of the liver. For a year he worked under General William Gorgas and Dr. S. T. Darling on anaemia caused by parasitic infections and especially on the lesions found in the intestinal tract in people suffering from these infections. He also studied the histology of the tissues in patients suffering from blackwater fever.

When he went to Johns Hopkins University as an assistant in the Department of Pathology, Whipple worked under William H. Welch on pigments related to liver necrosis caused by chloroform anaesthesia, his aim being to gather information about repair and regeneration of the liver cells. This problem was studied in the dog, and Whipple found that the liver cells had an almost limitless power of regeneration. He then became interested in jaundice, which is always associated with chloroform poisoning and injury to the liver. He studied the route by which the bile pigments pass into the blood and thus produce jaundice of various parts of the body and he found that the lympathic system was of little importance in transporting them. He then studied, by means of bile fistulas and other means, the bile pigments and their production outside the liver, and in this work he collaborated with C. W. Hooper.

After his appointment at the Hooper Foundation, Whipple continued his work with bile fistulas, and soon found that a better understanding of the production of haemoglobin was needed if the metabolism of bile pigments was to be understood. In collaboration with C. W. Hooper and Mrs. Robscheit-Robbins, he did experiments on short-term anaemia in dogs due to loss of blood, and further work was done on this subject and on diets consisting of liver in relation to the regeneration of blood. In Rochester, however, he decided to use anaemias due to blood loss which were uniformly sustained and were long maintained, and to study the effects on these of various factors in diets added to the rations. This work showed that the most effective addition to the diets was raw liver itself. For this work on the therapeutic value of liver in the treatment of pernicious anaemia he was awarded, together with George R. Minot and William P. Murphy, the Nobel Prize for Physiology or Medicine in 1934.

Whipple has, in addition to the researches just described, worked on tuberculosis, pancreatitis, chloroform poisoning in animals, the metabolism of pigments and iron, the constituents of the bile, and the regeneration of plasma protein, and he has studied protein metabolism by means of lysine labelled with ^{14}C, and also vitamin B_{12} labelled with ^{60}Co, and its distribution and functions in the body. He has also made studies of the stroma of red blood cells.

Among the many honours and distinctions he received are honorary doctorates of several American Universities as well as of the Universities of Athens and Glasgow; the Popular Science Monthly Gold Medal and Annual Award in 1930 (with Dr. Minot), and the William Wood Gerhard Gold Medal of the Pathological Society of Philadelphia, in 1934.

He is a Trustee of the Rockefeller Foundation. He is also a Corresponding Member of the Association of Physicians in Vienna and of the Royal Society of Physicians in Budapest, and of the European Society of Haematology, and a Foreign Corresponding Member of the British Medical Association. He is an Honorary Member of the Pathological Society of Great Britain and Ireland, and of the American Philosophical Society and the

Society of Experimental Biology and Medicine. He was, from 1936–1953, a member of the Board of Scientific Directors of the Rockefeller Institute, a member of the Board of Trustees of this Foundation from 1939–1953, Vice-Chairman of its Board of Trustees from 1953–1960, and in 1960 he was appointed Trustee Emeritus.

In 1914 Whipple married Katherine Ball Waring of Charleston, South Carolina. He has one son George Hoyt (b. 1917) and one daughter Barbara (b. 1921), and seven grandchildren.

GEORGE R. MINOT

The development of liver therapy in pernicious anemia

Nobel Lecture, December 12, 1934

The idea that something in food might be of advantage to patients with pernicious anemia was in my mind in 1912, when I was a house officer at the Massachusetts General Hospital, as is noted in certain case records there. Ever since my student days, when I had the opportunity, in my father's wards at the Massachusetts General Hospital, to distinguish from pernicious anemia two cases of chronic hereditary hemolytic jaundice, I have taken a deep interest in this disease. I watched many patients pass through relapses and remissions, and observed that, despite treatment with arsenic, blood transfusions, splenectomy, and other procedures, all eventually died. Prolonged observation permitted me to become aquainted with the multiple variations and many aspects of the disease, and to realize that from a few cases it was difficult to determine the effect of therapeutic procedures.

The study of the patients' diets was begun in 1915 in an attempt to determine if some sort of dietary deficiency could be found. The similarity of certain symptoms and signs of pernicious anemia to those in pellagra, sprue, and beriberi was appreciated, as was the fact that certain sorts of anemia were occasionally associated with a faulty diet. Elders, among others, suggested in 1922 that such a state of affairs existed in pernicious anemia. Furthermore, the almost constant occurrence of achlorhydria in pernicious anemia, which appears usually long before the anemia and remains in spite of liver therapy, led me to wonder if this disorder of the digestive system had something to do with the condition which might be in the nature of a dietary deficiency disease. Indeed, Fenwick, about 1880, suggested the possible primary role of the stomach, but it remained for Castle, in 1928, to demonstrate the part this organ plays in the causation of the disease.

The possibility of an excess of fats in the diet leading to excessive blood destruction was considered, but therapy with low-fat diets was futile. Later the effects of high-caloric feeding with an excess of protein, derived especially from meat, were studied at about the same time that other physicians, such as Barker, reported some benefit from this treatment. The results were not impressive but were perhaps suggestive.

Although Pepper in 1875 and Cohnheim in 1876 recognized that the bone marrow was abnormal, there was a prevailing opinion in the early part of this century that abnormal blood destruction played an important or primary role in the production of the disease. Nevertheless it was believed by many physicians, as I was taught, that the production of blood by the bone marrow was also deeply implicated. About 1919 the late Dr. James Homer Wright taught me to appreciate the character of the abnormality of the bone marrow in pernicious anemia, which led me to believe firmly that something was needed to make the primitive red cells that crowd the bone marrow in relapse grow to normal cells; and that it was of no particular value to aim treatment towards stopping what has been called excessive blood destruction in this disease. In 1922 Whipple suggested that in pernicious anemia there might be a scarcity of material from which the stroma of the red blood cells was formed, or that there existed a disease of the stroma-forming cells of the bone marrow. This concept fitted with the idea that there was a deficiency of something in the body and that dysfunction of pigments metabolism was resultant, or of secondary importance.

For centuries the concept that food bore a relationship to anemia had been vaguely expressed in the literature. It had been shown that liver and kidneys, rich in complete proteins, promoted the growth of animals, and that substances in liver could enhance cell division. It was likewise recognized that liver-feeding could benefit patients with sprue (Manson, 1883) and pellagra. These were among the reasons that led to the choice of liver as a substance likely to enhance blood formation. Of invaluable importance was Whipple's fundamental and classical work on hemoglobin regeneration by means of liver and other foods in anemia due to blood loss in dogs. He has now placed upon a secure quantitative basis the influence of food upon anemia.

A few patients were fed relatively small amounts of liver during 1924 and early 1925. Although these first patients did better than expected, the results permitted no more than speculations. Then Dr. Murphy joined in the work and we pursued the study of these and subsequent cases. Liver had been fed by Gibson and Howard and other individuals to pernicious anemia patients but without persistence or definite results. It seemed to us that to accomplish our object a large weighed amount of liver should be fed daily with regularity. Likewise to determine the effect it was considered essential that data should be obtained in a large number of cases to be appropriately compared with controls. By May, 1926, we had fed liver intensively and daily to 45 patients. In many of these patients symptomatic improvement was obvious within about a week. Soon they craved food, and color appeared in their faces. Tongue and digestive symptoms rapidly lessened. Within about 60 days the red blood cells counts had risen on the average from low levels to approximately normal. Dr. Murphy will describe to you in more detail the improvement that takes place. I wish especially to call to your attention the fact that an objective measure of the effects upon blood production was the chief basis of our conclusions that by feeding liver, significant improvement had been obtained. I refer especially to counts of new adult and young red blood cells (reticulocytes) appearing, as Peabody's studies demonstrated later, as a result of the maturation of the immature cells crowding the bone marrow.

The next step naturally was to attempt to determine the nature of the constituent in liver responsible for the effects, and to learn if an extract for therapeutic use could be obtained. Dr. Edwin J. Cohn, of the Department of Physical Chemistry in the Laboratories of Physiology of the Harvard Medical School, soon made a potent extract suitable for oral use. We tested on patients the preparations he prepared in an attempt to isolate the active principle. Although unsuccessful in this objective, as time passed we demonstrated (1929) that the potent material could be given intravenously, and produced maximal effects in very small quantities (0.15 g). These small experimental preparations were not practical for regular use, and it remained for Gänsslen in Germany to produce the first practical extract for parenteral therapy. Since then numerous individuals have prepared and studied such preparations. Extracts for parenteral use can be easily made by dissolving in water a powdered extract (Fraction G of Cohn) commonly used in America. The exact nature of the potent substance remains unknown, but especially from the studies of Cohn and West it appears to be a relatively small nitrogenous compound.

Extract given parenterally is at least 50 and perhaps 100 times as potent as extract given by mouth and it assures the individual of receiving into his body proper the material he lacks. It thus permits one to give large amounts easily: e.g., a few cubic centimeters of fluid a week, instead of about 1,500 to 3,000 grams of liver. The physician must know what a given amount of a given preparation may be expected to accomplish under usual circumstances. He must recognize that extraction causes loss of potency and that the potency of a good preparation is usually no more than 70 per cent of that of raw liver. Because one preparation is three times as concentrated as another, it by no means follows that the amount of potent material contained is three

times as great. It may actually be less potent. Active material is not confined
to liver. It has been found to occur in stomach tissue by Sturgis and Isaacs,
and is found in kidney, brain, placenta, and probably other organs. Castle
has shown that a reaction between an unidentified substance in normal
gastric juice and material in certain foods rich in the vitamin B complex
yields potent material. By utilization of the principle described by Castle,
Reimann has shown that liver may have its potency increased by incubation
with gastric juice. This has resulted in the practical use of combinations of
gastric and liver tissue.

Since the presentation of our original work, the regular beneficial results
of liver therapy have been confirmed in many parts of the world; for example, in Sweden by Strandell and in Copenhagen by Meulengracht. There
have been differences of opinion regarding the value of one or another preparation, but these differences can largely be explained if a proper comparison
of dosage is made. There are cases requiring much more potent material than
others and in some cases there is difficulty of absorption, making parenteral
therapy essential. As in other deficiency disorders there are factors, such as
infections, arteriosclerosis and serious damage to vital organs, which inhibit
the action of «liver extract» and when they are present it is often necessary
to give unusually large amounts of the active principle. The results of treatment will be essentially the same if, irrespective of its source for the given
case, enough potent material enters the body throughout life. Indeed, pernicious anemia, like other deficient states, should be treated on a quantitative
basis. Failure of liver therapy in a case diagnosed pernicious anemia implies
inadequate treatment, an incorrect diagnosis, or the existence of complication
sufficiently serious in itself to be disastrous for the patient.

The treatment should *not* consist in supplying enough material to remove
only one symptom, such as anemia, but enough both to supply indefinitely
all demands of the body for the substance and to fill it with an adequate
reserve supply. The disease affects the gastro-intestinal and neural systems as
well as the hemopoietic tissue. Probably more material is required to improve or inhibit the progress or development of neural lesions than to permit blood element to be maintained. With proper dosage, symptoms due
to neural lesions usually lessen, sometimes strikingly, as Dr. Murphy and
I originally noted. Unfavourable reports concerning the effects on the neural
system can be attributed to failure to realize that there is no standard dose
of «liver extract», and that there is great variation in the potency of different
preparations; also that sepsis and other complications can inhibit the action of

liver on the nervous tissue, as it does on blood formation. For proof of the effect of liver on neural lesions one must not so much study the degree of improvement, as seek for evidence of the arrest of the process. In my clinic this has been done especially by Dr. Strauss. There have been observed critically over from two to three years about 100 cases, including 26 cases with advanced combined system-disease, treated parenterally with liver extract. In no instance did a single objective neurologic sign become more marked nor did an abnormal sign develop. Abnormal signs sometimes decreased in intensity or became normal. There was thus objective arrest of the neural lesions in every case and subjective improvement in all. Indeed, certain patients who were originally unable to walk became sufficiently improved to return to their occupations. Failure to arrest the progress of neural lesions, like failure to restore the blood to normal and keep it there, usually means that more liver is required. Sometimes such severe infection or the like is present, that arrest cannot be achieved. One should not interpret parenteral therapy as acting in any fundamentally different manner from oral therapy. Certain cases treated for over eight years orally have had no progress of their neural lesions. Many cases can be adequately treated orally, but parenteral therapy is simple and permits suitable amounts of active principle easily and assuredly to enter the patient's body.

Liver therapy has been shown to be of value in other anemias than the idiopathic pernicious anemia originally described by Addison in 1849. There are other substances in liver and in certain liver extracts potent for pernicious anemia besides the principle active in pernicious anemia. Dr. Whipple has demonstrated a factor in liver aiding blood regeneration in «secondary» anemia. This factor can affect blood formation in certain clinical cases but as yet has received little critical study in man. Liver extract potent for pernicious anemia is of value in such other macrocytic anemias as arise in sprue (Bloomfield and Wycoff, Ashford and others), pregnancy (Wills, Strauss, and Castle), coeliac disease (Vaughan et al.) and the like. Partly because we recognized in pernicious anemia that, in addition to improvement of the blood, the alimentary tract and nervous systems were benefited, a still wider application has resulted. The possibility of controlling the gastro-intestinal symptoms in sprue by the use of parenteral therapy with liver extract has been demonstrated by Castle and Rhoads. For this purpose it is of great value. Likewise in pellagra the gastro-intestinal symptoms are responsive. The improvement of the skin lesions in this disease, which also occurs, suggests the possibilities of a broader application to other conditions.

It was suggested, in 1913, by Vogel and McCurdy, that determinations of the numbers of the young red blood cells (reticulocytes) might measure within a brief time the effect of therapy. Although I had studied these cells from my student days in many patients, it was not until Dr. Murphy's and my observations were well under way, that the significance of the reticulocyte reaction as an index of the effect of potent material was fully appreciated. It then became an important aid in the evaluation of therapy and in the subsequent development of effective liver extracts. It was soon recognized that, following the administration of effective doses of liver, the reticulocytes in pernicious anemia are increased for a few days, during which time their numbers follow a distinctive course. This reticulocyte reaction is orderly and simulates the curve for growth and death of organisms and their cells. Critical daily observation of these reactions and their proper interpretation thus gives useful information concerning both the state of the patient and the potency of the material administered.

A reaction that does not conform in character to that due to liver and in proper time relation to the administration of the substance under test, should always be looked upon with suspicion. The reticulocyte reaction induced by liver indicates a bone marrow reaction to a physiological need and not to a stimulant: a normal reaction to the existing anemia. A reticulocyte response does not necessarily mean that specific material which the body lacks has been supplied, for there are other substances and conditions that cause reticulocytosis in pernicious anemia, which do not regularly promote normal blood formation. For example, responses may be induced by potassium arsenite (Fowler's solution), but they usually differ very much from the physiological responses to liver. Non-specific responses, however, may closely resemble those due to supplying the deficient substance. One must be particularly critical of the nature of responses caused by substances given parenterally, because non-specific responses of various sorts can arise more readily from parenteral injection than from feeding.

It is important that the potency of products used in life-saving procedures be assured. The reticulocyte reaction is serviceable in yielding information regarding the strength of liver extracts and potent substitutes employed clinically. The factors which physiologically influence the reticulocyte reaction, also influence the rate of total red cell regeneration; but, as noted in 1927, the amount of material necessary to induce maximal reticulocyte rises is often less than the amount necessary for the maximal rate of increase of red cells. In testing products one must aim to give an amount expected to yield less than a maximal reticulocyte response, since otherwise considerable losses in potency may occur and remain undetected. The percentage of reticulocytes at the peak of their rise for a given red cell level has been used in comparative data. It is, of course, difficult to determine whether one agent is more effective than another, unless tests have been made in the same way. Very different types of curves for reticulocytes will be obtained from daily administration and from one relatively large parenteral injection, so that reticulocyte peaks cannot be compared directly. In either instance, however, the total number of young red cells poured forth during the reaction will be similar.

Because of the considerable variation in the reactivity of cases, more information can undoubtedly be obtained by comparative tests of known and unknown material in the same patient. This can be accomplished by observing the reticulocyte responses in successive ten-day periods of uniform *daily* administration of each substance. The dosage of the substance of known potency given first must be such that a submaximal reticulocyte response will result. If so much material has been given that a maximal reticulocyte response has occurred, an increase of potent material in the second period cannot induce a second reticulocyte rise. In a properly conducted test the occurrence of any orderly reticulocyte response to a second substance means that it is of greater potency than the first material given.

There is need for an animal or laboratory procedure to test the potency of products and to aid in determining the nature of the substance or substances effective in pernicious anemia. Numerous studies of this sort have been made; for example, by Vaughan on pigeons. Jacobson's recent report concerning the use of reticulocyte reactions in guinea pigs is suggestive of fruitful results.

In the early days, as has been noted, it was thought that pernicious anemia might be a dietary deficiency disease. The demonstration of the effectiveness of liver therapy indicated that this was very probably the case and thus provided the proper orientation towards a study of its cause. Castle in my clinic, from work entirely his own, has shown that it is a deficiency disease of a special sort, one which may be spoken of as a conditioned dietary deficiency disease. He and his associates have demonstrated that because of a specific defect of the gastric secretion the patient with pernicious anemia is unable to carry out an essential reaction with certain constituents of food. In the normal individual this reaction is necessary for the production of the supply of «liver extract» and thus for the prevention of the disease. The material is absorbed and stored in the liver and certain other tissues, so that when such animal tissues are fed to human beings, they receive the material needed in pernicious anemia without the necessity of the special reaction within their stomachs. In pernicious anemia it has been shown (Ivy and others) that «liver extract» is absent in the patient's liver. The factor in the normal gastric secretion responsible for the reaction has not been identified with any of its recognized constituents. Meulengracht has shown from what portion of the stomach the gastric factor is formed in pigs. Strauss and Castle have demonstrated that the food factor is associated with a number of natural sources of vitamin B, although, from the work of Wills, Lassen and others, it is probable that it is not a portion of the vitamin B complex.

As a logical consequence of this work Castle and his associates have postulated the significance of defects of the dietary factor, the gastric factor, and of difficulty in the absorption or utilization of substances promoting blood formation in the production of other types of macrocytic anemia responding to liver extract. Thus, in certain instances of the tropical macrocytic anemia of sprue (Castle and Rhoads) and in the anemia of pregnancy studied by Wills in India, a dietary defect seems of primary importance. The macrocytic anemia resulting from total ablation of the stomach or its destruction by cancer, like Addisonian pernicious anemia, is especially associated with loss of the gastric factor. In the macrocytic anemia of certain cases of late sprue, of coeliac disease and of rare instances of intestinal stenosis or multiple anastomoses, difficulty in absorption plays a major role. In the anemia of fish-tapeworm infestation these factors may also be involved. Theoretically there could occur a disorder of the internal metabolism of «liver extract», and an occasional case of macrocytic anemia might be explained on such an assumption. Today we realize also that the participation of these factors may be variable and temporary, as occurs in the pernicious anemia of pregnancy of the temperate zone (Strauss and Castle). During pregnancy the gastric factor may be absent, only to reappear after delivery. Macrocytic anemia in animals has been produced by Wills, and by Rhoads and Miller as a result of dietary defects, and Bence has shown that the liver of gastrectomized pigs becomes deficient in «liver extract».

In man, however, dietary deficiency is seldom confined strictly to one factor, nor are the results of disturbances of gastric secretion, of defects of intestinal absorption or of utilization necessarily concerned in only one type of metabolic process. Clearly such disturbances are involved in the production of iron deficiency resulting in certain types of hypochromic anemia.

We have noted that combined deficiency of iron and « liver extract» is not rare in the same individual. Other double and even multiple deficiencies occur. In one rare and striking case I have seen the tongue and skin lesions of pellagra subside upon the oral administration of yeast, the edema from protein deficiency vanish when beefsteak was fed and finally the macrocytic anemia disappear rapidly when liver extract was given orally.

Castle and his associates have shown that the gastric factor may return toward normal in pernicious anemia after treatment with liver extract. It is probable that the gastric reaction proceeds somewhat according to the law of mass action, so that very little of the intrinsic factor of the stomach might produce with a large amount of the extrinsic factor of the food, and vice versa, material for absorption sufficient to meet to a significant degree the demands of the body. Such a state of affairs could explain the responses in pernicious anemia obtained by feeding large amounts of autolyzed yeast-extract, as shown, for example, by Ungley. Here it is probable that traces of the intrinsic factor were present; and Isaacs, Goldhamer, and Sturgis have shown that traces, rather than complete absence, of this factor are apt to occur in Addisonian pernicious anemia. In this disease, however, it is almost always the gastric factor that is grossly deficient rather than the dietary factor; but both substances may be involved. We have seen patients who formerly had satisfactory diets, develop pernicious anemia lasting many months or a few years after partaking of distinctly undesirable diets. In such instances the poor diet probably precipitated the onset of the disease. It is thus not difficult, in the light of modern knowledge, to understand the probable causation of the « spontaneous» remissions and relapses in pernicious anemia. Probably in the natural course of the disease the gastric factor slowly declines but fluctuates in quantity from time to time. The extrinsic factor will also vary, and certain foods and rest may stimulate an increase of the intrinsic factor. Infection and severe gastritis could be responsible for temporary decrease of the gastric factor. The occasional patient who, after a relapse, remains without liver therapy in a state of complete remission for years, probably represents an instance of a temporary marked decline in the gastric factor, unless grossly deficient diet is shown to have existed.

Patients undergoing liver therapy for pernicious anemia sometimes, of their own accord, omit treatment and may remain in apparent good health for many months, but many more of such individuals soon find they are sick again. The former perhaps have accumulated a considerable reserve supply of « liver extract» in their bodies and have probably been enabled by treatment to manufacture some potent material from an ordinary diet. Sooner or later a very large number of these patients will relapse, if proper treatment is not resumed. The grave error in treatment is to prescribe too little liver extract or potent substitute. When there is doubt, more rather than less should be given. It is essential that the individual receive into his body indefinitely and with regularity enough potent material for his given case. The physician, however, must do more for his patient than prescribe a proper amount of liver, stomach, or the like; he should attend to all aspects of the case and not neglect attention to the individual's manifold problems of thought and action.

I have attempted to outline Dr. Murphy's and my contribution to the work for which the Caroline Institute has honored us. I have pointed out to you, also, some of the studies that have been made as the result of demonstrating with Dr. Murphy that liver feeding is dramatically effective for pernicious anemia patients. It seems to me that one may expect in the future more information to be obtained which, directly or indirectly, will follow as the result of these observations. Thus, upon the foundations laid by previous investigators, do medical art and science build a structure which will in its turn be the foundation of future knowledge.

Biography

George Richards Minot was born on December 2, 1885, at Boston, Massachusetts, U.S.A. His ancestor, George Minot, had migrated to America in 1630, from Saffron Walden, England. His father, James Jackson Minot, was a physician, and his mother was Elizabeth Whitney.

In his youth Minot was interested in butterflies and moths, and he published two articles on butterflies. He went to Harvard University and there took his A.B. degree in 1908, his M.D. in 1912, and gained an honorary degree of Sc.D. in 1928.

He did his hospital training at the Massachusetts General Hospital and then worked at Johns Hopkins Hospital and Medical School, under W. S. Thayer and W. H. Howell.

In 1915 he was appointed Assistant in Medicine at the Harvard Medical School and the Massachusetts General Hospital and was later appointed to a more senior post there.

In 1922 he became Physician-in-Chief of the Collis P. Huntington Memorial Hospital of Harvard University, and later was appointed to the Staff of the Peter Bent Brigham Hospital.

In 1928 he was elected Professor of Medicine at Harvard University and Director of the Thorndike Memorial Laboratory and Visiting Physician to the Boston City Hospital.

Minot early became, when he was a medical student, interested in the disorders of the blood with which his name is associated and he published during his life many papers on this and other subjects. Arthritis, cancer, dietary deficiencies, the part played by diet (vitamin B deficiency) in the production of so-called alcoholic polyneuritis and the social aspects of disease were among the subjects of his papers. Further he studied the coagulation of the blood, blood transfusion, the blood platelets and the reticulocytes as well as certain blood disorders, and he described an atypical familial haemorrhagic condition associated with prolonged anaemia. He also studied the condition of the blood in certain cases of industrial poisoning.

Among his other interests were leucaemia, disorders of the lymphatic tissues and polycythaemia, but his most important contributions to knowledge were made in his studies of anaemia. His name will always be associated with the therapy of pernicious anaemia, in which he first became interested in 1914, but it was not until later that he, like William P. Murphy, became impressed by the work of George Hoyt Whipple on the treatment of experimental forms of anaemia in dogs, and in 1926 he and Murphy described the effective treatment of pernicious anaemia by means of liver. For this work he and Murphy and Whipple were awarded, in 1934, the Nobel Prize for Physiology or Medicine. Subsequently, Minot, in collaboration with Edwin J. Cohn, extended this work by showing the efficacy of certain fractions of liver substance and he demonstrated the value of reticulocyte reactions in the evaluation of therapeutic procedures. He also added to knowledge of gastro-intestinal functions and of iron therapy for anaemia, and to knowledge of other aspects of this group of diseases.

Minot was member or fellow of numerous medical and allied organizations in his own country and abroad, and served as Editor of several medical publications. Among the many honours and distinctions he received, may be mentioned: the Cameron Prize in Practical Therapeutics of the University of Edinburgh, in 1930 (jointly with W. P. Murphy), the Popular Science Monthly Gold Medal and Annual Award for 1930 (jointly with G. H. Whipple), and the John Scott Medal of the City of Philadelphia.

On June 29, 1915, Minot married Marian Linzee Weld; there were two daughters and one son by this marriage.

After a long and busy life, during which he made many important contributions to medical knowledge, especially to that of diseases of the blood, Minot died, full of honours, in 1950.

WILLIAM P. MURPHY

Pernicious anemia

Nobel Lecture, December 12, 1934

During the twenty-year period following 1849, in which year Thomas Addison first described the diseased condition, which he designated as «idiopathic» anemia, reports of similar cases were published by such men as Barclay, Wilks, Bristowe, Lebert, Habershon, and others.

Further interest was aroused, both on the Continent and in America, by Biermer's discussion of a group of patients with severe anemia of varying etiology, in a paper published in 1872. As a designation for these cases he suggested «progressive pernicious» anemia, a name which became more generally used than «idiopathic» anemia, suggested by Addison. Biermer also called attention to the frequency of retinal hemorrhages and the occurrence of fever in his cases.

Even as early as 1878, Eichhorst published a 375-page monograph on Progressive Pernicious Anemia, and five years later Laache of Christiania published his monograph on anemia, which consisted of 256 printed pages together with many graphs and plates illustrating the blood changes. He described particularly the presence in the blood of large, deeply coloured corpuscles. Quincke had previously called attention to the variations in shape of the red blood corpuscles – the poikilocytosis.

Even these early papers had presented the clinical picture of pernicious anemia essentially as we see it today. Various theories as to the etiology were discussed but perhaps the one of most interest, when viewed in the light of knowledge available since the era of liver therapy, is that presented in 1880 by Fenwick in his book *Atrophy of the Stomach*. Fenwick wrote: «indeed most of the symptoms are not the immediate result of the atrophy of the stomach, but arise from the deficiency of the blood produced by it». And again: «it will be readily conceded that the anemia that accompanies atrophy of the stomach is the result of the imperfect secretion of the gastric juice consequent upon it».

Since the earliest use of liver in the treatment of pernicious anemia, however, new fields of observation have been made available both in the clinic and in the laboratory. We have been allowed the thrill of watching the patient through a few days of depression following the institution of liver therapy until remission occurs with its often sudden and almost unbelievable sense of well-being simultaneously with the maximum increase of the reticulocytes or new red blood cells. Then we have followed this remission through to completion, until the blood becomes normal, with a normal red blood cell level – that is 5,000,000 or more cells per cubic millimeter of blood. Perhaps even more dramatic has been the improvement in the disturbances of locomotion resulting from nerve damage. But all of this has been described in our early papers, so that further details need not here be recited.

Observations of the patients at intervals in the office or hospital blood clinic, and attention to the important details of treatment have made it possible for us to regularly maintain our patients in a state of economic efficiency and with reasonably good health. Forty-two of the forty-five patients originally treated and discussed in our first paper of 1926 have been kept under observation. Of this number, thirty-one, or approximately three fourths, are living and well after almost ten years of treatment. Eleven have died from various causes other than pernicious anemia.

The problem which, during the past few years, has particularly interested me, as a practitioner of medicine, has been the practical one of making treatment more bearable for the victim of pernicious anemia, who must necessarily continue treatment indefinitely in order to maintain a satisfactory state of health. For this purpose treatment must be simplified, its efficiency increased and its cost decreased. Definite progress has been made in this direction through the development of a liver extract for parenteral use. At the Peter Bent Brigham Hospital we have, for over three years, used an extract of uniform potency for intramuscular injection which is so concentrated that 3 cubic centimeters is prepared from 100 grams of liver, and represents the potency of fifty times the amount of liver from which it is prepared. Or, in other words, the injection intramuscularly of 3 cubic centimeters prepared from 100 grams of liver is equivalent in its effect to that of 5,000 grams of whole liver when taken perorally.

Because of the concentration of the extract and its uniform and high potency, it is possible to bring about improvement most rapidly and with confidence, to maintain a normal blood level and state of health with infrequent injections, perhaps three cubic centimeters every two to four weeks, and at a minimum of expense to the patient. A rough comparison of dosage and cost of treatment by the various means may be of interest. It is estimated that 5,000 grams of whole liver, 84 vials of liver extract for peroral use (prepared from 8,400 grams of liver) and 3 cubic centimeters or 1 vial of liver extract for intramuscular injection (prepared from 100 grams of liver) will have essentially similar effects. The liver may cost $ 5.50, the liver extract for peroral use $ 17.00 and the liver extract for intramuscular use $ 1.17, a striking difference.

Not only has this liver extract for intramuscular injection proved its value for pernicious anemia; it has also, which is perhaps of even greater importance, displayed its ability to stimulate the production of leucocytes. The fact that the granulocytes may be practically doubled in number within from six to eight hours after a single injection, recommends its use in many of those diseased states accompanied by granulocytopenia, such as pneumonia, influenza, agranulocytosis of Schultz, and even in some instances postoperatively.

Again, we find it effective in enhancing the rate of hemoglobin formation in the hypochromic anemias treated with iron. Perhaps its effects here is in producing the stroma or envelope in which the hemoglobin may be more quickly and normally stored.

So there will continue to be found more and more uses for this highly potent material and, with even greater concentration in the near future, it will further lighten for the patient the burden of continued treatment. During the next few years many important problems in the general field of the blood dyscrasias will also be solved as a direct result of the introduction of liver therapy for pernicious anemia.

Rather than enlarge further upon the details and results of the treatment of pernicious anemia, I shall now present, with your permission, a motion picture which will illustrate many points more clearly than I could discuss them here.

Biography

William Parry Murphy was born on February 6, 1892, at Stoughton, Wisconsin, U.S.A. He is the son of Thomas Francis Murphy and Rose Anna Parry, his father being a congregational minister with various pastorates in Wisconsin and Oregon. William Parry was educated at the public schools of Wisconsin and Oregon and at the University of Oregon, where he took his A.B. degree in 1914.

For the next two years he taught physics and mathematics at the high schools of Oregon, and then spent one year at the University of Oregon Medical School at Portland, where he also acted as a laboratory assistant in the Department of Anatomy. He then attended a summer course at the Rush Medical School in Chicago and was later awarded the William Stanislaus Murphy Fellowship at Harvard Medical School, Boston. He held this Fellowship for three years and graduated as a Doctor of Medicine in 1922.

Two years as House Officer at the Rhode Island Hospital followed and he then became Assistant Resident Physician at the Peter Bent Brigham Hospital under Professor Henry A. Christian. This appointment he held for eighteen months and then he was appointed Junior Associate in Medicine at this hospital.

In 1924 he was appointed Assistant in Medicine at Harvard, and from 1928 until 1935 he was Instructor in Medicine there. From 1935 until 1938 he was Associate in Medicine at Harvard and from 1948 until 1958 Lecturer in Medicine, becoming in 1958 Senior Associate in Medicine, and subsequently Emeritus Lecturer in that subject.

In 1923 Murphy practised medicine for a time and subsequently engaged in research on diabetes mellitus and on diseases of the blood. Murphy's work on pernicious and other forms of anaemia was outstanding. For the treatment of pernicious and hypochromic anaemia and for granulocytopenia he used intramuscular injections of extract of liver and he was associated with George Richards Minot and George Hoyt Whipple in work on pernicious anaemia and the treatment of it by means of a diet of uncooked liver. For this work he was awarded, together with George Richards Minot and

George Hoyt Whipple, the Nobel Prize for Physiology or Medicine for 1934. He wrote *Anemia in Practice: Pernicious Anemia* (1939).

He has been consulting haematologist to several hospitals, and he now lives at Brooklyn, Mass., U.S.A. Among his many distinctions and honours are the Cameron Prize of the University of Edinburgh, together with George Richards Minot for their work on pernicious anaemia (1930), the Bronze Medal of the American Medical Association for an exhibit demonstrating his methods of treating anaemias with liver extract (1934), the First Rank of Decoration-Commander of the Order of the White Rose, Finland (1934), and the National Order of Merit, Carlos J. Finlay, Official, Cuba (1952).

He is member of numerous medical and allied societies at home and abroad, including the Deutsche Akademie der Naturforscher Leopoldina.

Murphy married Pearl Harriett Adams on September 10, 1919, and they have one son, Dr. William P. Murphy, Jr. Their only daughter, Priscilla Adams, died in 1936.

Physiology or Medicine 1935

HANS SPEMANN

«for his discovery of the organizer effect in embryonic development»

Physiology or Medicine 1935

Presentation Speech by Professor G. Häggquist, member of the Staff of
Professors of the Royal Caroline Institute

Your Majesty, Your Royal Highnesses, Ladies and Gentlemen.

When the Staff of Professors of the Caroline Institute decided that Professor Hans Spemann should be considered pre-eminently for this year's Nobel Prize it was the first time that a representative of that branch of physiology which is called developmental mechanics was to receive this award.

Developmental mechanics seeks to establish the inner causal connection between the developmental processes. Wilhelm Roux founded this branch of science at the end of the 80's of the last century. Although Roux himself, Driesch, and many others have enriched our knowledge with interesting facts, it was really Spemann and his school who first established developmental mechanics as a current branch of science which has revealed laws and relationships which encompass the entire biological world.

In his technical work, Spemann can be called a micro-surgeon. His instruments are simple: glass rods drawn to a point, glass tubes which can be used as fine pipettes, or loops of children's hair. His experimental material consisted of the eggs of newts and frogs. An egg cell of this kind is a little ball of living matter with a diameter of $1-1\frac{1}{2}$ mm. Normally after fertilization it develops by continued segmentation until it changes into a small hollow sphere whose wall consists of small cells. Subsequently this hollow sphere invaginates rather as if you were to take a burst rubber ball and squeeze it together in the hand; only the difference is that the walls grow together so that the orifice of the now double-walled sphere will be small and cleft-shaped. After that, a further layer of cells grows between the two walls of the sphere. These three layers are called *ectoderm*, *mesoderm*, and *entoderm* from outside inwards respectively. The orifice is called the *blastopore*. Then in front of this blastopore there arise from the ectoderm the primordia of the brain and spinal cord. Beneath the primordial brain an invagination of the ectoderm against the entoderm is formed, later to become the mouth. The mesoderm will form the skeleton (in the first place the dorsal strip, then the notochord) and muscles. The entoderm forms gut.

Much thought has been given to the nature of the forces and causality regulating this development. It is at this point that Spemann's researches begin. He used eggs of various animal species which differ in colour, and with his simple instruments transplanted small pieces of tissue in different stages of development. By this means he was able to establish that, for example, a cell mass normally destined to become ventral epidermis – Spemann calls it presumptive ventral epidermis – could develop into nerve tissue if it were put in the place where the spinal cord was to develop. Hence, the course of development of these cells was not laid down in advance or it could – if such was the case – be altered by transplantation; so that the transplanted portion adjusted itself to its new environment. When Spemann then transplanted the anterior lip of the blastopore of an embryo into the ventral side of another embryo it grew a new brain and spinal cord. This brain and spinal cord did not arise from the transplanted cell material, but from the presumptive ventral epidermis whose course of development was thus altered by the presence of the blastopore. From this Spemann could ascertain that the blastopore had an organizing influence on its environment. The cell material which was grafted into the ventral epidermis and caused the development of the new spinal cord was actually of the kind that, developing normally, would have given rise to the notochord. Further experiments showed that it is the notochord primordia which organize the development of the primordial spinal cord, while, on the other hand, the mesoderm in the head causes the development of a primordial brain. Near this arise the so-called optic vesicles which are the origin of the retina of the eye. Where these approach the ectoderm of the head they organize the development of the lens of the eye. Or, to take another example: the anterior end of the primordial gut (the oesophagus) organizes the development of a primordial mouth and primordial teeth inside it. Thus, we now see how cell masses originally undifferentiated have the course of their development laid down by the influence of rudiments of organs formed earlier. Thereafter, a cell mass such as this can assume the role of organizer in relation to its environment.

In this way we begin to understand how the laws of development work. We begin to perceive why a primordial head arises at the anterior end of the embryo, why a brain always arises in the head and never anywhere else, or why the mouth always has its place below the primordial brain and never elsewhere.

When the main principles of normal development become clear we may indeed hope that we shall soon come to understand abnormal developmental processes and how malformations arise. In his experiments Spemann has already succeeded in producing individuals with « situs inversus »: those peculiar malformations in which the relationship of the organs to left and right is the opposite of that in a normal individual. Such cases are also known among human beings: people in whom the heart is mainly on the right side, the stomach on the right, the main mass of the liver as well as the appendix are on the left side, to name only a few organs. Perhaps – and as doctors we hope so – Spemann's researches may also lead to a better understanding of the development of those strange and fateful structures known as tumours. For these can actually be regarded as the result of a disorganization of normal development and of normal conditions within the tissue.

Be that as it may. Even if our hopes on this score are not to be fulfilled, nevertheless Spemann has revealed conditions in the developmental process which are of deep significance. A mountain of difficulties rears up before him who seeks to wrest from Nature, secrets connected with the origin and development of a new individual. Spemann has opened this mountain and has brought to light rich treasures of knowledge Moreover, a group of disciples has followed him, who can carry his thoughts forward and continue the work at such time as the master himself grows weary. As evidence of the great esteem in which they hold Spemann's merits the members of the Staff of Professors of the Caroline Institute have awarded him this year's Nobel Prize for Physiology or Medicine.

Herr Geheimrat. You are a student of Theodor Boveri, and occupy the Chair once held by Professor August Weismann. These are two names of vast reputation evoking feelings of gratitude and admiration in anyone engaged in biological research. They are, however, also names imposing on the student and successor responsibilities for carrying on a great tradition. You, Herr Geheimrat, have been successful in upholding this proud scientific tradition. You have, with new tools, continued where Weismann and Boveri had to stop, and have paved new ways in biology. August Weismann managed, although ignorant of Mendel's observations, to outline the significance of the nucleus as bearer of heredity; Boveri laid, together with Oscar Hertwig, the foundation of our knowledge of the fertilization phenomena; and you, Herr Geheimrat, have discovered secret forces regulating the early development of the fertilized egg. You have also created a school of scientists from whom Science can expect further valuable contributions.

As a result of this you have occupied a place in the first rank of great cultural personalities in which your country is so rich.

As a token of its great appreciation of your scientific achievements, the Staff of Professors of the Caroline Institute has decided to confer upon you the Nobel Prize in Physiology or Medicine for this year. I ask you to receive the prize from the hands of His Majesty the King.

HANS SPEMANN

The organizer-effect in embryonic development

Nobel Lecture, December 12, 1935

The experiments which finally led to the discovery of the phenomena which are now designated as «organizer-effect» were prompted by a question which actually goes back to the beginnings of developmental mechanics, indeed to the beginnings of the history of evolution in general. How does that harmonious interlocking of separate processes come about which makes up the complete process of development? Do they go on side by side independently of each other (by «self-differentiation», Roux), but from the very beginning so in equilibrium that they form the highly complicated end product of the complete organism, or is their influence on each other one of mutual stimulation, advancement or limitation?

These question, various answers to which constitute the theories of preformation or epigenesis, were lifted out of the realm of speculation up into that of an exact science when first Wilhelm Roux and then Hans Driesch used experimental methods in their research into development. The first experiments consisted in separating the individual parts of the embryo from each other and culturing them in isolation. This would show what each part was capable of by itself, while at the same time showing how far the developmental processes depending on them were dependent on or independent of each other.

In this way Roux was able after taking a frog's egg, pricking and destroying one of its two blastomeres, to obtain half an embryo from the other. Driesch, on the other hand, took a sea-urchin's egg, separated one segmental cell from the other and obtained a smaller but complete embryo. Further experiments showed that the differing results depended not on the material but on the method. The completely isolated segmental cell which has been reduced by half can grow into a whole in the case not only of the sea-urchin's egg, but also of amphibian's egg. This growth is inhibited if the dead cell is left attached; when this happens, the cell grows in accordance with its original determination, forming, first at least, half an embryo.

Even in those early days of research into developmental mechanics a second method of enquiry into this same question was discovered – that of «embryonic transplantation». Gustav Born observed that portions of young larval amphibians united if their freshly cut edges happened to come into contact with each other. He followed up this phenomenon and found that the individual portions were capable of self-differentiation to an astonishing degree.

It was from these premises that I began my experiments. They were all carried out on young amphibian embryos, mostly those of the common striped newt (*Triton taeniatus*). To make these experiments intelligible to the non-specialist it will be necessary in the first place to describe the main features in the normal development of these eggs.

Development begins immediately after fertilization, with a fairly protracted period of cell division which is called segmentation on account of the furrowing which appears on the surface. By the formation of an inner cavity or blastocoele, the blastocyst or blastula comes into being. Its lower, vegetative half (the thick floor of the blastocyst) consists of large cells rich in yolk, while the upper, animal half (the thin roof) is made up of numerous small cells poorer in yolk. Between the two is the marginal zone – a ring of medium-sized cells.

Next begins a very complicated and in many ways puzzling process: the so-called gastrulation. The end result of it is that all the material of the marginal zone and of the vegetative half of the blastula becomes invaginated and is thus covered over by animal material. Then along the line of invagination, i.e. the primitive orifice or blastopore, runs the outer layer of cells or ectoderm into the two invaginated layers, the mesoderm (originating from the marginal zone), and entoderm (corresponding to the vegetative half of the blastula, rich in yolk).

With this the primordia of the most important organs, the skin and central nervous system, vertebral column and musculature, gut and body cavity have in the main achieved their final dispositions. Their visible differentiation occupies the next phase of development.

The primordium of the central nervous system originates in the ectoderm of the dorsal surface, starting from the blastopore and coming forward as a thickened plate shaped like a shield with its anterior half broader than its posterior. This is the neural plate, and its lateral margins rise up as the neural folds. The neural folds are brought closer to each other and fused together so that the neural plate becomes a tube – the neural tube. This becomes separated from the epidermis and sinks below the surface. Its front end, which is thicker and originated in the broader anterior part of the neural plate will become brain; its thinner posterior part will become spinal cord.

The neural plate lies over mesoderm. When the plate forms the neural tube, separates off and sinks below the surface, the mesoderm divides into five longitudinal strips lying side by side. The median strip is destined to be the axial skeleton or notochord. To the right and left of it is a row of mesodermal blocks or somites. These in turn are flanked on either side by the lateral plates from which arises the primordium of the coelum.

Finally, the entoderm first forms a broad open gutter, which is shaped like a trough. Its margins then bend inward towards the middle, and, along the mid-line – that is, just beneath the notochord – it completes the intestinal tube.

All these processes which, given a favourable temperature, go forward surprisingly quickly depend essentially not on the production of new material from the embryo substance but on the rearrangement of what is already there. It is therefore possible, and W. Vogt did this to perfection by means of staining, to show in the blastula or early gastrula, as it were, a topography of the rudiments of the presumptive organs.

In the face of this sort of topographical map we are again confronted with the question whether there is a real diversity in these parts which corresponds to the pattern of the presumptive rudiments in the early gastrula; whether they are more or less predestined, i.e. «determined», for their subsequent fate or whether they are still indifferent and do not have their ultimate determination impressed on them until later.

The first answer to this question was given by experiments in isolation. Thus, if the bisection is not made as early as between the two cells after the first segmentation but later, even at the blastula stage, or at that of the very young gastrula, you can still get twins. So up to this stage the cell material must still be to a large degree indifferent and capable of being used for various purposes in constructing the body. This becomes especially clear when the bisection is made in such a way that it separates the ventral half of the embryo from the dorsal half. Even then the latter half can develop into a miniature embryo of normal proportions. Here the new allocation of the material becomes perfectly clear. According to the evidence of our topographical map, the dorsal half contains almost all the material for the neural plate, i.e. much too much for a half-sized embryo; on the other hand, it lacks all of the presumptive epidermis. This latter must therefore be made good by material from the former.

Now if presumptive neural plate and presumptive epidermis are interchangeable, they must therefore also be interchangeable without prejudicing further normal development. Embryonic transplantation at this early stage must therefore produce different consequences than it would if performed in the later stages in which Gustav Born experimented.

It was on these thoughts and on the development of a way to facilitate the manipulation of these uncommonly fragile young embryos and operation upon them that the success of the new experiments rested.

The first experiment consisted in exchanging a portion of presumptive epidermis and neural plate between two embryos of the same age, each being at the beginning of gastrulation. The grafts took so smoothly and development proceeded so normally that their margins left no trace except that the grafted tissue itself was distinguishable for a while by means of its natural pigmentation, or by artificial vital staining. From this it was obvious that, as we had expected, the portions were interchangeable – that is to say, presumptive epidermis could become neural plate and presumptive neural plate could become epidermis.

From this we can infer not only the very indifferent nature of the cells at this early stage of development; the result allows the much more important conclusion that the transplanted portion must in its new environment be subjected to some kind of influence which determines its subsequent development.

It is here that the analytical superiority of this experiment is shown over the previous ones, whereby use was made of the regulation power of the embryo. For it was now possible to examine all the parts of the embryo separately for their active and re-active induction capacity, and also to vary the age and species of the implant with great latitude.

At the same time this opens important fresh possibilities: first of all in the matter of procedures. The interchangeability may be undertaken not only between embryos of the same species but also between those of different species, e.g. between embryos of *Triton taeniatus* which have a fair amount of pigmentation and those of *Triton cristatus* which have little or none. This allows us to distinghuish the implant more or less clearly for a very long time even in sections and often to define its limits in terms of its cells. Let me describe a case of this kind in more detail.

A portion of presumptive neural plate was removed from an embryo *Triton taeniatus* at the beginning of gastrulation and exchanged with a portion of presumptive epidermis from a *Triton cristatus* embryo of the same age. The embryo in which the host was taeniatus later showed anteriorly and to the left in the neutral plate a smoothly grafted oblong area of white cristatus tissue which developed further into parts of the brain and eye. The other embryo with cristatus as the host showed on the right-hand side in the epidermis of the gill area a long streak of dark taeniatus tissue which developed further as epidermis and formed the covering of the outer gills. Since the portions have been exchanged, and since one portion is now settled where the other came from, we can see at once from sections that brain substance has come from presumptive epidermis, and epidermis has come from presumptive brain substance.

Because the implant in this « heteroplastic » transplantation remains distinguishable for a fairly long time it is possible to test the interchangeability of those parts of the embryo which develop inwards during gastrulation. We can, for example, establish whether the exchange is feasible not only as between one and the same layer of cells but also as between two different layers.

By and large this is in fact the case. So O. Mangold was able to show that mesodermal organs such as notochord, somites and pronephric ducts could arise from presumptive ectoderm by suitable transplantation at the beginning of gastrulation.

Now, when random samples were taken from the whole surface of the gastrula and transplanted in this way in an indifferent place it became apparent that a limited area, namely the region of the upper and lateral blastopore lip did not conform. A portion of this kind, transplanted in an indifferent place in another embryo of the same age did not develop according to its new environment but rather persisted in the course previously entered upon and constrained its environment to follow it. It invaginates altogether as if it were still in its old place, builds up part of the axial organs and completes itself out of the mesodermal environment. Above all, it induces in the overlying ectoderm a neural plate which closes to a tube, in favourable cases bulges out into optical vesicles and adds lenses and auditory vesicles.

First carried out at my instigation by Hilde Mangold, this experiment shows, therefore, that there is an area in the embryo whose parts, when transplanted into an indifferent part of another embryo, there organize the primordia for a secondary embryo. These parts were therefore given the name of « organizers » and the region of the embryos in which they are gathered together at the beginning of gastrulation was called the « centre of organization ». H. Bautzmann has defined the limits of this area by systematic probing outwards and has found that it coincides more or less with the area of the presumptive notochord-mesoderm which invaginates later.

From these two facts – the development of an indifferent piece in conformity with its location and the inductive effect of an organizer – several series of experiments proceeded, connected with obvious questions. We will just touch on a few of them.

Since at first the organizer becomes invaginated, that is, completes the gastrulation it has begun, so that material in the neighbourhood can be included in the process, one might suppose that it is this process itself which causes further determination of the parts it has affected. But this is, to say the least, extremely unlikely, because the induction of neural plate takes place even though it has not itself been invaginated. This can be proved by a method which is highly significant for the whole progress of research. That is to say, those parts of the embryo which are being examined for their inductive capacity can be made to bypass the active invagination and can be made effective by inserting them in the blastocoele through a small slit in the roof of the blastula or young gastrula which quickly heals over. The gastrulation does not suffer any essential disturbance from this and while it goes on, the blastocoele disappears and the piece we are examining comes to lie directly under the ectoderm and there shows what it is capable of. Thus a portion of the upper marginal zone of the blastula or early gastrula, or else a piece of the roof of the archenteron of the mature gastrula was planted in the blastocoele of a young gastrula and so brought beneath the ectoderm from the beginning; it was demonstrated that these portions were able to induce neural plate.

Now, these methods made it also possible to examine for their inductive capacity pieces which could not be embodied in the host embryo by any other means, either because they differed too much in age and origin or else because they were no longer living, or even not of living origin. We will have a look at these experiments next.

It had already been demonstrated in my early experiments that host and donor did not need to be exactly the same age in order to be able to work together. It was O. Mangold in particular who followed up this question and made the important discovery that the inductive reaction capacity is strictly limited in time while the inductive action capacity remains for a long time, far beyond the stage necessary for normal development.

This is true not only, as H. Bautzmann showed, for the notochord which normally induces in the earlier stages, but strangely enough also for a portion of embryo in which there would otherwise be no question if this kind of induction, viz. the neural plate. Both O. Mangold and I found simultaneously but independently, and starting from different lines of enquiry, that it can induce after transplantation. To this, O. Mangold added the important statement that the inductive capacity of this tissue persists into late stages, until there is a functioning brain in the hatched out larva.

Associated with this is the question whether and how far the inductive influence is specific in nature. Also, and this is connected with the other question, what role the action and reaction system plays in bringing about the highly complicated product of development. I had already expressed the opinion earlier that the inductive stimulus does not prescribe the specific character but releases that already inherent in the reaction system. The inductive potential already adduced of parts which have far exceeded the stage of observed normal effectiveness also points in the same direction. Still more is this true of the more recent experiments by Holtfreter which prove the extensive diffusion of factors which are able to induce a neural plate in the ectoderm of the young gastrula. So pretty well the whole animal kingdom from tapeworms to human beings was examined by the implantation method and shown to be capable of induction.

However, this does not only make obvious the largely unspecific character of the inducting agent; it also seems probable that it is chemical in nature. It was always thought to be so from the beginning. To make quite sure, experiments had to be made in which the inductor had been destroyed in various ways—by desiccation, freezing, or boiling. We got no clearly positive result from these first experiments; not until later similar ones by Holtfreter. It became apparent that this kind of treatment did not destroy the capacity of the inductors and, further, quite paradoxically, that this can in fact call forth such capacity in non-inductors.

The first experiment with a chemically treated inductor was carried out by Else Wehmeier and proved that an inductor immersed in 96% alcohol for $3\frac{1}{2}$ minutes did not lose its capacity.

After this, the chemical analysis was tackled in various quarters: in Germany by F. G. Fischer and E. Wehmeier, later with H. Lehmann, L. Jühling, and K. Hultzsch; in England by J. Needham, D. M. Needham, and C. H. Waddington. From the large number of separate results which still seem to be coming in I should like to draw attention to one only which is of the utmost importance in this connection. Chemically simple substances as, for example, synthetic oleic acid can nevertheless induce a complicated and in a certain sense complete structure such as a neural plate which will close over into a neural tube. Again, that would therefore indicate, as do some of the results from abnormal inductors, that most of the complication is based in the structure of the reaction system, and that the inductor has only a triggering and in some circumstances directing effect. Whether and, if so, how far and in what way such «unorganized inductors» (for it would be a contradiction in terms to speak here of «organizers») determine the direction is at the moment one of the most interesting but also most difficult questions.

But this broaches a new complex of questions which goes right back to the first induction experiments. It had already turned out in Hilde Mangold's experiments that the induced embryonic primordia were in the main arranged in the same direction as the primary ones and on a level with them. This seemed to emerge either from a general structural plan of the embryo or else from an influence of the primary embryonic primordia.

To investigate the former phenomenon, the similarity of direction of the constituents of the two embryos, two different experiments were set up. Upper blastopore lip still engaged in invagination was implanted in a different orientation in relation to the host embryo – crosswise and opposite to the orientation of the later primary primordia. With crosswise implantation it was shown that the invaginating cells of the graft were carried along by the gastrulating movements of the host and that thus the substratum was laid down along the long axis of the embryo. With opposite implantation the cells of the graft migrating against the stream get jammed but are not deflected. A controlling structure of the embryo, therefore, only works in so far as it determines the direction of the gastrulation movements both of the host embryo and the graft. It becomes even more obvious when a piece of the roof of the archenteron is planted in the blastocoele. The graft does not lie fixed in the cell formation of the host embryo so it can rather keep its original position and the induced secondary embryo primordia can be either crosswise or entirely opposite to those of the primary.

Of even greater interest, perhaps, is the result of the experiments which were to explain how the secondary primordia of the embryo were on the same level. For example, it can be seen that the auditory vesicles of both lie in nearly the same cross section of the embryo. In order to find out the cause of this regional determination or at least to establish its position the implantation was varied in two ways. To understand this we must remember one simple fact about development. In the course of gastrulation the invaginating material is rolled inwards around the upper lip of the blastopore. Thus, the material first invaginated lies farthest towards the front underneath the subsequent brain, while material invaginating later underlies the future spinal cord. Now it could be that the substratum of the head also determines the brain character of the anterior end of the neural plate («head-organizer») and the substratum of the trunk area determines the character of the spinal cord («trunk-organizer»). In order to test this, a portion of upper blastopore lip at the beginning of gastrulation (head-organizer) and one from an advanced and mature gastrula (trunk-organizer) were transplanted in the *same* place in an early gastrula, i.e. at the site where the lower blastopore lip would later develop; this was done also at *different* sites – in the head and trunk areas. It was shown that in fact something like a head- and trunk-organizer does exist, since the former is able to induce a brain also in the trunk region. It was shown moreover that the level in the embryo at which the induction takes place co-determines its nature, since at the head level even a trunk-organizer can induce a brain.

We have already indicated above that this last could have two different reasons. It could be that the disposition for building the head surrounds the whole embryo at head level in a broad circular band. But it could equally well be that a regional differentiating influence is exerted by the primary embryo primordia which co-determines the shape of the secondary embryo. In the region of the primary brain, respectively its primordia, there would be a «brain area» in which neural substance which had been stimulated by induction would develop into brain.

On the basis of definite facts established by experiment, Holtfreter has decided against the first and in favour of the second possibility. Moreover he has in addition discovered some more extremely interesting examples of these «embryonic areas». As we have seen, inducing tissues retain their induction capacity for a long time, and far beyond the stage of development required in the normal course. That being so, in a normal embryo neural substance would have to be induced afresh in the epidermis which lies over the neural tube or the somites, unless that tissue had already exceeded its ephemeral period of reaction capacity. We could therefore infer, what Holtfreter discovered in a different enquiry, that a young portion still capable of reaction would in fact behave differently in this site. And it really is true that in particles of ectoderm from early gastrula implanted superficially at different levels in older gastrula a great variety of inherent potencies is activated. It depends on the region, so that in an anterior area, brain with optic and aural vesicles is induced, while further back, notochord and pronephric ducts are induced, and further back still, little tails. That shows that even the older embryo is still riddled with «embryonic areas» which do not normally come to light but can be detected at any time by indicators rich in potencies.

These inductions between parts of different ages do not complete the embryo by replacing what has been taken away; they are not «complementary» (O. Mangold) as in the case of a graft of the same age from an exactly similar site. Rather do the induced parts develop according to site only in a general sense, through «autonomic» induction; they are produced in excess and have a certain independence (O. Mangold).

A still further series of questions and experiments arose out of the first induction experiments and we will just touch on these in conclusion. As said earlier the induction effect is also possible with heteroplastic transplantation, i.e. between embryos of different species. For example, presumptive brain of a *Triton taeniatus* embryo can be made into epidermis in the gill area of a *Triton cristatus*. But the outer gills covered by it will have taeniatus properties; that is to say, they will be similar not to those of the species which has caused their development (instead of that of brain) but will resemble that of the species from which the implant originates. Potencies are not transferred to the «gill area» of the host; it is merely that those potencies relevant to its location are awakened. And in heteroplastic transplantation these diverge somewhat from those of the host. If an exchange between samples of different genus or even between systematic groups remote from each other (xenoplastic) were possible and followed by induction effects, very valuable conclusions could be expected.

In this respect there is another question that must be dealt with which cropped up during those first experiments: whether in fact the induced organ is laid down part for part or as a whole. From the example of the outer gills we were not able to answer the question, but we could do so from two other organs – the lens and the balancers.

In the Triton, as with most amphibia, the lens of the eye arises as a sequel to the optic cup and its size depends strictly on it. Thus, if the optic cup diminishes in size so does the lens. So it follows that the smaller eye of the *Triton taeniatus* has a smaller lens than the larger eye of *Triton cristatus* at the same stage of development. E. Rotmann now interchanged presumptive lens epidermis with presumptive ventral epidermis in each of the two species at the beginning of gastrulation. The lenses which are formed at a certain moment thereafter follow the size and degree of development of the donor. This can be seen very clearly in the constricted lens primordia with early fibre development; but even quite early stages show lens growth in the epidermis which in one case is too large for the optic cup and in the other case too small. The lens potencies therefore react in the field that activates them not only qualitatively but also quantitatively in accordance with the heredity of the species to which they belong. The lens potencies are not stimulated by the optic cup to the extent within which, with its drawn-in retina layer, it comes into contact with the epidermis. Rather is the lens more or less put in hand as a whole with the epidermis.

The balancers behave in the same way in a further completely analogous experiment of Rotmann's. In its structure and in its angle to the head it is similar to the species from which the transplanted ectoderm is derived and not to the other from which the induction has proceeded.

Added to this problem of uniformity according to species there is another in those cases of xenoplastic transplantation in which organs of different morphological significance are situated in the same region. This is so, for instance, when the ectoderm of the presumptive mouth region is exchanged between the embryos of Urodela and Anura. In the newt larva, lateral to the head and beneath the eyes are two balancers, while the tadpole has beneath the mouth near the ventral mid-line two lower suction cups. Moreover, the newt has real teeth in its mouth which both in origin and structure are comparable to our own teeth. The tadpole's mouth, on the other hand, is furnished with horny jaws and little horny processes. These are quite different in origin and structure from real teeth and indeed have nothing to do with them morphologically. It has been an old dream of mine to substitute for the presumptive mouth region of a newt the foreign ectoderm which comes from a frog early in gastrulation, since I wanted to find out what kind of « armoury » the mouth would form then. This experiment has now been successfully carried out several times since then, and also the other way round. It was first performed at my instigation and in my Institute by O. Schotté, later by Holtfreter, O. Mangold, and E. Rotmann with results we expected but hardly dared hope for. In the mouth region of a Triton larva there arose from transplanted Anura ectoderm of the early gastrula, suction cups and horny jaws; in a tadpole, balancers arose from Urodela ectoderm. When the foreign implant was so narrow that it left the place of origin of the characteristic organs wholly or partly free, these could then themselves develop alongside.

After these results we can say with all certainty of the inducing stimulus that as regards *what* arises, it must be of a very special nature; but as to *how* it arises, it must be of a very general character. We have, however, no idea at all how the «mouth area» releases potencies of the «mouth structures», even when they are of an entirely different species.

Biography

Hans Spemann was born on June 27, 1869, at Stuttgart. He was the eldest son of the publisher, Wilhelm Spemann. From 1878 until 1888 he went to the Eberhard-Ludwig School at Stuttgart and when he left school in 1888 he spent a year in his father's publishing business.

From 1889–1890 he did his military service and then, after a period as a retail bookseller, he entered, in 1891, the University of Heidelberg. There, until he took his preliminary examination in 1893, he studied medicine, and was especially attracted by the work of the comparative anatomist there, Carl Gegenbaur.

During the winter of 1893–1894 he studied at the University of Munich, where he became more closely acquainted with August Pauly – a fact of great importance to him. From the spring of 1894 to the end of 1908, he worked in the Zoological Institute at the University of Würzburg. In 1895 he took his degree in zoology, botany, and physics (subjects to serve his anatomical studies), having worked under Theodor Boveri, Julius Sachs, and Wilhelm Röntgen, all of whom had the greatest influence on his scientific development.

In 1898 he qualified as a lecturer in zoology at the University of Würzburg, and in 1908 he was asked to become Professor of Zoology and Comparative Anatomy at Rostock, and in 1914 he became Associate Director of the Kaiser Wilhelm Institute of Biology at Berlin-Dahlem. In 1919 he was appointed Professor of Zoology at the University of Freiburg-im-Breisgau, in succession to Hans Doflein, a post which he held until he retired and became Emeritus Professor in 1935.

Spemann's name will always be associated with his work on experimental embryology. He made himself a master of micro-surgical technique and, working on the relatively large eggs of amphibians he discovered in 1924, together with Hilde Mangold, the existence of an area in the embryo, the portions of which, upon transplantation into an indifferent part of a second embryo there organized (induced) secondary embryonic primordia. The name « organizer centre » or « organizer » was therefore given by him to those parts. For this discovery of the organizer effect in embryonic development, he was awarded the Nobel Prize in 1935.

Later Spemann showed that different parts of the organization centre produce different parts of the embryo. The anterior parts of it tend to produce parts of the head, and the posterior parts of it parts of the tail. Further, tail organizers, when they are grafted into the head region of another embryo, may produce heads instead of tails, the reason being that they are influenced by the head organizer in their new environment.

Earlier Spemann had transplanted the optic cups of new embryos into the outermost layer of the region of the abdomen and had found that they induced the production, in this new situation, of a lens of the eye. This was interpreted as being evidence of the existence of secondary organizers which operate after the induction exerted by the primary organizer has been completed.

By these and other experiments of a similar kind Spemann laid the foundations of the theory of embryonic induction by organizers, which led later to biochemical studies of this process and the ultimate development of the modern science of experimental morphogenesis. He described his researches in his book *Embryonic Development and Induction* (1938).

Spemann died at Freiburg on September 9, 1941.

Physiology or Medicine 1936

Sir HENRY HALLETT DALE

OTTO LOEWI

«for their discoveries relating to chemical transmission of nerve impulses»

Physiology or Medicine 1936

Presentation Speech by Professor G. Liljestrand, member of the Staff of Professors of the Royal Caroline Institute

Your Majesty, Your Royal Highnesses, Ladies and Gentlemen.

In the second book of his famous work on the history of Rome, Livy has described how Menenius Agrippa, sent out by the Senate to attempt to bring about a reconciliation with the Plebeians who were on strike, told them the fable of the revolt of the limbs against the stomach, stressing the necessity of the cooperation of all parts in the interest of the whole. This cooperation, or « consensus partium », described here in a simple form, is the main objective of physiological research. To a large extent it is brought about by the body fluids and especially by the blood. These not only effect the necessary distribution of the supply from outside, but also the removal of waste products; the intensive research done today on internal secretions has also shown how important it is that the various hormones should be distributed by this means, from the organs in which they are produced to other parts of the body. Characteristic of the whole pattern of this cooperation, either humoral or chemical, is the fact that it is established relatively slowly but extends over a considerable time. Simultaneously also, another mechanism is set up, through the development of the nervous system, which permits the exchange of rapid messages and their swift transposal into action. Occasionally such messages are sent out through an act of will which brings the skeletal muscles into action. But our inner organs also are under the influence of the nervous system. The heart beats which are accelerated by work and mental emotion, the pupils which contract when light enters the eye, and the gastro-intestinal canal which, through its movements, dispatches food according to its kind are examples of how activity adapts itself to the influence of certain nerves which, in these instances, are not under the command of the will. In this portion of the nervous system, then, there is a kind of self-government and it is therefore known as the autonomic nervous system. This is composed of two main parts, which are both of equal importance, but which to a certain extent represent conflicting interests. Taking the heart as an example, one section, the so-called sympathetic system, conveys those impulses which accelerate beating, while the parasympathetic system on the other hand conveys those which have a slowing-down effect.

If external work has to be performed, or if dangers threaten, then the sympathetic part of the autonomic nervous system takes over the direction and develops increased activity. The heart pumps more blood, the muscles are put into a state of defence and receive an extra supply of fuel, while, at the same time, there is a momentary cessation of activity in a variety of other places, for instance in the movements of the intestinal canal. In contrast, different activity occurs in the parasympathetic system as local conditions require, for instance in the function of a single organ.

It was generally thought that impulses in the nerves act directly on the muscles or glands bringing about a change in their activity. But as early as 1904, Elliott presented a different interpretation. From the medulla of the adrenal glands, which, as embryonic development shows, is related with the sympathetic nervous system, a substance can be produced, i.e. adrenaline, the effect of which is remarkably similar to that produced by increased activity in the sympathetic system. Elliott therefore supposed that the impulses in the sympathetic nerves produced a release of adrenaline in the nerve endings which would then be the real vehicles of the stimulation effect. Ten years later, Dale published a comprehensive investigation of another substance, acetylcholine, for which he found a corresponding conformity with the effect of the parasympathetic stimulation. As, however, at that time acetylcholine had not been met with in the body, there was not sufficient basis for a discussion as to whether it normally transmitted impulses.

While the idea that nerve stimulation could be brought about by the release of certain substances was not entirely new, it is nevertheless thanks to Loewi that the idea was brought from the realm of unproven hypotheses on to the firm ground of certain experience. He first used a heart with its nerve trunk, removed from a frog or toad, connecting up the heart chamber with a small glass container in which was a small quantity of a suitable nutrient fluid. If the nerve trunk was stimulated by electrical means, the number and strength of heart beats altered according to circumstances – there are, namely, in the nerve trunk fibres from both the sympathetic and the parasympathetic systems. If after such stimulation Loewi transferred the fluid which had been pumped in and out of the heart into another similarly prepared heart, he found that the fluid itself had taken on properties capable of producing changes in the activity of the organ corresponding to those which had earlier been produced by the nerve stimulation. Through this very simple but ingenious experiment it was proved that the nerve stimulus can release substances having the action characteristic for the nerve stimulation and further observations left no doubt whatever that the nerve stimulus itself was passed on to the organ by chemical means.

Painstaking work now began with the object of determining the nature of the substances concerned – it was soon apparent that different substances were involved in the stimulation of the two different kinds of nerves. This task would appear hopeless considering the incredibly small quantities in which the substances are released. Chemical methods alone were of no avail. But Loewi carried out instead a model analysis, using those activities which were obtained in the living organism under changing conditions. With the sympathicus substance he was able to prove in this way that, in a series of important points such as destruction through oxidation and under the effects of certain kinds of irradiation, as well as in regard to its action, it corresponded absolutely with adrenaline. As regards the parasympathetic substance, the task was more difficult on account of its rapid breaking down in the presence of blood and tissue – this supports the contention made previously that the parasympathetic nerves act locally whereas the action of the sympathetic nerves is more widespread. Loewi and Navratil discovered that the breaking down could be prevented by the addition of the vegetable base physostigmine and this made it possible to work out a method which later made the detection of the substance very much easier. After considerable work, Loewi was able to determine the nature of this substance too, and to prove that the parasympathetic substance was identical with acetylcholine.

Loewi's discoveries have successfully withstood the searching test of re-examination. Numerous investigations have shown also that the release of the two substances mentioned is in no way restricted to the nervous system of the heart. Many scientists, notably Cannon, after comprehensive tests, have discovered that adrenaline, or a very similar substance, appears after stimulation of other sympathetic channels. And Engelhart, a colleague of Loewi's, proved the existence of acetylcholine in the anterior chamber of the eye as a result of contraction of the pupil with the entry of light. Corresponding observations were made for many other organs, among others by Dale and his collaborators. Further support for the view that acetylcholine plays a part in the body under physiological conditions was obtained when Dale and Dudley were able to prepare from the body small quantities of this substance.

In recent years Dale and his distinguished collaborators have been able to add to our knowledge of the chemical transmission of stimuli in two extremely important points. In his earlier investigations with acetylcholine Dale was able to observe an effect on the nerve ganglia themselves or on the ganglia of the autonomic nervous system in which a kind of change-over takes place. He was then led to ask whether there could be a conduction of impulses from one nerve cell to another through the agency of acetylcholine. Using an elegant method described by the Russian Kibjakov, Feldberg and Gaddum were able to prove with Dale that acetylcholine appears in the nerve ganglia after stimulation of the connecting nerves. One can appreciate how sensitive the methods used must be when it is realized that only one hundred thousandth (1/100,000) of a milligram of acetylcholine a minute was produced under favourable conditions. The role of acetylcholine as a transmittor is not, however, restricted to the autonomic nervous system. Dale and his pupils proved with great skill that it plays a role in the production of muscular contractions on the part of the motor nerves. On the one hand, the appearance of the substance in connection with the transmission of impulses was confirmed, on the other hand it was also proved that under suitable experimental conditions infinitely small quantities of acetylcholine produced muscle contractions of a corresponding nature.

In understanding the effects of a series of different substances on the organism, the discovery of the chemical transmission of nerve stimuli represents a revolution. A simple and natural explanation is found for the strange conformity between the effect of adrenaline and acetylcholine on the one hand and the stimulation of the sympathetic and parasympathetic systems on the other hand; and the same applies for different substances having a more or less similar effect. But one now has a different point of view with regard to the effect of other substances as well, for example the vegetable bases atropine and physostigmine. Certainly, the observations made have a fundamental significance for our interpretation of the physiological processes of the nervous system, where, in the light of chemical transmission, various so-called summation and inhibitory phenomena can be better understood. Certain observations made during recent years point to practical consequences which will be of value in combating a number of pathological conditions. The importance of any discovery, however, does not only lie in the fact that it brings clarity and understanding to a number of observations not previously understood; it also poses quite new problems and leads research into new channels. The intensive work which is at present being carried out in different laboratories on questions connected with these observations proves convincingly what a stimulating effect the fresh ideas connected with the transmission of nerve stimuli have already had.

Sir Henry Dale, Professor Otto Loewi. The Royal Caroline Institute has decided to award to both of you jointly, this year's Nobel Prize for Physiology or Medicine for your discoveries in respect of the chemical transmission of nerve action. You, Professor Loewi, first succeeded in establishing proof of such transmission and in determining the nature of the effective substances. This work was, in part, built up on earlier research to which you, Sir Henry, made an essential contribution. The results were consolidated and complemented in many important respects by you and your collaborators. You and your school have also greatly extended the range of the new conception by later discoveries. Through these various discoveries, which have stimulated research in innumerable parts of the world, therefore demonstrating once again the international character of science, pharmacology has been very considerably influenced, and physiology or medicine enriched to a high degree.

On behalf of the Staff of Professors, I express to you our heartiest congratulations and hope that it may be granted to you to take part in further research into this new territory for a long time to come. With this hope, I have the honour to ask you to receive the Nobel Prize for Physiology or Medicine from the hands of His Majesty the King.

HENRY H. DALE

Some recent extensions of the chemical transmission of the effects of nerve impulses

Nobel Lecture, December 12, 1936

The transmission of the effects of nerve impulses, by the release of chemical agents, first became an experimental reality in 1921. In that year Otto Loewi published the first of the series[1, 2, 3, 4, 5, 6] of papers from his laboratory, which, in the years from 1921–1926, established all the principal character-istics of this newly revealed mechanism, so far as it applied to the peripheral transmission of effects from autonomic nerves to the effector units innervated by them. Of the general history of this discovery, of the speculations which preceded it, and of its more recent developments in detail in many labora-tories, as regards one aspect of it particularly by Cannon and his co-workers in Boston, you have heard from Professor Loewi himself. I propose to deal with a wider application of this conception of chemical transmission, which has resulted from researches carried out during the past three years in my own laboratory, by a number of able investigators – J. H. Gaddum, W. Feld-berg, A. Vartiainen, Marthe Vogt, G. L. Brown, Z. M. Bacq. These investi-gations have made it possible to suggest that a fundamentally similar chem-ical mechanism is concerned in the transmission of excitatory effects at the synapses in all autonomic ganglia, and at the motor nerve endings in ordi-nary, voluntary muscle.

You will see that, according to this relatively new evidence, a chemical mechanism of transmission is concerned, not only with the effects of auton-omic nerves, but with the whole of the efferent activities of the peripheral nervous system, whether voluntary or involuntary in function. This exten-sion of the principle of chemical transmission has come as a surprise to many; the relative ease, with which the evidence justifying it can be obtained, has been surprising to ourselves. But the basic conception, which encouraged us to undertake experiments in this direction, was no novelty to me; and for its origin I must ask you to look briefly at some experiments which I had already made and published in 1914[7]. My chemical collaborator at that time, Dr. Ewins[8], had isolated the substance responsible for a characteristic activity which I had detected in certain ergot extracts, and it had proved to be acetylcholine, the very intense activity of which had been observed by Reid Hunt[9] already in 1906. Since we had found this substance in nature, and it was no longer merely a synthetic curiosity, it seemed to me of interest to explore its activity in greater detail. I was thus able to describe it as having two apparently distinct types of action. Through what I termed its « muscarine » action, it reproduced at the periphery all the effects of parasympathetic nerves, with a fidelity which, as I indicated, was comparable to that with which adrenaline had been shown, some ten years earlier, to reproduce those of true sympathetic nerves. All these peripheral muscarine actions, these parasympathomimetic effects of acetylcholine, were very readily abolished by atropine. When they were thus suppressed, another type of action was revealed, which I termed the « nicotine » action, because it closely resembled the action of that alkaloid in its intense stimulant effect on all autonomic ganglion cells, and, as later appeared, on voluntary muscle fibres. I am tempted here to quote some words which I wrote in that paper, in 1914.

« It is clear, then, that the distinction between *muscarine* and *nicotine* activity cannot be made with absolute sharpness... Nor is there any evidence enabling us to regard one group of the molecule as responsible for the one type of action, and another for the other. One can merely conclude that there is some degree of biochemical similarity between the ganglion cells of the whole involuntary system and the terminations of voluntary nerve fibres in striated muscle, on the one hand, and the mechanism connected with the peripheral terminations of craniosacral involuntary (i.e. parasympathetic) nerves on the other. »

In the same paper I had speculated on the possible occurrence of acetylcholine in the animal body, and on its physiological significance if it should be found there; and had pointed out the extraordinary evanescence of its action, suggesting that an esterase probably contributed to its rapid removal from the blood.

When, therefore, some seven years later, Loewi described his beautiful experiments, showing that stimulation of the vagus nerve produced its inhibitor effects on the frog's heart by the liberation of a chemical substance; and when his successive papers provided cumulative evidence of the similarity of this substance to acetylcholine, including its extreme liability to destruction by an esterase, which Loewi extracted from the heart muscle; I believe that I was more ready than most of my contemporaries for immediate acceptance of the evidence for this « Vagusstoff », and more eager, almost, than Professor Loewi himself, to assume its identity with acetylcholine. There was wanting, it seemed to me, only one item of evidence to justify certainty as to the nature of this substance, namely, a proof that acetylcholine itself, and not merely some choline ester of closely similar properties, was an actual constituent of the animal body.

Professor Loewi has already mentioned the extraction and identification of acetylcholine, as a natural constituent of a mammalian organ, by my late and deeply lamented colleague, H. W. Dudley, and myself[10], in 1929. He has also dealt with the general rule that parasympathetic effects are transmitted by acetylcholine, and true sympathetic effects by what his own most recent experiments appear definitely to identify as adrenaline. He has mentioned also the important exceptions to that rule. In view of such exceptions, it seemed to me desirable to have a terminology enabling us to refer to a nerve fibre in terms of the chemical transmission of its effects, without reference to its anatomical origin; and, on this functional basis, I[11] proposed to refer to nerve fibres and their impulses as « cholinergic » or « adrenergic », as the case might be. Such a functional terminology seemed to me the more important, in view of the evidence which was already coming from our experiments, that acetylcholine had a much wider function as a transmitter of nervous excitation, than that concerned with the post-ganglionic fibres of the autonomic system, and their effects on involuntary muscle and gland cells. For all such effects of acetylcholine, directly analogous to those which Loewi discovered in relation to the heart vagus, were covered by what I had termed the « muscarine » action of acetylcholine, and were all very readily suppressed by atropine. But there remained, as yet without any corresponding physiological significance, the other type of action of acetylcholine, so similar in distribution to that of nicotine, which had come to my notice nearly twenty years earlier. Was it credible, I asked myself, that this sensitiveness of ganglion cells, and of voluntary muscle fibres, to the substance now known to be the transmitter of peripheral parasympathetic effects, was entirely without physiological meaning? I could not believe it. At the same time, it had to be recognized that the transmission of nervous excitation at ganglionic synapses, and at motor nerve endings in voluntary muscle, was a phenomenon of a different order from any of those in connexion with which the intervention of a chemical transmitter had hitherto been demonstrated, or even considered. Acetylcholine, released at the peripheral endings of the vagus or the chorda tympani, could be pictured as reaching the heart

cells or those of the salivary gland by diffusion, and there inhibiting an automatic rhythm, or exciting glandular secretion. At a ganglionic synapse or a
motor ending on a voluntary muscle fibre, on the other hand, the evidence
was clear, that a single impulse, reaching the end of the preganglionic or
motor nerve fibre, caused the passage from the ganglion cell, along its postganglionic axon, of a single nerve impulse, and no more; or caused the
passage, from the motor end plate of the muscle fibre, of a single wave of
excitation, of propagated contraction, and no more. In both cases, the phenomenon had the appearance of a direct, unbroken conduction, to ganglion
cell or muscle fibre, of the same propagated wave of physico-chemical disturbance as had constituted the preganglionic or the motor nerve impulse,
with only a slight, almost negligible retardation in its passage across the
ganglionic synapse or the neuromuscular junction. And, indeed, such continuity of the conduction, in both cases, had generally been assumed, and,
in the case of the neuromuscular conduction, in particular, had been implicit
in the interpretation of a great body of detailed evidence, which the ingenui-
ty and the labours of two generations of physiologists had produced.

   Could the stimulating action of acetylcholine on ganglion cells and on
muscle fibres, its «nicotine» actions, be pictured as intervening in these rapid
and strictly limited transmissions of excitation across ganglionic and neuromuscular synapses? We could only imagine such intervention, if we could
think of acetylcholine as appearing and disappearing in a manner entirely
different from that involved in its transmission of peripheral parasympathetic effects. We must suppose that an impulse, arriving at the ending of a
preganglionic or a voluntary motor nerve fibre, releases with a flashlike
suddenness a small charge of acetylcholine, in immediate contact with the
ganglion cell or the motor end plate of the muscle fibre. We must suppose
that this sudden rise in concentration of acetylcholine stimulates the ganglion cell to the discharge of a postganglionic impulse, or initiates a propagated wave of excitation along the muscle fibre. And we must suppose,
further, that the acetylcholine then disappears with a suddenness comparable
to that of its liberation, so that it has vanished by the end of the brief refractory period of the ganglion cell or the muscle fibre, which is thus left fully
responsive to another discharge of acetylcholine, by another nerve impulse.
Such a sequence of events seems to involve two things. The first is a depot,
closely related to the preganglionic or motor nerve ending[12], in which acetylcholine may be held in some association which prevents its action and
protects it from destruction, and from which it can be immediately liberated

by the arrival of a nerve impulse. Professor Loewi has mentioned the evidence for such storage of acetylcholine, waiting for liberation, at parasympathetic nerve endings; and Brown and Feldberg[13], in my laboratory, have obtained evidence that nearly the whole of the acetylcholine, obtainable by extraction from a normal sympathetic ganglion, disappears when the preganglionic nerve fibres are caused to degenerate by section; so that its maintenance is, in fact, dependent on the integrity of the preganglionic nerve endings. The second thing required, by the suggested action of acetylcholine in transmitting the kind of excitation we are discussing, is a mechanism for its very rapid removal, so that it disappears completely within the few milliseconds of the refractory period of muscle fibre or ganglion cell. One naturally thinks of the specific cholinesterase, first detected by Loewi in heart muscle, and since found to be widely distributed in the blood and tissues. Even when obtained in solution this potent enzyme destroys acetylcholine with a quite remarkable rapidity; and if we could suppose it to be concentrated on surfaces at preganglionic or motor nerve endings, in immediate relation to the site of liberation and action of acetylcholine, it might furnish an adequate mechanism for the complete destruction of this substance, even during the very brief interval of the refractory period. Here, again, I am permitted to make preliminary mention of experiments which Dr. Franz Brücke, who earlier worked under Prof. Loewi, is even now making in my laboratory, and which have already given uniform evidence that a large amount of cholinesterase is present in a sympathetic ganglion, and that this, like the acetylcholine obtainable from such a ganglion, disappears largely when the preganglionic fibres, and their endings in the ganglion, are caused to degenerate. We have evidence, then, that both the reserve of acetylcholine, and the esterase required for its destruction, are in fact associated with the preganglionic nerve endings, as our hypothesis demands. I am departing, however, too far from the true historical order, and presenting recent and confirmatory details of evidence, before I have described the initial observations, which opened this new field to our experimental exploration.

Although from the time when it first became clear that Loewi's Vagusstoff was acetylcholine, I had begun to consider the possible significance of its « nicotine » actions, it was long before the possibility of its intervention as transmitter at ganglionic synapses, or at voluntary motor nerve endings, seemed to be accessible to investigation. Experiments on the ganglion came first in order. Chang and Gaddum[14] had found, confirming an earlier observation by Witanowski, that sympathetic ganglia were rich in acetylcholine. Feldberg, just before he returned to my laboratory for a stay of some years, had observed, with Minz and Tsudzimura[15], that the effects of splanchnic nerve stimulation are transmitted to the cells of the suprarenal medulla by the release of acetylcholine in that tissue. Now these medullary cells are morphological analogues of sympathetic ganglion cells, and Feldberg, continuing this study in my laboratory, found that this stimulating action of acetylcholine on the suprarenal medulla belonged to the « nicotine » side of its actions. Clearly we had to extend these observations to the ganglion; and a method of perfusing the superior cervical ganglion of the cat, then recently described by Kibjakov[16], made the experiment possible. Feldberg and Gaddum[17], though unable to reproduce effects obtained by Kibjakow with pure Locke's solution, found that, when eserine was added to the fluid perfusing the ganglion, stimulation of the preganglionic fibres regularly caused the appearance of acetylcholine in the venous effluent. It could be identified by its characteristic instability, and by the fact that its activity matched the same known concentration of acetylcholine in a series of different physiological tests, covering both « muscarine » and « nicotine » actions. It appeared in the venous fluid in relatively high concentrations, so strong, indeed, that reinjection of the fluid into the arterial side of the perfusion caused, on occasion, a direct stimulation of the ganglion cells. It was clear that, if the liberation took place actually at the synapses, the acetylcholine liberated by each preganglionic impulse, in small dose, indeed, but in much higher concentration than that in which it reached the venous effluent, *must* act as a stimulus to the corresponding ganglion cells. Feldberg and Vartiainen[18] then showed that it was, in fact, only the arrival of preganglionic impulses at synapses which caused the acetylcholine to appear. They showed, further, that the ganglion cells might be paralysed by nicotine or curarine, so that they would no longer respond to preganglionic stimulation or to the injection of acetylcholine, but that such treatment did not, in the least, diminish the output of acetylcholine caused by the arrival of preganglionic impulses at the synapses. There was, in this respect, a complete analogy with the paralysing effect of atropine on the action of the heart vagus, which, as Loewi and Navratil had shown many years before, stops the action of acetylcholine on the heart, but does not affect its liberation by the vagus impulses.

My colleagues have added other chapters of interest to this story of chemical transmission at the synapses in the ganglion. I may just mention Brown and Feldberg's[13] observation that potassium ions, the mobilization of which is so intimately connected with the nervous impulse, will liberate acetylcholine from its depot in the ganglion, in a manner closely recalling the effect of preganglionic impulses; and their more recent finding[19] that, with prolonged preganglionic stimulation, the ganglion sheds into the fluid perfusing it several times as much acetylcholine as can be obtained from a similar, unstimulated ganglion by artificial extraction. The effects of eserine, on the transmission of excitation in the ganglion, are complicated by a paralysing action of this alkaloid on the ganglion cells, and still need further elucidation. I can more usefully pass to our recent work on voluntary muscle, in which such effects are much clearer.

The difficulty facing us in the case of the voluntary muscle was largely a quantitative one. In a sympathetic ganglion, the synaptic junctions, at which the acetylcholine is released by the incident preganglionic impulses, form a large part of the small amount of tissue perfused. In a voluntary muscle the bulk of tissue, supplied by a rich network of capillary blood vessels, is relatively enormous in relation to the motor nerve endings, of which only one is present on each muscle fibre. The volume of perfusion fluid necessary to maintain functional activity is, therefore, relatively very large, in relation to the amount of acetylcholine which the scattered motor nerve endings can be expected to yield when impulses reach them. With the skilled and patient co-operation of Dr. Feldberg and Miss Vogt[20], however, it was possible to overcome these difficulties, and to demonstrate that, when only the voluntary motor fibres to a muscle are stimulated, to the complete exclusion of the autonomic and sensory components of the mixed nerve, acetylcholine passes into the Locke's solution, containing a small proportion of eserine, with which the muscle is perfused. If, by calculation, we estimate the amount of acetylcholine thus obtained from the effect of a single motor impulse, arriving at a single nerve ending, the quantity is of the same order as that similarly estimated for a single preganglionic impulse and a single ganglion cell; in both cases 10^{-15} gram, which corresponds to about three million molecules of acetylcholine. We found that, if the muscle was denervated by degeneration, direct stimulation, though evoking vigorous contractions, produced no trace of acetylcholine. If, on the other hand, the muscle was completely paralysed to the effects of nerve impulses by curarine, stimulation of its motor nerve fibres caused the usual output of acetylcholine, though the muscle remained completely passive. Again there is a complete analogy with Loewi's observations on the heart vagus and atropine.

With this demonstration, that acetylcholine was liberated at the endings of motor nerve fibres in voluntary muscle, in immediate relation to the motor end plates of the muscle fibres, only one side of our problem had been solved. Acetylcholine, injected into the vessels of a ganglion, could be shown to stimulate the ganglion cells to the discharge of postganglionic impulses. In the case of normal voluntary muscle, on the other hand, the evidence before us suggested only that certain muscles of frogs, reptiles and birds responded to the application of acetylcholine, not by quick, propagated contractions like those evoked by motor nerve impulses, but by slow, persistent contractures, of low tension. As for the normal muscles of mammals, on which our evidence of acetylcholine liberation had been obtained, these were supposed, on evidence provided by myself among others, to give no response at all to acetylcholine, except in large doses, and then only irregularly. The denervated mammalian muscle was known to be highly sensitive to acetylcholine, but the evidence, again from myself among others, suggested that its response was of the nature of a contracture, and not of a quick, propagated contraction.

Considering the manner in which acetylcholine must reach the motor end plates of the muscle fibres, if it were indeed the transmitter of motor nerve excitation – that it must appear with a flash-like suddenness, in high concentration, simultaneously at every nerve ending – we concluded that the ordinary method of injecting acetylcholine, so that it reached the muscle by slow diffusion from the general circulation, could not possibly reproduce this abrupt appearance at the points responsive to its action. We attempted a nearer approach to these supposed conditions of its natural release, by a method which enabled us, after a brief interruption of the arterial blood supply, to inject a small dose of acetylcholine, in a small volume of saline solution, directly and rapidly into the empty blood vessels of the muscle[21]. The responses which we thus obtained were of an entirely different kind from any which had previously been recorded. A dose of about 2 γ of acetylcholine, thus injected at close range into the vessels of a cat's gastrocnemius, produced a contraction with a maximal tension equal to that of the twitch produced by a maximal motor nerve volley, and of a rapidity but little less than that of the motor nerve twitch. We have direct evidence that only a small part of the acetylcholine so injected actually reaches the muscle end plates by diffusion from the vessels; and we argued that, in any case, it could not reach them simultaneously, but only in rapid succession; so that the response, in spite of its superficial resemblance to a rather slow twitch, must actually be a brief, asynchronous tetanus. My colleague, G.L. Brown, using a strictly localized electrical lead from the muscle, involving only a few fibres, has obtained clear evidence that the response has, indeed, that nature. It is a brief burst of unsynchronized and repetitive responses of the individual muscle fibres; but these individual responses are, without doubt, quick, propagated contractions, and there is no semblance of contracture about the phenomenon. Unlike the response of the denervated muscle to acetylcholine, this quick response of normal mammalian muscle is suppressed with great ease by curarine.

At this point I must briefly refer to some observations made only in the past few weeks, and still in progress. The normal mammalian muscle had seemed to present us initially with the greatest difficulty, being supposed not to react to acetylcholine at all. This difficulty being removed by a more adequate technique, we had to face the fact that the function of acetylcholine, as transmitter of voluntary motor nerve impulses, could not be confined to the case of mammalian muscle. The muscle of the frog, the classical object of innumerable studies of neuromuscular conduction, had been found to respond to acetylcholine, indeed, but only by contractures of low tension, and not by propagated contractions comparable to those evoked by nerve volleys. Here again, we reflected that the method which had been used for the application of acetylcholine, the immersion of the excised muscle in a suitable dilution of the substance, could hardly be expected to reproduce that rapidity of access to the appropriate points on the fibres, which its simultaneous liberation at all nerve endings would achieve. The patient skill of my colleague, G. L. Brown, has now made it possible to apply acetylcholine to the frog's muscle by direct injection of a small dose into its empty blood vessels, in a manner quite analogous to that which produced such significant results in the mammalian muscle. If 1 γ of acetylcholine, for example, dissolved in 0.1 cc of Ringer's solution, is thus injected suddenly into the artery supplying the frog's gastrocnemius, the surface of the muscle, covered with its glistening aponeurosis, shows immediately the ripple and shimmer of innumerable, unsynchronized contractions, propagated along the fibres and fascicles of the muscle; at the height of the effect a tension of several hundred grams is developed; and the electrical record gives decisive evidence that this response is an irregular, asynchronous tetanus, and not a contracture. With larger doses this tetanus is cut short and extinguished by the contracture – the only effect of acetylcholine on frog's muscle which earlier work had recognized.

From the study of the mammalian muscle we have also obtained what seems to be clear evidence concerning a mechanism by which acetylcholine, suddenly liberated at the nerve ending to transmit the excitatory effect of a motor impulse to the muscle fibre, may, with a comparable suddenness, be removed completely during the refractory period. If this removal is due, as we have suggested, to the destructive action of cholinesterase, concentrated on surfaces at the nerve ending, we should expect that eserine, with its depressant effect on the action of the cholinesterase, discovered by Loewi and Navratil, would delay the disappearance, from the neighbourhood of the motor end plates of the muscle fibres, of the acetylcholine liberated by a single nerve volley, and would thereby modify the response of the muscle. The effect is easy to demonstrate. Eserine causes, in fact, a great increase of the maximum tension attained by the contraction of the muscle in response to a maximal nerve volley. The all-or-none principle forbade us to suppose that such a potentiated response was a single twitch; and the electrical records showed that it was, indeed, repetitive, and had the nature of a brief, diminishing tetanus[21].

The eserine has so depressed the action of the esterase at the nerve endings, that the acetylcholine liberated by a single nerve volley lingers there, and reexcites the muscle at each emergence from successive refractory periods, until the concentration falls at last below the stimulation threshold. Bacq and Brown[22] have more recently extended these observations to a series on artificial eserine analogues, and have found that the potentiating action on the response of mammalian muscle to single nerve volleys is, in fact, proportional, in the different compounds of the series, to the anticholinesterase action, as independently determined.

There are many other aspects of these phenomena, some of them still under active investigation in my laboratory. I must be content today to have presented the main headings of the evidence, which, as it seems to me, is forcing upon us the conclusion, in spite of the preconceptions which made the idea initially so difficult to entertain, that acetylcholine does actually intervene as a chemical transmitter of excitation, in the rapid and individu alized transmission at ganglionic synapses and at the motor endings in voluntary muscle; that, in the terminology which I have proposed, the preganglionic fibres of the autonomic system, and the motor nerve fibres to voluntary muscle, are also « cholinergic ».

You will see that we are thus led to the conclusion that nearly all the efferent neurones of the whole peripheral nervous system are cholinergic; only the postganglionic fibres of the true sympathetic system are adrenergic, and not even all of these. As I have earlier pointed out, on more than one occasion[12, 23], before the evidence for the cholinergic function of voluntary motor nerves was nearly as strong as it has now become, this new classification of nerve fibres, by chemical function, renders at once intelligible the formerly puzzling evidence as to the functional compatibility of different types of nerve fibre, in replacing one another in experimental regeneration. The whole of the evidence of such replacement, obtained by Langley and Anderson early in the present century, can now be summarized by the simple statement that any cholinergic fibres can replace any other cholinergic fibres, and that adrenergic fibres can replace adrenergic fibres, but that no fibre can be functionally replaced by one which employs a different chemical transmitter. The chemical function, as I have expressed it, seems to be characteristic of the neurone, and unchangeable. In that connexion, particular interest appears to me to attach to the recent observations of Wybauw[24], which seem to provide clear evidence that the antidromic vasodilatation, generally believed to be produced through peripheral axon branches from sensory fibres, also employs a cholinergic mechanism. If this is substantiated, and if my suggestion holds good that the chemical mechanism is characteristic of the neurone, the question at once presents itself, whether at the other ending of the same sensory neurone, in a central synapse, the same cholinergic transmission of excitation will be found.

Hitherto the evidence concerning a chemical transmission in the central nervous system, of the type which we have found prevailing at all peripheral synapses, is scattered and insufficiently uniform in its indications. The basal ganglia of the brain are peculiarly rich in acetylcholine, the presence of which must presumably have some significance; and suggestive effects of eserine and of acetylcholine, injected into the ventricles of the brain, have been described. I take the view, however, that we need a much larger array of well-authenticated facts, before we begin to theorize. It is here, especially, that we need to proceed with caution; if the principle of chemical transmission is ultimately to find a further extension to the interneuronal transmission in the brain itself, it is by patient testing of the groundwork of experimental fact, at each new step, that a safe and steady advance will be achieved. The possible importance of such an extension, even for practical medicine and therapeutics, could hardly be over-estimated. Hitherto the conception of chemical transmission at nerve endings and neuronal synapses, originating in Loewi's discovery, and with the extension that the work of my colleagues has been able to give to it, can claim one practical result, in the specific, though alas only short, alleviation of the condition of myasthenia gravis, by eserine and its synthetic analogues.

1. O. Loewi, *Pflügers Arch. Ges. Physiol.*, 189 (1921) 239.
2. O. Loewi, *Pflügers Arch. Ges. Physiol.*, 193 (1921) 201.
3. O. Loewi, *Pflügers Arch. Ges. Phys.*, 203 (1924) 408.
4. O. Loewi and E. Navratil, *Pflügers Arch. Ges. Physiol.*, 206 (1924) 123.
5. O. Loewi and E. Navratil, *Pflügers Arch. Ges. Physiol.*, 214 (1926) 678.
6. O. Loewi and E. Navratil, *Pflügers Arch. Ges. Physiol.*, 214 (1926) 689.
7. H. H. Dale, *J. Pharmacol.*, 6 (1914) 147.
8. A. J. Ewins, *Biochem. J.*, 8 (1914) 44.
9. R. Hunt and M. Taveau, *Brit. Med. J.*, 2 (1906) 1788.
10. H. H. Dale and H. W. Dudley, *J. Physiol.*, 68 (1929) 97.
11. H. H. Dale, *J. Physiol.*, 80 (1933) 10 P.
12. H. H. Dale, *Dixon Memorial Lecture, Proc. Roy. Soc. Med.*, 28 (1935) (*Sect. Therapeutics and Pharmacology*, pp. 15–28).
13. G. L. Brown and W. Feldberg, *J. Physiol.*, 86 (1936) 290.
14. H. C. Chang and J. H. Gaddum, *J. Physiol.*, 79 (1933) 255.
15. W. Feldberg, B. Minz, and H. Tsudzimura, *J. Physiol.*, 81 (1934) 286.
16. A. V. Kibjakov, *Pflügers Arch. Ges. Physiol.*, 232 (1933) 432.
17. W. Feldberg and J. H. Gaddum, *J. Physiol.*, 81 (1934) 305.
18. W. Feldberg and A. Vartiainen, *J. Physiol.*, 83 (1934) 103.
19. G. L. Brown and W. Feldberg, *J. Physiol.*, 86 (1936) 40 P.
20. H. H. Dale, W. Feldberg, and M. Vogt, *J. Physiol.*, 86 (1936) 353.
21. G. L. Brown, H. H. Dale, and W. Feldberg, *J. Physiol.*, 87 (1936) 394.
22. Z. M. Bacq and G. L. Brown, *J. Physiol.*, 89 (1937) 45.
23. H. H. Dale, *Nothnagel Lecture, No. 4., Vienna*: Urban and Schwarzenberg (1935).
24. L. Wybauw, *Compt. Rend. Soc. Biol.*, 123 (1936) 524.

Biography

Henry Hallett Dale was born in London on June 9, 1875. He attended Leys School, Cambridge, and in 1894 he entered Trinity College with a scholarship. He graduated through the Natural Sciences Tripos, specializing in physiology and zoology. From 1898 to 1900 he was a Coutts-Trotter Student in Physiology at Trinity College, working then under J. N. Langley. In 1900 he gained a scholarship and entered St. Bartholomew's Hospital, London, for the clinical part of the medical course. He qualified as B.Ch., Cambridge in 1903 and became M.D. in 1909. Meanwhile, he had been awarded the George Henry Lewes Studentship in Physiology and he used it to carry out research under Professor Starling at University College, London. It was here that he met his lifelong friend, Otto Loewi. During 1903, he spent four months with Paul Ehrlich in Frankfurt before returning to University College as Sharpey Scholar. He held this post for only six months before he took an appointment as pharmacologist at the Wellcome Physiological Research Laboratories in 1904. He became Director of these laboratories in 1906, working for some six years with the chemical cooperation of George Barger.

In 1914, Dale was appointed Director of the Department of Biochemistry and Pharmacology at the National Institute for Medical Research in London, becoming in 1928 Director of this Institute; and he served in this capacity until his retirement in 1942 when he became Professor of Chemistry and a Director of the Davy-Faraday Laboratory at the Royal Institution, London. Since 1946, he has devoted his knowledge and energies to the administration of the Wellcome Trust for the support of medical research and medical scholarships. He has been a Trustee since 1936 and served as Chairman of the Board from 1938 until 1960. He was elected a Fellow of the Royal Society in 1914 and served as Secretary from 1925 to 1935. During World War II, Sir Henry served on several Advisory Committees to His Majesty's Government. He was knighted in 1932 and appointed to the Order of Merit in 1944.

Sir Henry's researches have involved a painstaking investigation of the pharmacology of ergot alkaloids and a study of the effects of incidental bases of a simpler nature, such as tyramine and histamine. He discovered the oxytocic action of pituitary extracts, and his continued work on the action of histamine led to studies on anaphylaxis and on conditions of shock. He identified acetylcholine as a constituent of certain ergot extracts, and an analysis of its action served as a basis for later researches, extending the application of Loewi's discoveries, which have been recognized in the joint award of the Nobel Prize for 1935, given on account of the discoveries relating to chemical transmission of nerve impulses. In addition to numerous articles in medical and scientific journals which record his work, Sir Henry is the author of *Adventures in Physiology* (1953), and *An Autumn Gleaning* (1954).

Sir Henry was President of the Royal Society (1940–1945), President of the British Association (1947), and President of the Royal Society of Medicine (1948–1950). He has received many public honours including the G.B.E. (Knight Grand Cross, Order of the British Empire) in 1948, Medal of Freedom (Silver Palm), U.S.A., in 1947, the Grand-Croix de l'Ordre de la Couronne (Belgium) in 1950, and l'Ordre pour le Mérite (Western Germany) in 1955. The Royal and Copley Medals of the Royal Society, the Gold Albert Medal of the Royal Society of Arts, the Baly Medal of the Royal College of Physicians (London), the Cameron Prize (Edinburgh) and the Schmiedeberg plaquette from the German Pharmacological Society are among the many awards he has gained and, in addition, he has been awarded Fellowships of numerous learned societies and institutions throughout the world, including the Royal Society of Edinburgh and Trinity College, Cambridge. He is also a Foreign Associate of the National Academy of Sciences (Washington), Académie de Médecine (Paris) and l'Académie Royale de Belgique, as well as Academies in Denmark, Germany, Italy, Rumania, Spain, Sweden, U.S.A. (New York). He is the recipient of over twenty honorary degrees, and amongst the many lectures he has given are the Nothnagel Lecture (Vienna) and the Pilgrim Trust Lecture to the National Academy of Sciences, Philadelphia.

Sir Henry married Ellen Harriet Hallett, his first cousin, in 1904. Their eldest daughter, Alison Sarah, is married to Lord Todd, Nobel Laureate in Chemistry, 1957.

The chemical transmission of nerve action

Nobel Lecture, December 12, 1936

Natural or artificial stimulation of nerves gives rise to a process of progressive excitation in them, leading to a response in the effector organ of the nerves concerned.

Up until the year 1921 it was not known how the stimulation of a nerve influenced the effector organ's function, in other words, in what way the stimulation was transmitted to the effector organ from the nerve-ending. In general it was thought that it came about through direct transmission of the stimulation wave from the nerve fibre to the effector organ. But the possibility of transmission by chemical means had also been considered and experiments had been conducted on these lines. As a result of his own experiments, Howell[1] had come to believe that vagus stimulation released potassium in the heart and that this was the cause of the resultant effect, and Bayliss[2] discussed the possibility, in view of the similarity in action of the so-called vagomimetic substances and chorda stimulation, that this stimulation might be caused by the production of such substances. Although these data were known to me, my attention was only drawn years after my discovery to the fact that earlier (in 1904 to be exact) Elliott[3], in the last paragraph of a short note, suggested the possibility that the stimulation of sympathetic nerves might be brought about by the release of adrenaline, and that Dixon[4] had already communicated experiments in an inaccessible site to test whether, during vagus stimulation, a substance was released which contributed to the stimulation reaction.

In the year 1921 I was successful for the first time[5] in obtaining certain proof that by stimulation of the nerves in a frog's heart substances were released which to some extent passed into the heart fluid and, when transferred with this into a test heart, caused it to react in exactly the same way as the stimulation of the corresponding nerves. In this way it was proved that the nerves do not act directly upon the heart, but rather that the direct result of nerve stimulation is the release of chemical substances and that it is these which bring directly about characteristic changes of function in the heart.

It was, of course, possible right from the start that this mechanism which I described at the time as «humoral transference», but which is now known as «chemical» transference as the result of a well-founded suggestion by H. H. Dale, does not represent an isolated phenomenon but a special condition which also appears elsewhere. We shall soon see that this supposition was justified. But before I go into that I should like to characterize in more detail the substances which are released by nerve stimulation and produce the effect. First of all, I must mention my distinguished collaborators E. Navratil, W. Witanowski, and E. Engelhart, and thank them.

Let me begin with the transfer medium of the reaction in vagus stimulation which I have called «vagus substance». We were able to determine that its effect is inhibited by atropine[5] and very quickly disappears[6]. In looking for a substance with both these characteristics, I found that out of a series of the known vagomimetic substances, muscarine, pilocarpin, choline, and acetylcholine, only the last-named possessed them[7]. We were then able to establish further that the rapid disappearance of the action of the vagus substance and acetylcholine (Ac.Ch.) through the breaking down of these substances was caused by the action of an esterase in the heart[6], which had already been postulated by Dale[8]. I was able to show furthermore that the action of this esterase could be specifically inhibited through minimum concentrations of eserine[7]. This discovery was important not only because, for the first time, the operational mechanism of an alkaloid had been revealed, but especially because the discovery enabled the theory of the chemical transference of nerve stimulation to be developed for the first time. On the one hand, this eserine action provided a means of revealing the minimal quantities of Ac.Ch. being released by nerve stimulation which would otherwise, because of their rapid destructibility, have remained undisclosed. On the other hand, we are able, in cases where for any reason it is technically impossible or difficult to prove directly the release of Ac.Ch. in nerve stimulation, to draw the conclusion indirectly from the increase in effect of nerve stimulation after previous eserination that the nerve stimulation is being produced by the release of Ac.Ch. And now we must return to the characterization of the vagus substance.

The vagus substance behaves identically with Ac.Ch. not only in regard to its reaction to atropine, and to its destructibility with esterase but also concerning all other characteristics. As Dale and Dudley[9] were able to produce it directly from the organs, there can be no more doubt that the vagus substance is Ac.Ch. and in future I shall refer to it as such.

As regards the character of the substance which is released through stimu-

lation of the sympathetic nerves of the heart and other organs, I was able
to show earlier that it shares many properties with adrenaline; both, for
example, are destroyed by alkali[20] and by fluorescence and ultraviolet light[6],
the activity of both is abolished by ergotamine[21]; on the other hand, as
Cannon and Rosenblueth[10] have shown, it is raised by small and in themselves ineffective quantities of cocaine, the adrenaline-sensitizing action of
which Fröhlich and I[11] found some 25 years ago.

Like the effect of adrenaline, an equal effective strength of the sympathicus
substance declines very slowly in the heart, much more slowly incidentally
than might have been expected in view of the rapid oxidizability of adrenaline
or sympathicus substance in vitro. The cause of this, as revealed by Dr. Ralph
Smith of Ann Arbor and me in a series of specially conducted experiments
not as yet published, turns out to be the giving off of substances from the
heart which inhibit adrenaline oxidation. There must, of course, be some
physiological purpose in the fact that individual devices exist, on the one
hand to remove the acetylcholine as quickly as possible and the adrenaline,
on the other hand, as slowly as possible. And now we must return to the
chemical nature of the sympathicus substance.

Although for some time it had been considered probable after all we had
seen that the sympathicus substance was adrenaline, I was only able to give
direct proof of it this year. Gaddum and Schild[13], on the basis of a statement
by Paget, investigated the significance of a green fluorescence visible in
ultraviolet light which pointed to adrenaline in the presence of O_2 and alkali,
and found that this appears to a high degree specific for adrenaline. I was
now able to show that not only the heart extract, but also the heart fluid,
shows this reaction after accelerated periods of stimulation[12]. Accordingly I
consider it proved that the sympathicus substance is adrenaline.

Now I must briefly consider the question of to what extent the neurochemical mechanism, that is to say the chemical transference of nerve stimulation, is important other than to the heart.

Firstly, Rylant[14] and others were able to show that with warm-blooded
animals too, vagus stimulation released Ac.Ch. which was responsible for
the resultant stimulation reaction. I must mention in this connection that
my collaborator Engelhart[15] was able to show, in accordance with the well-
known fact that the heart vagus in warm-blooded animals ends at the au-
ricular/ventricular boundary, that here considerably more Ac.Ch. was to be
found before and after stimulation in the auricle than in the ventricle, whereas in a frog's heart, where the vagus extends over the ventricle as well, the

distribution of Ac.Ch. over auricle and ventricle is even. As the heart vagus belongs to the parasympathetic system, the question had to be examined whether and to what extent the neurochemical mechanism applied here. The first investigation on this point also came from my Institute, from Engelhart[16], who was able to prove the release of Ac.Ch. as a result of stimulation of the oculomotor nerve. The total result of the many different, resultant investigations on various organs can be summarized by saying that up until now no single case is known in which the effect of the stimulation of the parasympathetic nerves was not caused by the release of Ac.Ch.

As, to my mind, a lecture should concern itself not only with results, but also with still open questions, I must touch on the following: As all activity caused by the application of Ac.Ch. can be halted by atropine, one might expect that wherever Ac.Ch. is released as a result of nerve stimulation, the effect could everywhere be halted by atropine. This, however, is not so. Contractions of the bladder after stimulation of the pelvic nerve, dilation of the vessels of the salivary gland after stimulation of the chorda nerve still occur even after atropinization. And here we must mention the following strange observation by V.E.Henderson[17]: he found that after preliminary atropinization,[35] vagus stimulation in the intestine produced no increase of tonus, but an increase of peristaltic contractions. The reason for these remarkable exceptions has so far escaped us.

The neurochemical mechanism is everywhere apparent in the field of activity of the parasympathetic system, as in the sympathetic system. But we have Dale[18] and his collaborators to thank for the recognition that the stimulation of certain nerve fibres which belong anatomically to the sympathetic system lead to the release, not of adrenaline, as in the overwhelmingly large number of cases, but of Ac.Ch.

To sum up then, it may be said that the neurochemical mechanism applies in the stimulation of all autonomic nerves.

But it also embraces a much wider area. We owe this knowledge in the main to the basic investigations of Dale. There is no need, therefore, for me to go further into this in my lecture.

We now have to discuss the important question of whether the nerve stimulation influences only the function of the effector organ by the release of nerve substances, as I will call the chemical transmitters for the sake of briefness, or whether it perhaps exerts another influence as well.

Here we shall be well advised to take as a starting-point the mechanism of action of atropine or ergotamine. With Navratil[19] I was able to show (and this finding was confirmed many times over) that these alkaloids do not, as had been thought previously, attack and incapacitate the nerves themselves. We were able to show this by demonstrating that even after using atropine and ergotamine, nerve stimulation still released nerve substances. This shows that atropine and ergotamine do not impair the function of the nerves, which is a liberating one, that is to say, they do not paralyse the nerves, but exert an antagonistic influence on the action of the substances produced. By recognizing that after previous application of atropine or ergotamine the stimulation of the respective nerves is known to have no effect at all upon the effector organ, it has been proved that nerve stimulation has no other effect but to release nerve substances. What other kind of function can remain for the nerve if the action of the substance released coincides absolutely with the effect of the nerve stimulation? Although what follows is self-explanatory, I still think it desirable to state it expressly: in all cases in which the neurochemical mechanism occurs, the nerves only control function to the extent of the release of the substance: the place where this occurs is in the effector organ of the nerve. From then onwards, the released substance exerts control: the functioning organ is, therefore, *its* effector organ exclusively.

And now we must consider in which directions our knowledge of the physiological process has been extended, beyond what we have already said, by the discovery of the neurochemical mechanism.

There will be no cause for argument if we see the most importance in the fact that at last a clear answer has been found to the age-old question as to the nature of the stimulus-transfer from nerve to effector organ.

Next in importance appears to me to be the explanation of the nature of the peripheral inhibition. Up until now, it appeared quite inconceivable that the stimulation of a nerve could lead to inhibition in the effector organ. With the proof that this inhibition comes about because the nerve releases a function-inhibiting substance, the reason for it becomes clear. At the same time, however, something else is proved which seems to me to be of great importance: the release of a substance by the nerves is the expression of a positive function, an activation. This proves that the *direct* effect of the stimulation of all nerves, whether activating or inhibitory, represents a promotion of function, for this is what the release of the substance does.

Today, because we know how it happens, this solution strikes us as self-evident. For, since the process of stimulation is, to a certain degree, unspecific and furthermore interference in stimulus frequencies which certainly form the basis of some inhibitory manifestations in the animal region of the central nervous system cannot, in the case of peripherally inhibitable organs, be regarded as the cause of inhibition, I see no other possibility, at least in general, as to how nerve stimulation can lead to inhibitions of the effector organ at all than by chemical means; in other words, the chemical mechanism is the only conceivable way.

So much for the field of activity and the importance of the neurochemical mechanism.

After this description which touches upon the general nature only of the neurochemical mechanism, we will now consider more exactly its finer mechanism.

First of all the question arises: where are the substances released by nerve stimulation localized, or, in other words, where is the point of attack of the nerve stimulation? A priori, two possibilities exist: the substances are released in the nerve endings or in the effector organ. Investigations of this question carried out so far are concerned only with Ac.Ch.

For the time being we shall only draw upon findings which concern the Ac.Ch. content of organs after nerve degeneration.

As far back as 30 years ago, Anderson[22] observed the following: after degenerative division of the oculomotor nerve, light stimulation was for a long time without effect, regardless of whether the eye had been eserinized or not. There followed a period when light stimulus was still ineffective to the uneserinized eye, but not to the eserinized eye. At this moment, as could be shown, a weak regeneration of the oculomotor nerve had begun. In Anderson's time it was not possible to give an adequate explanation of these findings. Today, when we know that oculomotor stimulation releases Ac.Ch., the action of eserine is revealed as being simply to increase the effect of the Ac.Ch. by inhibiting that of the esterase, and Anderson's results become absolutely clear. With degeneration of the oculomotor nerve the Ac.Ch. disappears. Eserine then also becomes ineffective. With the start of regeneration of the oculomotor nerve the Ac.Ch. appears again, but in too small quantities to cause miosis with light stimulus alone, i.e. without the increased activity provided by eserine. Thus Anderson's experiments provide the first proof that the existence of Ac.Ch. in the eye is dependent upon the nerves. Later Engelhart[16] in my own Institute produced this proof in a direct manner. With direct Ac.Ch. determination he found that after degeneration of the oculomotor nerve in corpus ciliare and iris, the Ac.Ch., present in considerable quantities in preserved nerves, completely disappears. This shows that, in many organs at any rate, the Ac.Ch. content and its maintenance is connected with the presence of the nerve. There are two possible explanations for the disappearance of the Ac.Ch. after nerve degeneration. Either the Ac.Ch. is a part of the nerve and disappears then naturally with its degeneration, or it belongs to the effector organ. Then we should have to assume that the formation and maintenance of the Ac.Ch. amount in the effector organ was, in some mysterious and trophic manner, dependent upon the nerve, so that it would disappear with its degeneration. Should the Ac.Ch. be a product of the effector organ and not the nerve ending, then, according to Dale, it would have to disappear, after degeneration, through some kind of atrophy. This hypothesis would then require a further sub-hypothesis, that of separate and specific transmission system in the effector organ quite unlike any other. This assumption would be necessary, because, after oculomotor nerve degeneration, the effector organs, corpus ciliare and iris do not degenerate, and yet the Ac.Ch. disappears. The influence of the oculomotor nerve degeneration must, in that case, only extend to the mysterious transmission system. In respect of these difficulties alone, a far likelier assumption is that the Ac.Ch. which is released by nerve stimulation belongs to the neurone itself, or more exactly to the nerve ending. There is in my opinion, in at least one instance, compelling proof for the correctness of this supposition.

In Dale's Institute, Feldberg and Gaddum[23] have shown that stimulation of the preganglionic sympathetic fibres in the neck releases Ac.Ch. in the sup.cerv. ganglion, which itself stimulates the ganglion, so that progressive stimulation is set up in the postganglionic fibres. In elegant experiments directed towards the question of the localization of the release of Ac.Ch. in the ganglion, Feldberg and Vartiainen[24] were recently able to prove that it was released neither by the preganglionic fibres nor by the ganglion cells themselves, the only direct effector organ. They concluded, therefore, that the Ac.Ch. was produced in the synapse. Synapse is not an anatomical but a purely functional concept. It indicates the spot where the nerve ending comes into contact with the cell, and has been adopted by histologists only in this sense. If, therefore, it can be proved that Ac.Ch. is formed in the « synapse », it can only, in my opinion, be in the preganglionic nerve ending or in the ganglion cell. As the ganglion cell can be ruled out, as Feldberg and Vartiainen have shown, there only remains, it appears to me, the nerve ending as the site of release. Although proof of this has so far only been obtained directly in the case of preganglionic sympathetic endings, there is, nevertheless, much to make us think that in other places as well the nerve substances are released in the nerve endings themselves. We know that in many organs by no means each single, functioning unit is accorded a nerve fibre. At most, according to Stöhr, one occurs for every hundred capillaries. When the nerve is stimulated, however, all react. In these cases, how does the nerve substance diffuse to those regions without nerves? I believe that the nerve ending is here the liberation centre. This supposition is supported when we consider that when the autonomous nerves are stimulated the two same substances are always released in very different organs having a quite different chemical structure and accordingly undergoing quite different chemical changes. If the substances were not being released in the nerve endings, but peripherally of them, then we should again have to assume the presence of some mysterious mechanism capable of transferring the stimulation of the nerve ending to the supposed peripheral position where the substance would be released; in which case, the discovery of the neurochemical mechanism would not, in my opinion, represent any important progress.

We come now to the next question concerning this delicate mechanism. So far we have only spoken of the release of the substance from the nerve ending. This is only to say that a free nerve substance emerges from the nerve ending. But it is important for an understanding of the nature of nerve function to know what exactly we should imagine is implied by this release. A priori the following possibilities exist: either the substances are not present in the nerve ending when the nerves are in a state of rest and are only formed by nerve stimulation and, once formed, diffuse, or they are already present in the state of rest, but can only diffuse after stimulation. As regards the formation of nerve substances through the nerves, it is certain that this can be done. Even Witanowski[25] in his day found Ac.Ch. in the vagus, in the sympathicus and in the sympathetic ganglia. The last two findings were confirmed by Chang and Gaddum.[26] As Ac.Ch. is not present in the blood, it cannot diffuse from there, and neither, on account of its ready destructibility, could it diffuse from elsewhere in the nerves and ganglia. The same applies for adrenaline. Recently we have succeeded in showing the presence of adrenaline in a frog's brain in a state of rest or even anaesthetized, and also in the upper cervical ganglia of cattle. It was characterized by its effect upon the heart which was similar to that of adrenaline, through the neutralizing of this effect by ergotaminization and also by its destructibility through fluorescent light. These findings, therefore, confirm that the nerve substances are formed by the nerve and are present even in a state of rest. Whether the nerve, when stimulated, produces further substance as well is another still undecided question which we are not touching upon here. However interesting in itself the answer to this question may be, it does not appear to me to be of essential importance, since the basic effect of nerve stimulation is the release of the substances. There are two possibilities as regards the processes of release and diffusion: either the substances are present in a free and diffusible state in the nerve ending, but the nerve ending when in a state of rest is impermeable and only made permeable to them after stimulation, when they become diffusible and effective, or, the substances in the resting state are in some way combined and indiffusible and only the stimulation releases the combination and thereby makes them diffusible and effective. If the first possibility were to apply, then we must not find the Ac.Ch. at all, since, as has been shown, esterase is found everywhere in the nerves and this, as we shall soon see, destroys the free Ac.Ch. But we do find it in the nerve. This fact alone suffices to show that it is not present in a free, diffusible state in the nerve ending. In addition, Bergami[28] recently found, in confirmation of earlier experiments by Calabro[27], that Ac.Ch. only issues from the free end of severed nerves if the nerve is stimulated. In this case, the release cannot, of course, be attributed to any change in the state of permeability brought about by stimulation, since the free nerve ending has no membrane. The second possibility which I mentioned earlier must apply, namely that the Ac.Ch. in the unstimulated nerves is bound in some way and thereby protected from the assault of the esterase. In fact, it is present in such quantity in hearts where there is no vagus stimulation, that in a freely diffusible state it would be more than sufficient to stop the heart altogether. On its own it is ineffectual and is protected against the action of the esterase, in contrast to when it is in a diffusible state.

In experiments directed towards the study of this question Engelhart and I[29] found the following: If one determines the initial value of Ac.Ch. in a heart section, leaving the remaining portion of the heart intact for a few hours, as much Ac.Ch. is found in it afterwards as in the beginning. Dale and Dudley, incidentally, found the same in the case of the spleen. In an organ in a state of rest, therefore, the Ac.Ch. is protected against the esterase. But if free (that is to say diffusible) Ac.Ch. is added to a heart in a state of rest, it is destroyed. All this goes to show that obviously, as Dale also assumed, the Ac.Ch. is present in the organ in a state of rest in some kind of loose, non-diffusible combination, and for that reason it is non-susceptible to attack by esterase and non-effective. Such combinations we know do very often occur in an organism. The so-called « vehicle function » of the blood implies in fact no more than the ability of the blood's component parts to bind substances and, when necessary, to release them. But the binding must in any case be a very loose one, as after destroying the structure, for instance by mincing the organ, the Ac.Ch. is very quickly destroyed by esterase. Nerve stimulation would accordingly appear to have the effect of releasing from this combination the Ac.Ch. which has been proved to be present in the nerve.

The same applies also for the nerve substance in the sympathetic system, adrenaline. As I was able to show this year[12], the heart contains 1γ to 2γ per gram, which corresponds to a concentration of $1:1$ million to $1:500,000$. Whereas adrenaline added to the heart will already be effective in a concentration of $1:100$ million to the maximum, the concentration of 100–200 times more adrenaline in a heart in a state of rest will be without effect. Therefore it also must be present in some kind of inactive combination in the heart. This fact also seems to me to be of importance in the possible interpretation of certain other findings. It is known that in many organs the adrenaline action is quickly over. Up until now this has been explained by the speedy oxidation of adrenaline. This is certainly the case for pure adrenaline solutions *in vitro*. *In vivo*, on the other hand, adrenaline is not only not easily oxidized, but all the organs contain substances – among them, as has been proved, amino acids – which, even in minimal quantities, have a direct inhibiting effect upon the oxidation of adrenaline. How then does this rapid cessation of activity come about? It may, in part, be due to counteractions. In some cases, however, the disappearance of activity could be due to rapid transference of the adrenaline into an ineffective linkage as is to be found in the heart.

Now let us return from this digression to the subject of the release of the nerve substances. This occurs very quickly and the action of the released nerve substance is very rapid also, although between release and effect the diffusion process has also to be set in motion. The time interval varies in length in different cases, but is in part certainly dependent upon the distance of the releasing nerve ending from the effector cell. According to Brown and Eccles[30] this is 80–100 σ in the case of the heart, but only 2 σ in the ganglionic synapse. This must mean that release coincides with stimulation. Dale is able to explain quite easily the fact that the effect reaches the ganglion cell almost without any time lapse by the fact that the release in the nerve ending occurs directly with contact with the ganglion cell, whereas in the heart, where incidentally the first contraction after vagus stimulation is smaller, a certain time is required for diffusion to the effector cells. As in the case of release and effect, the speed with which the substance and with it the effect disappears, varies in different objects. The discovery of the chemical mechanism of the effect of vagus stimulation in the heart was only possible because in this case the destruction of the Ac.Ch. occurs so slowly that the substance had time to diffuse, in sufficient quantity to be active, into the heart; in the ganglia on the other hand, the destruction occurs so rapidly that the Ac.Ch. in the perfusion fluid is only demonstrable after preliminary eserination. The differences in time between freeing and disappearance in both cases are easily understandable if we consider the quite different purposes which the nerve stimulation serves in both these cases.

And now, finally, we come to the localization of the point of attack of the nerve substances.

As long as it was not known that the autonomic nerves, when stimulated, release substances which condition the successful effect of the nerve stimulation, it was assumed in general, in consideration of the fact that the action picture of the so-called vago- and sympathico-mimetic substances is identical with the stimulating of the corresponding nerves, and, further, with the fact that it was believed that the alkaloids, atropine, and ergotamine, which inhibit the action of the substances, paralyse the corresponding nerves, that the vago- and sympathico-mimetic substances stimulate the nerves somewhere peripherally. But as they are effective even after nerve degeneration, it was assumed, with justification at the time, that a non-degenerative myoneural junction was the point of attack. Today, now that we know that the nerves do release nerve substances, this view is no longer tenable. The nerve substances, considered as vago- or sympathico-mimetic substances, would have to act like these, that is to say, they would have to stimulate the myoneural junction and release substances, etc. on their own. In this case there would be no kind of effect upon the effector organ. Quite apart from this, the supposition that the nerve substances stimulate the nerve somewhere is quite superfluous by the proof shown above, that the alkaloids atropine and ergotamine which inhibit the activity of the vago- and sympathico-mimetic substances, do not, as was supposed, paralyse the nerves, but are simply antagonistic to the substances. If all this is evidence against the nerve as point of attack, it has also been proved that Ac.Ch. and adrenaline are also effective in the absence of nerves. Ac.Ch., for instance, dilates vessels which are not parasympathetically innervated. Adrenaline increases the activity of the still nerveless embryonic heart and stimulates the arrectores pilorum, which, according to Stöhr, are also nerveless, etc. Therefore, the point of attack of the nerve substances must be some part of the effector organ itself, probably chemical or chemico-physical in character and not morphological.

As Dale has proved, we can no longer say that the nerve substances reproduce the action picture of the nerves but rather it is a fact that the nerves reproduce the action picture of the substances, since they release these and thus lead to effective action. That the activity caused by any one nerve substance appears principally at the spot where it is released, that is to say, that in that particular spot the cells are receptive to its action, is a local phenomenon of the specific sensitivity to certain chemical substances which is met with everywhere in the living organism and which is one of the foundations of its function and, therefore, of its very existence and which can only be understood teleologically and not causally; think, for example, of the finely graduated, specific sensitivity of the respiratory centre to CO_2.

Up until now we have discussed only the effect of the nerve substances on the organ in which they are released through nerve stimulus. Are they only active there, or in other distant organs too? We have already mentioned that a part of the released substance diffuses into the blood or into some other perfusing fluid. This could present the possibility of its action being extended to other more distant organs. What is the position here? Given special conditions, which I would like to characterize as pathological, this could happen. It has been proved that when the breaking up of the Ac.Ch. by an esterase, is inhibited by eserine, the Ac.Ch. penetrates with the blood to other organs in sufficient quantities to cause activity. Furthermore, Cannon[31] by preliminary sensitizing of organs through denervation, or cocainization, made them so hypersensitive to the sympathicus substance that they reacted to its release in any organ. In the same way as in these experimentally induced disturbances, it could also happen perhaps that in cases of illness, the release of surplus quantities of substance or incomplete destruction may interrupt the normal release and destruction, leading to hypersensitivity of organs and the appearance of effect at a distance. It would be very desirable if in future clinicians would give consideration to these relationships with a view to explaining certain symptoms and groups of symptoms which until now, partly without sufficient foundation, have been considered as purely reflex. Under normal conditions, however, the effect of the nerve substance would be limited to the organ in which it is released. The hormones are there to exert a general control, that is to say not a localized chemical one, on the organs.

In conclusion a word or two on the question of how the neurochemical mechanism fits into the connecting pattern of cells. With the discovery that its influence comes about through substances which are released by the nervous system itself, we have the first proof that the nervous system is not only an effector organ for chemical influences from outside, and not only a participant in general metabolism, but that it has itself a specific chemical influence upon happenings in the organism. On closer examination this is not surprising.

In nerve-free multicellular organisms, the relationships of the cells to each other can only be of a chemical nature. In multicellular organisms with nerve systems, the nerve cells only represent cells like any others, but they have extensions suited to the purpose which they serve, namely the nerves. Accordingly it is perhaps only natural that the relationships between the nervous system and other organs should be qualitatively of the same kind as that between the non-nervous organs among themselves, that is to say, of a chemical nature.

1. W. H. Howell, *Am. J. Physiol.*, 21 (1908) 51.
2. W. M. Bayliss, *Principles of General Physiology*, 3rd ed., 1920, p. 344.
3. T. R. Elliott, *J. Physiol.*, 31 (1904) 20 P.
4. W. E. Dixon, *Med. Mag.*, 16 (1907) 454.
5. O. Loewi, *Pflügers Arch. Ges. Physiol.*, 189 (1921) 239.
6. O. Loewi and E. Navratil, *Pflügers Arch. Ges. Physiol.*, 214 (1926) 678.
7. O. Loewi and E. Navratil, *Pflügers Arch. Ges. Physiol.*, 214 (1926) 689.
8. H. H. Dale, *J. Pharmacol.*, 68 (1924) 107.
9. H. H. Dale and H. W. Dudley, *J. Physiol.*, 68 (1929) 97.
10. W. B. Cannon and A. Rosenblueth, *Am. J. Physiol.*, 99 (1932) 396.
11. A. Fröhlich and O. Loewi, *Arch. Exptl. Pathol. Pharmakol.*, 62 (1910) 159.
12. O. Loewi, *Pflügers Arch. Ges. Physiol.*, 237 (1936) 504.
13. J. H. Gaddum and H. Schild, *J. Physiol.*, 80 (1934) 9 P.
14. P. Rylant, *Compt. Rend. Soc. Biol.*, 96 (1927) 1054.
15. E. Engelhart, *Pflügers Arch. Ges. Physiol.*, 225 (1930) 722.
16. E. Engelhart, *Pflügers Arch. Ges. Physiol.*, 227 (1931) 220.
17. V. E. Henderson, *Arch. Intern. Pharmacodyn.*, 27 (1922) 205.
18. H. H. Dale, *J. Physiol.*, 80 (1933) 10 P.
19. O. Loewi and E. Navratil, *Pflügers Arch. Ges. Physiol.*, 206 (1924) 123.
    E. Navratil, *Pflügers Arch. Ges. Physiol.*, 217 (1927) 610.
20. O. Loewi, *Pflügers Arch. Ges. Physiol.*, 193 (1921) 201.

21. O. Loewi, *Pflügers Arch. Ges. Physiol.*, 203 (1924) 408.
22. H. K. Anderson, *J. Physiol.*, 33 (1905) 156, 414.
23. W. Feldberg and J. H. Gaddum, *J. Physiol.*, 81 (1934) 305.
24. W. Feldberg and A. Vartiainen, *J. Physiol.*, 83 (1934) 103.
25. W. R. Witanowski, *Pflügers Arch. Ges. Physiol.*, 208 (1925) 694.
26. H. C. Chang and J. H. Gaddum, *J. Physiol.*, 79 (1933) 255.
27. Q. Calabro, *Riv. Biol.*, 19 (1935).
28. G. Bergami, *Klin. Wochschr.*, 15 (1936) 1030.
29. E. Engelhart and O. Loewi, *Arch. Intern. Pharmacodyn.*, 38 (1930) 287.
30. G. L. Brown and J. C. Eccles, *J. Physiol.*, 82 (1934) 211.
31. W. B. Cannon and Z. M. Bacq, *Am. J. Physiol.*, 96 (1931) 392.

Biography

Otto Loewi was born on June 3, 1873, in Frankfurt-am-Main, Germany, the son of Jacob Loewi, a merchant, and Anna Willstätter.

After having attended the humanistic Gymnasium (grammar school) in his native town, he entered in 1891 the Universities of Munich and Strassburg (at that time part of Germany) as a medical student. Apart from his attendance at the inspiring anatomy courses of Gustav Schwalbe, however, he seldom went to the medical lectures, being more inclined towards those held at the philosophical faculty. Only in the summer of 1893 did he seriously prepare for his « Physicum », the first medical examination, which he just managed to pass. It was not until the autumn of 1894 that his indifference to medicine suddenly gave way to almost enthusiastic interest. In 1896 he took his doctor's degree at Strassburg University, his thesis dealing with a subject suggested by Professor Oswald Schmiedeberg, the famous « Father of Pharmacology ». Also responsible for his medical education were: Bernhard Naunyn, distinguished clinician and experimental pathologist, Oscar Minkowski, and Adolph Magnus-Levy.

After his graduation he followed a course in inorganic analytical chemistry with Martin Freund, in Frankfurt, and afterwards spent a few months working in the biochemical institute of Franz Hofmeister in Strassburg. During 1897–1898 he was assistant to Carl von Noorden, clinician at the City Hospital in Frankfurt. Soon, however, after seeing the high mortality in countless cases of far-advanced tuberculosis and pneumonia, left without any treatment because of lack of therapy, he decided to drop his intention to become a clinician and instead to carry out research in basic medical science, in particular pharmacology. In 1898 he succeeded in becoming an assistant of Professor Hans Horst Meyer, the renowned pharmacologist at the University of Marburg-an-der-Lahn, from 1904 Professor of Pharmacology in Vienna. In 1905 Loewi became Associate Professor at Meyer's laboratory, and in 1909 he was appointed to the Chair of Pharmacology in Graz.

During his first years in Marburg, Loewi's studies were in the field of metabolism. As a result of his work on the action of phlorhizin, a glucoside provoking glycosuria, and another one on nuclein metabolism in man, he was appointed «Privatdozent» (Lecturer) in 1900. Two years later he published his paper «Über Eiweisssynthese im Tierkörper» (On protein synthesis in the animal body), proving that animals are able to rebuild their proteins from their degradation products, the amino acids – an essential discovery with regard to nutrition.

That same year he also published the first part of a series of papers about experimental contributions to the physiology and pharmacology of kidney function.

In 1902 Loewi also spent some months in Starling's laboratory, in London, where he also worked with W. M. Bayliss, Starling's brother-in-law. And it was in this laboratory that he first met his lifelong friend Henry Dale, who was later to share the Nobel Prize with him.

After his return to Marburg in 1902 Loewi continued to study the function of the kidney and the mechanism of the action of diuretics. On his arrival in Vienna in 1905 he again took up the problems connected with carbohydrate metabolism. He proved thereby that preference for fructose rather than glucose is not only characteristic of pancreatectomized dogs, as earlier demonstrated by Minkowski, but also of dogs deprived of their glycogen by other means, e.g. by phosphorus poisoning. He also proved that the heart in contrast to the liver, cannot utilize fructose. And finally that epinephrine injections into rabbits completely depleted of their liver glycogen by starvation brought the glycogen back to almost normal values in spite of continued starvation. His other investigations in Vienna, done jointly with Alfred Fröhlich, dealt with the vegetative nervous system (stimulated by the discovery made by Gaskell and Langley of the existence of two divisions of this nervous system, and also as a result of his coming into contact with T. R. Elliott in Cambridge, where the latter was conducting his final experiments on the action of epinephrine). His classic paper in this field was published in 1905, the best-known result of these studies being the observation that small doses of cocaine potentiate the responses of sympathetically innervated organs to epinephrine and sympathetic nerve stimulation.

It was as Professor in Graz that Loewi cultivated his gifts as a lecturer. A number of his associates during this period came from the U.S.A. Loewi continued his studies of carbohydrate metabolism, investigating among other things the conditions responsible for epinephrine hyperglycaemia.

In 1921 Loewi discovered the chemical transmission of nerve impulses the research of which was greatly developed by him and his co-workers in the

years following, culminating ultimately in his demonstation that the parasympathetic substance («Vagusstoff») is acetylcholine and that a substance
closely related to adrenaline played a corresponding role at the sympathetic
nerve endings. It was for these researches that he received the Nobel Prize in
1936, jointly with Sir Henry Dale. This and other discoveries in the fields of
chemistry, physics, and pharmacology have since then led to a complete
renewal of the concepts of the sympathetic nervous system.

When the Germans invaded Austria in 1938, Loewi was forced to leave
his homeland. (But only after he had been compelled to instruct the Swedish
bank in Stockholm to transfer the Nobel Prize money to a prescribed Nazi-
controlled bank.)

After spending some time as Visiting Professor at the Université Libre in
Brussels, and at the Nuffield Institute, Oxford, Loewi accepted an invitation
to join the College of Medicine, New York University, as Research Professor
of Pharmacology, and to work in George Wallace's Laboratory. He arrived
in the United States in 1940. In America Loewi came into close contact with
many outstanding biologists from all over the world and here he found much
inspiration for his work.

Loewi held honorary degrees from New York University, Yale Univer-
sity, and from the Universities of Graz and Frankfurt. He was recipient of
the Physiology Prize of the Royal Academy of Sciences of Bologna, of the
Lieben Prize of the Academy of Vienna, and of the Cameron Prize of the
University of Edinburgh (1944). He was an Honorary Member of the
Physiological Society (London), of the Harvey Society (New York), and
of the Società Italiana di Biologia Sperimentale; he was also Corresponding
Member of the Society of Physicians in Vienna, of the Viennese Biological
Society, and of the Society for the Advancement of Natural Sciences in
Marburg-an-der-Lahn; and was Member of the Deutsche Akademie der
Naturforscher Leopoldina, in Halle. In 1954 he was appointed Foreign Mem-
ber of the Royal Society.

Since his schooldays, Loewi showed keen interest in the humanities. He
always enjoyed music, architecture, and painting, and in his younger years
seldom missed an opportunity to visit museums and exhibitions.

In 1908 he married Guida Goldschmiedt, daughter of Dr. Guido Goldschmiedt, then Professor of Chemistry in Prague, and later in Vienna. They
had three sons, Hans, Victor, Guido; and one daughter, Anna. Professor
Loewi became an American citizen in 1946. He died December 25, 1961.

Physiology or Medicine 1937

ALBERT SZENT-GYÖRGYI VON NAGYRAPOLT

«for his discoveries in connection with the biological combustion processes, with special reference to vitamin C and the catalysis of fumaric acid»

Physiology or Medicine 1937

Presentation Speech by Professor E. Hammarsten, member of the Staff of Professors of the Royal Caroline Institute

Your Majesty, Your Royal Highnesses, Ladies and Gentlemen.

The Staff of Professors of the Caroline Institute, pursuant to the task devolving upon them by the terms of the will of Alfred Nobel, have awarded the Prize for Physiology or Medicine for the year 1937 to Professor Albert von Szent-Györgyi, *in recognition of his discoveries concerning the biological combustion processes* with especial reference to vitamin C and to the fumaric acid catalyst. The wording of the above sentence indicates that the mechanism of biological oxidation has been investigated beyond the great discoveries in this field made by Otto Warburg, Heinrich Wieland, and their successors. Their systems of catalysts for oxidation have been shown to be dependent on Szent-Györgyi's new catalysts.

It was generally known before that combustion liberates energy in living cells which can be employed there without loss – directly for the building up of new substances – for storage or for the building of functioning cell structures. The building up of living organisms then is dependent in essential respects on combustion, which is guided by catalyst systems. Thus catalyst systems are conditional for the building up of living organisms. Consequently in the unknown period during which organic life originated, the formation of these and other catalyst systems must have preceded the completion of the living animal organisms.

Preferably I should wish to confine my remarks to the new conquistador from Szeged. The survey is however of higher importance on this occasion, and moreover the course of the events is dramatically concentrated. Each one of the three has conquered new ground by intuitive daring and skill. Szent-Györgyi's greatest achievement has intimately linked up the accomplishments of the two others and of their successors, giving us for the first time a picture of a coherent oxidation process – of the interplay of three catalyst systems and the oxidation thereby in metabolism.

Warburg, who always stood alone with some few faithful co-workers, is the foremost pioneer, and he had to overcome the greatest difficulties. At this day there is none who can any longer throw doubt on his discoveries, but that was not so, when in 1931 underestimated by the majority, he was awarded the Nobel Prize by the Caroline Institute. He has shown that the inert oxygen, with which the red blood corpuscles are fully loaded, is taken up from them by a catalyst system to which many red pigments belong (for brevity's sake called «the red system»). These are related to the red blood pigment. They contain as active groups (for the most part) iron and specific proteins. In this system the oxygen combines with the iron during varying periods of time. In the case of the most rapid catalysts, it combines with the iron, is converted into a lively, reactively disposed form and is delivered – all at a speed that gives a flowing stream of active oxygen from the catalyst system. One thought that this active oxygen oxidized directly. That is not the way however. On the contrary, the active oxygen meets *hydrogen* – but that is another story, belonging to Szent-Györgyi's great discovery. The manner in which the life-giving active oxygen's dramatic encounter in the darkness of the cells ensues, had been unknown ever since the morning of time until, in 1933, Szent-Györgyi carried out some experiments which proved to be the prelude to the revelation of the secret.

For the moment I will leave oxygen, and direct the attention to the first, apparently unimportant, experiments carried out by Wieland. These led him to the conception of an idea, which was destined to carry him on to the disclosure of an extensive part of the mechanism of oxidation. A large number of investigators were soon attracted by Wieland's opinion. This seemed to be at variance with the oxygen activation – at any rate that was the view of a majority. This apparent inconsistency was never considered by Szent-Györgyi, nor by Warburg.

Wieland had observed that palladium is capable of absorbing hydrogen from certain organic compounds, which means their partial combustion or oxidation. Through the cooperation of many investigators the presence was revealed of extensive metal-free catalyst systems, the effect of which was shown to consist in the removal of hydrogen from metabolic substances, in agreement with Wieland's concept. These catalysts were given a name in common: *dehydrogenases* (hydrogen-removers, hydrogen-absorbers, or hydrogen-transporters) and the idea was held pretty generally that the hydrogen activated by this system would be capable of reacting directly with the inert oxygen molecules. Hydrogen superoxide was supposed to form an intermediate product. That is not the highway of oxidation however. On the contrary, the hydrogen first meets Szent-Györgyi's catalyst system from a different side to the one where the activated oxygen flows into it from the «red system». That again is another story, which also belongs to Szent-Györgyi's great discovery. From 1925 onwards he had been investigating a number of hydrogen-absorbers. Previously to anyone else he formed the view of these as members of a catalyst system in the service of oxidation (in other quarters loosely conceived of as being auxiliary catalysts of some kind for fermentation). He was also occupied with experiments on a yellow substance, termed *flave* by him, while his investigation regarding vitamin C was being completed, and conducted on to the isolation of that substance, enabling him later to insert it in the catalyst system of certain hydrogen-removers. Vitamin C and another substance, containing sulphur as a hydrogen-removing group and defined by Sir Frederick Gowland Hopkins and others, were however until 1934 the only substances belonging to the hydrogen-transporters in the oxidation-chain that had been isolated. The rank that they possess as catalysts is dependent on the velocity of the hydrogen-transportation and the degree of the activation of the hydrogen – problems that still await a satisfactory solution. On the other hand, Hugo Theorell succeeded in 1934 in isolating, in Warburg's laboratory, the first really rapid hydrogen-transporter, called «the yellow enzyme». He could also show that it was a phosphoric-acid ester of vitamin B_2, linked to a specific protein. Warburg and Christian, in 1935, defined the nature of the active group in two other dehydrogenases, colourless and metal-free (co-ferment and co-zymase), which had long frustrated the attempts of other investigators. One of them was the catalyst that Szent-Györgyi had placed in this section of the oxidation-chain.

The magnificent series of Szent-Györgyi's discoveries commenced in 1933. They were carried out and pursued at Szeged with extraordinary rapidity and precision. His clear vision for essentials induced him, in spite of his isolation of *ascorbic acid* and of his identification of it with the so-termed vitamin C – a feat that was justly hailed with enthusiasm – to hand over to others for the time being the tempting pursuit of the further development of that discovery, and to devote the whole of his energy to the problem of combustion, notwithstanding the difficulties it presented. Many investigators had been working at the so-called plant acids in the muscular system, and had observed their capacity for intensifying oxidation in that tissue. The readiest explanation however of how that came about, viz. that they are easily combusted themselves, simply did not fit in with Szent-Györgyi's intuitive perception. By elaborating reliable methods of analysis for the substances in question, and by means of consistent experiments, he and his co-workers proved *that the plant acids were not consumed by combustion, were not ordinary nutrient substances, but were on the contrary themselves active groups of catalysts which served to maintain the combustion without themselves suffering any diminution thereby.* The process involves a peregrination of hydrogen more intricate than the adventurous journeys of Odysseus, though more rapid. Hydrogen is released out of the metabolic substances, probably through co-operation between Szent-Györgyi's and Warburg's co-ferment and Theorell's yellow enzyme, and encounters the plant acids, entering in that way Szent-Györgyi's system. These acids transfer the system into the order: oxalacetic acid, malic acid, fumaric acid, and succinic acid, then, in the form of active hydrogen, to encounter the active oxygen from « the red system» and form water and free energy – a series of providentially subdued explosions which I alluded to before as a dramatic encounter. The plant acids act as catalysts by cooperation with specific proteins, and the effect of the yellow enzyme probably extends some way into this Szent-Györgyi's intermediate system.

Thus, the oxygen-activation in the red iron system and hydrogen-transfer from nutrients by the yellow metal-free system along with co-agents have been united by Szent-Györgyi through the discovery of this intermediate system. The interplay of « the red system»'s cytochrome-group and the yellow enzyme might probably also, according to Theorell, proceed directly. The flaws are numerous, but not of a character to constitute any essential breach in the highway of the oxidation-chain. Numerous ramifications of the latter however already begin to be discernible.

It is of especially great importance that at least two vitamins – C and B_2, and possibly B_1, and P – are in cooperation in the oxidation chain and are catalysts, illustrating the way in which these vitamins act in the organism. It may be that development in the near future will reveal the importance for our organism of copper concerning oxidation and of vitamin C with certain followers in plants, viz. oxidating enzyme, and oxidizable and reducible substances (Szent-Györgyi's flavonoles, termed vitamin P), which are capable of forming a sensitively attuned system with the vitamin, hydrogen-superoxide and proteins, or parts of them, with active and activating sulphur in the molecule. The sulphur of the alchemists of old, out of which everything was to radiate, is destined to experience a renaissance.

Professor Albert von Szent-Györgyi.   As a representative for the Caroline Institute on this occasion, I am commissioned to give expression to our high estimation of your researches.

You never swerved from your unyielding purpose to study the primary and fundamental processes of biological oxidation. Entering upon this difficult field of biochemical research you soon became a pioneer by interpreting the position and real function of co-ferment as an important link in the chain of dehydrogenating catalysts. Not even your important discoveries regarding vitamin C could deter you from following a certain strain of thought. I am deducing now from a close observation of your work that you were drawing distinctions in your mind at this occasion between your interesting discovery of ascorbic acid and the bare possibility of some other audacious plans of yours coming true. At this early stage they must have involved the investigation of the fundamental mechanism of connecting hydrogen activation with that of oxygen activation. Your intuitive mind decided in favour of the possibility of success, and you won through. In the year 1933 the first signs became visible for outsiders, and from then on the pace set by you and your co-workers at Szeged was astonishing, and your results were fundamentally new and highly important. In the midst of fervent research work with most promising aspects you are the discoverer and idealist to the mind of Alfred Nobel.

I ask you, Professor Szent-Györgyi, to receive the prize from the hands of His Majesty, our gracious King.

Oxidation, energy transfer, and vitamins

Nobel Lecture, December 11, 1937

A living cell requires energy not only for all its functions, but also for the maintenance of its structure. Without energy life would be extinguished instantaneously, and the cellular fabric would collapse. The source of this energy is the sun's radiation. Energy from the sun's rays is trapped by green plants, and converted into a bound form, invested in a chemical reaction. It can easily be observed that, when sunlight falls on green plants, they liberate oxygen from carbon dioxide, and store up carbon, bound to the elements of water, as carbohydrate:

$$\text{Energy} + n\,CO_2\,(H_2O) = n\,O_2 + C_nH_{2n}O_n \qquad (1)$$

The radiant energy is now locked up in this carbohydrate molecule. This molecule is our food and the plant's foodstuff. When energy is required, the above reaction takes place in the reverse direction, i.e. the carbohydrate is again combined with oxygen to form carbon dioxide, oxidized, and energy released thereby:

$$n\,O_2 + C_nH_{2n}O_n = n\,CO_2\,(H_2O) + \text{Energy} \qquad (2)$$

According to our earlier views, carbon and carbon dioxide played the central role in this process. Supposedly, radiant energy was used to break down *carbon* dioxide. On oxidation *carbo*hydrate was again combined with oxygen to form *carbon* dioxide.

Investigations during the last few decades have brought hydrogen instead of carbon, and instead of CO_2, water, the mother of all life, into the foreground. It is becoming increasingly probable that radiant energy is used primarily to break water down into its elements, while CO_2 serves only to fix the elusive hydrogen thus released:

$$\text{Energy} + 2n\,H_2O = 4n\,H + n\,O_2 \qquad (3)$$
$$4n\,H + n\,CO_2 = C_nH_{2n}O_n + n\,H_2O \qquad (4)$$

While this concept of energy fixation was still being developed, the importance of hydrogen in the reversal of this process, whereby energy is liberated by oxidation, had already been confirmed by H. Wieland's experiments. This could be represented as follows:

$$C_nH_{2n}O_n + n\,H_2O = 4n\,H + n\,CO_2 \qquad (5)$$
$$4n\,H + n\,O_2 = 2n\,H_2O + \text{Energy} \qquad (6)$$

This way of representing it is meant to bring out the fact that our body really only knowns *one* fuel, hydrogen. The foodstuff, carbohydrate, is essentially a packet of hydrogen, a hydrogen supplier, a hydrogen donor, and the main event during its combustion is the splitting off of hydrogen. So the combustion of hydrogen is the real energy-supplying reaction. To the elucidation of reaction (6), which seems so simple, I have devoted all my energy for the last fifteen years.

When I first ventured into this territory, the foundations had already been laid by the two pioneers H. Wieland and O. Warburg, and Wieland's teaching had been applied by Th. Thunberg to the realm of animal physiology. Wieland and Thunberg showed, with regard to foodstuffs, how the first step in oxidation is the « activation » of hydrogen, whereby the bonds linking it to the food molecule are loosened, and hydrogen prepared for splitting off. But at the same time oxygen is also, as Warburg showed, activated for the reaction by an enzyme. The hydrogen-activating enzymes are called dehydrases or dehydrogenases. Warburg called his oxygen-activating catalyst, « respiratory enzyme ».

These concepts of Wieland and Warburg were apparently contradictory, and my first task was to show that the two processes are complementary to one another, and that in muscle cells *activated oxygen* oxidizes *activated hydrogen*.

This picture was enriched by the English worker D. Keilin. He showed that activated oxygen does not oxidize activated hydrogen directly, but that a dye, cytochrome, is interposed between them. In keeping with this function, the « respiratory enzyme » is now also called « cytochrome oxidase ».

About ten years ago, when I tried to construct this system of respiration artificially and added together the respiratory enzyme with cytochrome and some foodstuff together with its dehydrogenase, I could justifiably expect that this system would use up oxygen and oxidize the food. But the system remained inactive. So there had to be other links missing, and I set off in search of them. To start with, I found that the dehydrogenation of certain donors is linked to the presence of a co-enzyme. Analysis of this co-enzyme showed it to be a nucleotide, identical with v. Euler's co-zymase, which H. v. Euler and R. Nilsson had already shown to accelerate the process of dehydration.

As a result of Warburg's investigations, this co-dehydrogenase has recently come very much into the foreground. Warburg showed that it contains a pyridine base, and that it accepts hydrogen directly from food when the latter is dehydrogenated. It is therefore, the primary H-acceptor.

While working on the isolation of the co-enzyme with Banga, I found a remarkable dye, which showed clearly by its reversible oxidation that it, too, played a part in the respiration.

We called this new dye *cytoflav*. Later Warburg showed that this substance exercised its function in combination with a protein. He called this protein complex of the dye, «yellow enzyme». R. Kuhn, to whom we owe the structural analysis of the dye, called the dye *lactoflavin* and, with Györgyi and Wagner-Jauregg, showed it to be identical with vitamin B_2.

But the respiratory system stayed inactive even after the addition of both these new components, co-dehydrogenase and yellow enzyme.

With the help of my loyal collaborators, especially Annau, Banga, Gözsy, Laki, and Straub, I succeeded over the last few years in showing that the C_4-dicarboxylic acids together with their activators, were involved as links in this chain of oxidation, and that with their addition the system was now complete, showing an oxygen uptake corresponding to normal respiration.

My time is too short to permit me to go into the details of the demonstration and the countless measurements which led to this conclusion. I will only describe the end result of this work in a few words. This is as follows: the C_4-dicarboxylic acids and their activators which Thunberg discovered are interposed between cytochrome and the activation of hydrogen as intermediate hydrogen-carriers. In the case of carbohydrate, hydrogen from the food is first taken up by oxaloacetic acid, which is absorbed onto the protoplasmic protein, the so-called malic dehydrogenase, and thereby activated. By taking up two hydrogen atoms, oxaloacetic acid is changed into malic acid. This malic acid now passes on the H-atoms, and thus reverts to oxaloacetic acid, which can again take up new H-atoms.

The H-atoms released by malic acid are taken up by fumaric acid, which is similarly activated by the plasma protein, the so-called succinic dehydrogenase. The uptake of two H-atoms converts the fumarate to succinate, to succinic acid. The two H-atoms of succinic acid are then oxidized away by the cytochrome. Finally the cytochrome is oxidized by the respiratory enzyme, and the respiratory enzyme by oxygen.

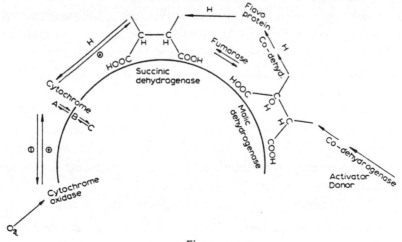

Fumarase

Succinic dehydrogenase $\quad$ H$_2$O $\quad$ Malic dehydrogenase $\qquad$ Dehydrogenase

O$_2$

Warburg's respiratory enzyme $\rightarrow$ $++\rightleftarrows+++$ Cytochrome A B C $\rightleftarrows$

COOH	COOH		COOH	COOH
HCH	CH		HCH	HCH
HCH	CH	2H	HCOH	CH
COOH	COOH		COOH	COOH

$\rightleftarrows$ $\quad$ 2H $\quad$ $\rightleftarrows$ $\quad$ $\leftarrow$2H $\quad$ Donor

Succinic acid $\qquad$ Fumaric acid $\qquad$ Malic acid $\qquad$ Oxaloaceitc acid

Fig. 1

The function of the C$_4$-dicarboxylic acids is not to be pictured as consisting of a certain amount of C$_4$-dicarboxylic acid in the cell which is alternately oxidized and reduced. Fig. 2 corresponds more to the real situation. The protoplasmic surface, which is represented by the semi-circle, has single molecules of oxaloacetate and fumarate attached to it as prosthetic groups. These fixed, activated dicarboxylic molecules then temporarily bind the hydrogen from the food.

The co-dehydrogenases and the yellow enzymes also take part in this system. I have attempted in Fig. 2 to add them in at the right place.

Fig. 2

This diagram, which will probably still undergo many more modifications, states that the « foodstuff » – H-donor – starts by passing its hydrogen, which has been activated by dehydrase, to the co-dehydrogenase. The co-enzyme passes it to the oxaloacetic acid*. The malic acid then passes it on again to a co-enzyme, which passes the hydrogen to the yellow enzyme. The yellow enzyme passes the hydrogen to the fumarate. The succinate so produced is then oxidized by cytochrome, the cytochrome by respiratory enzyme, the respiratory enzyme by oxygen.

So the reaction $2 H + O \rightarrow H_2O$, which seems such a simple one, breaks down into a long series of separate reactions. With each new step, with each transfer between substances, the hydrogen loses some of its energy, finally combining with oxygen in its lowest-energy compound. So each hydrogen atom is gradually oxidized in a long series of reactions, and its energy released in stages.

This oxidation of hydrogen in stages seems to be one of the basic principles of biological oxidation. The reason for it is probably mainly that the cell would not be able to harness and transfer to other processes the large amount of energy which would be released by direct oxidation. The cell needs small change if it is to be able to pay for its functions without losing too much in the process. So it oxidizes the H-atom by stages, converting the large banknote into small change.

I myself was led into the territory of oxidation some 15 years ago by a false supposition. I was interested in the function of the adrenal cortex. When this organ ceases to function, life itself ceases (Addison's disease). But before life ceases, a brown pigmentation makes its appearance in the individual, as happens with certain fruits: apples, pears, bananas, etc., which, as they decay, also turn brown. As a result of investigations by Palladin, the great Russian botanist, it was known that this brown discolouration was related to the damaged oxidation mechanism. I myself was (and still am) convinced, that, with regard to basic functions, as oxidation may be regarded representative of them, there are in principle no fundamental differences between animals and plants. So I set out to study the oxidation system in the potato, which, if damaged, causes the plant to turn brown. I did this in the hope of discovering, through these studies, the key to the understanding of adrenal function.

It was already known that the plants which turn brown when damaged –

---

* One cannot exclude the possibility that the yellow enzyme may also mediate the transfer of hydrogen between carbohydrate and oxaloacetic acid.

about half of all plants – contain a polyphenol, generally a pyrocatechol derivative, together with an enzyme, polyphenoloxidase, which oxidizes polyphenol with the help of oxygen. The current interpretation of the mode of action of this oxidase was a confused one. I succeeded in showing that the situation was simply this, that the oxidase oxidizes the polyphenol to quinone with oxygen. In the intact plant the quinone is reduced back again with hydrogen made available from the foodstuff. Phenol therefore acts as a hydrogen-carrier between oxygen and the H-donor, and we are here again faced with a probably still imperfectly understood system for the stepwise combustion of hydrogen. In the damaged plant, reduction of quinone cannot keep pace with the mounting oxidation of the phenol, and quinones remain unreduced and form pigments.

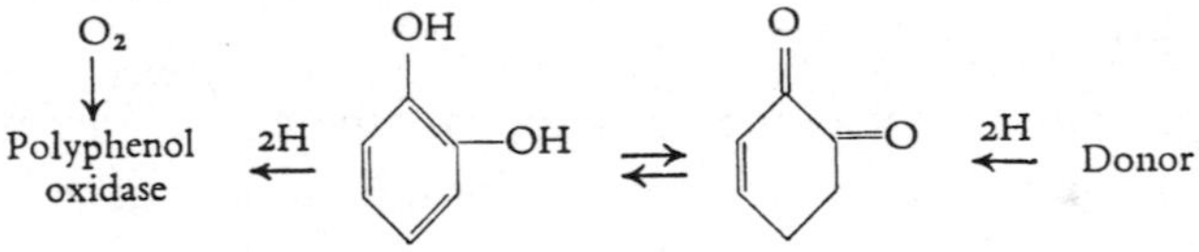

Fig. 3

However, this system gave me no information about adrenal function. So I turned to the plants which do not turn brown when they die, and therefore had to contain an oxidation system with a different structure. All that was known of these plants was that they contained a very active peroxidase. This peroxidase is able to activate peroxide. In the presence of this enzyme, peroxide can oxidize various aromatic substances to coloured pigments. This reaction does not occur without peroxidase. For example, if benzidine is added to a peroxide in the presence of peroxidase, a deep-blue colour appears immediately, which is caused by the oxidation of the benzidine. This reaction, which also serves to indicate the enzyme's presence, does not occur without peroxidase.

But if, for this reaction, I simply used some juice which had been squeezed from these plants instead of a purified peroxidase, and added benzidine and peroxide, the blue pigment appeared, but only after a small delay of about a second. Analysis of this delay showed that it was due to the presence of a powerful reducing substance, which reduced the oxidized benzidine again, until it had itself been used up.

There was great excitement in my little basement room in Groningen, when I found that the adrenal cortex contained a similar reducing substance in relatively large quantities.

Both my means and my knowledge of chemistry were inadequate for investigating the substance more closely. But thanks to the invitation from F. G. Hopkins and the help of the *Rockefeller Foundation*, I was able ten years ago to transfer my workshop to Cambridge, where for the first time I was able to pay more serious attention to chemistry. Soon I succeeded in isolating the substance in question from adrenals and various plants, and in showing that it corresponded to the formula $C_6H_8O_6$ and was related to the carbohydrates. This last circumstance induced me to apply to Prof. W. N. Haworth, who immediately recognized the chemical interest of the substance and asked me for a larger quantity to permit analysis of its structure. Unfortunately it appeared that the only material suitable for preparation on a large scale was adrenal gland. All my efforts to find a suitable plant raw material remained unsuccessful, and adrenals were not available in large quantities in England.

Prof. Krogh tried to help me, generously sending me adrenals from Copenhagen by plane. But unfortunately the substance perished in transit. Then the Mayo Foundation and Prof. Kendall came to my help on a large scale, and made it possible for me to work, regardless of expense, on the material from large American slaughter-houses. The result of a year's work was 25 g of a crystalline substance, which was given the name « hexuronic acid ». I shared this amount of the substance with Prof. Haworth. He undertook to investigate the exact structural formula of the substance. I used the other half of my preparation to gain a deeper understanding of the substance's function. The substance could not replace the adrenals, but caused the disappearance of pigmentation in patients with Addison's disease.

Unfortunately it turned out that the amount of substance was inadequate for finding out its chemical constitution. Through lack of means the preparation could not be repeated, and no cheaper material was found from which the acid could have been obtained in larger quantities.

From the beginning I had suspected that the substance was identical with vitamin C. But my unsettled way of life was not suited for vitamin experiments, concerning which I had not had any experience either. In 1930 I gave up this way of life, and settled down in my own country at the University of Szeged. Fate, too, soon sent me a first-rate young American collaborator, J. L. Svirbely, who had experience in vitamin research, but besides this experience brought only the conviction that my hexuronic acid was not identical with vitamin C. In the autumn of 1931 our first experiments were completed, and showed unmistakably that hexuronic acid was powerfully anti-scorbutic, and that the anti-scorbutic acitvity of plant juices corresponded to their hexuronic acid content. We did not publish our results till the following year after repeating our experiments. At this time Tillmans was already directing attention to the connection between the reducing strength and the vitamin activity of plant juices. At the same time King and Waugh also reported crystals obtained from lemon juice, which were active anti-scorbutically and resembled our hexuronic acid.

Suddenly the long-ignored hexuronic acid moved into the limelight, and there was an urgent need for larger amounts of the substance, so that on the one hand its structural analysis could be continued and on the other its vitamin nature confirmed. However, in the course of our vitamin experiments we had used up the last remnants of our substance, and we had no chance of preparing the substance from adrenals, and, as already mentioned, every other material was unsuitable for large-scale work.

My town, Szeged, is the centre of the Hungarian paprika industry. Since this fruit travels badly, I had not had the chance of trying it earlier. The sight of this healthy fruit inspired me one evening with a last hope, and that same night investigation revealed that this fruit represented an unbelievably rich source of hexuronic acid, which, with Haworth, I re-baptized ascorbic acid. Supported on a large scale by the American *Josiah Macy Jr. Foundation*, it was still possible by making use of the paprika season, which was then drawing to a close, to produce more than half a kilogram, and the following year more than three kilogram of crystalline ascorbic acid. I shared out this substance among all the investigators who wanted to work on it. I also had the privilege of providing my two prize-winning colleagues P. Karrer and W. N. Haworth with abundant material, and making its structural analysis possible for them. I myself produced with Varga the mono-acetone derivative of ascorbic acid, which forms magnificent crystals, from which, after repeated dissolving and recrystallization, ascorbic acid can be separated again with undiminished activity. This was the first proof that ascorbic acid was identical with vitamin C, and that the substance's activity

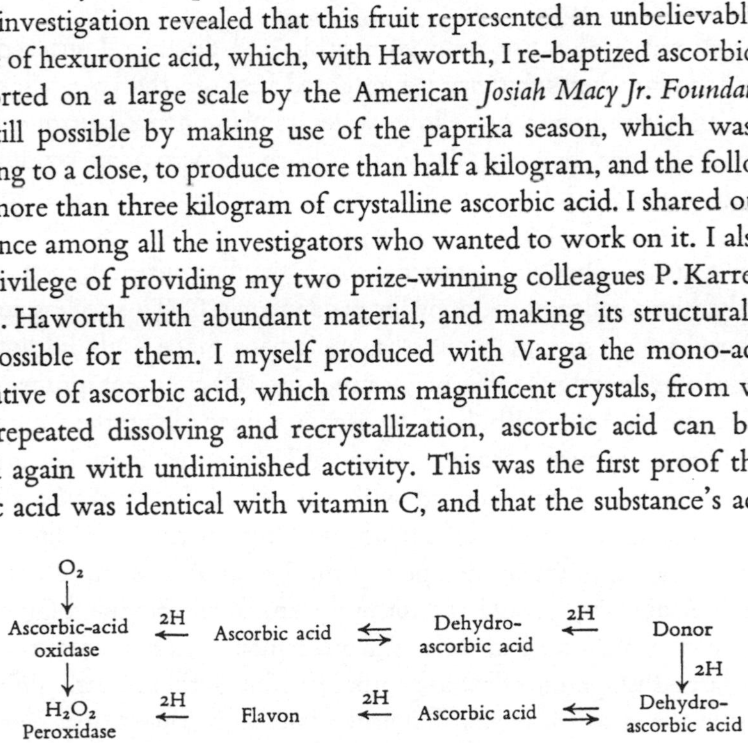

Fig. 4

was not due to an impurity. I do not wish to linger any more over this well-known story, which developed in such a dramatic fashion. Thanks to international collaboration, in the unbelievably short space of two years the mysterious vitamin C had become a cheap, synthetic product.

Returning to the processes of oxidation, I now tried to analyse further the system of respiration in plants, in which ascorbic acid and peroxidase played an important part. I had already found in Rochester that the peroxidase plants contain an enzyme which reversibly oxidizes ascorbic acid with two valencies in the presence of oxygen. Further analysis showed that here again a system of respiration was in question, in which hydrogen was oxidized by stages. I would like, in the interests of brevity, to summarize the end result of these experiments, which I carried out with St. Huszák.

Ascorbic acid oxidase oxidizes the acid with oxygen to reversible dehydro-ascorbic acid, whereby the oxygen unites with the two labile H-atoms from the acid to form hydrogen peroxide. This peroxide reacts with peroxidase and oxidizes a second molecule of ascorbic acid. Both these molecules of dehydro-ascorbic acid again take up hydrogen from the foodstuff, possibly by means of SH-groups.

But peroxidase does not oxidize ascorbic acid directly. I succeeded in showing that another substance is interposed between the two, which belongs to the large group of yellow, water-soluble phenol-benzol-γ-pyran plant dyes (flavone, flavonol, flavanone). Here the peroxidase oxidizes the phenol group to the quinone, which then oxidizes the ascorbic acid directly, taking up both its H-atoms.

At the time that I had just detected the rich vitamin content of the paprika, I was asked by a colleague of mine for pure vitamin C. This colleague himself suffered from a serious haemorrhagic diathesis. Since I still did not have enough of this crystalline substance at my disposal then, I sent him paprikas. My colleague was cured. But later we tried in vain to obtain the same therapeutic effect with pure vitamin C. Guided by my earlier studies into the peroxidase system, I investigated with my friend St. Rusznyák and his collaborators Armentano and Bentsáth the effect of the other link in the chain, the flavones. Certain members of this group of substances, the flavanone hesperidin (Fig. 5) and the formerly unknown eriodictyolglycoside, a mixture of which we had isolated from lemons and named citrin, now had the same therapeutic effect as paprika itself. It is still too early on in our experience for us to make any definitive statements. But it does seem that these substances possess great biological activity. They influence most obviously the capillary blood vessels, whose permeability and resistance suffer gravely in many disease states. These dyes are able to restore the state of affairs to normal, and to judge by the first experiences, it seems that these substances will enrich the doctor's inventory with a really useful new weapon for him to fight illnesses with. Our experiments made it probable that certain members of this group possess vitamin-like properties. For this reason I called the substance vitamin P. Unfortunately these vitamin-like properties have not yet been successfully demonstrated in a completely irreproachable and reproducible fashion.

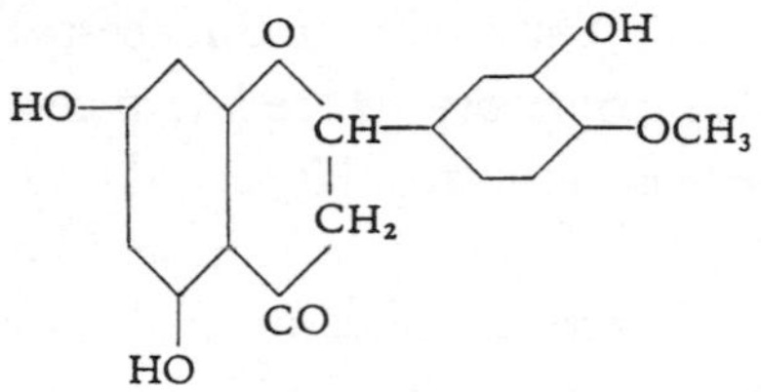

Fig. 5

Ladies and Gentlemen, I have tried to sketch out for you a rapid picture of my work. When I myself look back, I am always only aware of the distressing puniness of my efforts, as compared with the magnitude of Nature and of my problems. One circumstance, however, fills me always with the greatest happiness and gratitude, when I look back on my own struggles. From the moment I seized my staff, a novice in search of knowledge, and left my devastated fatherland to tread the wanderer's path – which has not been without its privations – as an unknown and penniless novice, from that moment to the present one, I always felt myself to belong to a great, international, spiritual family. Always and everywhere I found helping hands, friendship, cooperation and international solidarity. I owe it solely to this spirit of our science that I did not succumb, and that my endeavours are now crowned with the highest human recognition, the award of the Nobel Prize. This Nobel Prize, too, is but a fruit of this spirit, of this pan-human solidarity. I can but hope, my heart filled with gratitude, that this spirit may be preserved and that it may spread its bounteous rays beyond the limits of our knowledge, over the whole of humanity.

Biography

Albert von Szent-Györgyi was born in Budapest on September 16, 1893, the son of Nicolaus von Szent-Györgyi, a great landed proprietor and Josefine, whose father, Joseph Lenhossék, and brother Michael were both Professors of Anatomy in the University of Budapest. He matriculated in 1911 and entered his uncle's laboratory where he studied until the outbreak of World War I when he was mobilized. He served on the Italian and Russian fronts, gaining the Silver Medal for Valour, and he was discharged in 1917 after being wounded in action. He completed his studies in Budapest and then worked successively with the pharmacologist, G. Mansfeld at Pozsony, with Armin von Tschermak at Prague, where he studied electro-physiology, and with L. Michaelis in Berlin, before he went to Hamburg for a two-year course in physical chemistry at the Institute for Tropical Hygiene.

In 1920 he became an assistant at the University Institute of Pharmocology in Leiden and from 1922 to 1926 he worked with H. J. Hamburger at the Physiology Institute, Groningen, The Netherlands. In 1927 he went to Cambridge as a Rockefeller Fellow, working under F. G. Hopkins, and spent one year at the Mayo Foundation, Rochester, Minnesota, before returning to Cambridge. In 1930 he obtained the Chair of Medical Chemistry at the University of Szeged and in 1935 he also took the Chair in Organic Chemistry. At the end of World War II, he took the Chair of Medical Chemistry at Budapest and in 1947 he left Hungary to settle in the United States where he is Director of Research, Institute of Muscle Research, Woods Hole, Massachusetts.

Szent-Györgyi's early researches at Groningen concerned the chemistry of cell respiration. He described the interdependence of oxygen and hydrogen activation and made his first observations on co-dehydrases and the poly-phenol oxidase systems of plants. He also demonstrated the existence of a reducing substance in plant and animal tissues. At Cambridge and during his early spell in the United States, he isolated from adrenals this reducing substance, which is now known as ascorbic acid. Returning to Cambridge in 1929, he later described the pharmacological activity of the nucleotides with Drury.

On his return to Hungary, he noted the anti-scorbutic activity of ascorbic acid and discovered that paprika (*capsicum annuum*) was a rich source of vitamin C. His persistent studies of biological oxidation led to the recognition of the catalytic function of the C_4-dicarboxylic acids, the discovery of «cytoflav» (flavin) and a recognition of the biological activity and probable vitamin nature of flavanone (vitamin P).

In 1938 he commenced work on muscle research and quickly discovered the proteins actin and myosin and their complex. This led to a reproduction of the fundamental reaction of muscle contraction which formed the foundation of muscle research in the following decades. The preservation of biological material in glycerine, which has had extensive application including agricultural use in the preservation of sperm, has resulted from his more recent work. He has also developed the use of rabbit psoas muscle as an experimental material, published theories on the problems of energetics and investigated the regulation of growth and cell membrane potential, and the hormonal function of the thymus gland.

Szent-Györgyi, a member of many scientific societies, is a Past President of the Academy of Sciences, Budapest, and a Vice-President of the National Academy, Budapest. He was Visiting Professor, Harvard University in 1936 and Franchi Professor, University of Liège, 1938. He received the Cameron Prize (Edinburgh) in 1946 and the Lasker Award in 1954. His many publications include *Oxidation, Fermentation, Vitamins, Health and Disease* (1939); *Muscular Contraction* (1947); *The Nature of Life* (1947); *Contraction in Body and Heart Muscle* (1953); and *Bioenergetics* (1957).

Szent-Györgyi married Cornelia Demény, daughter of the Hungarian Postmaster-General, in 1917. During the 1930's he was actively anti-Nazi and during World War II he became a Swedish citizen – he was given extensive help by the Swedish Embassy in Budapest. In 1941, he married Màrta Borbiro, a co-worker at Woods Hole: they have one daughter.

He is interested in sport of all kinds, his favourites being sailing and alpinism.

Physiology or Medicine 1938

CORNEILLE JEAN FRANÇOIS HEYMANS

«for the discovery of the role played by the sinus and aortic mechanisms in the regulation of respiration»

Physiology or Medicine 1938

The following account of Heymans's work is by Professor G. Liljestrand, member of the Staff of Professors of the Royal Caroline Institute

For over a century now it has been known that respiration in vertebrates in-cluding man is regulated from a small area in the medulla, known as the respiratory centre. From this centre nervous impulses of variable strength travel along the spinal cord and motor nerves, and reach the respiratory muscles. These muscles then come into play to produce respiratory move-ments. It is a well-known fact that respiratory movements can be intention-ally modified, particularly during speech or singing, but respiration can also be influenced in different ways by mechanisms without conscious volition. For instance, entering a cold bath stops respiration for a few moments, pain tends to increase respiration. A sudden expansion of the lungs stops inspira-tion and induces expiration. Similarly, when air is withdrawn from the lungs by suction, expiration is halted and inspiration is induced. These facts were revealed by Hering and Breuer, and demonstrate the way in which reflexes influence respiration. Following the centripetal nervous pathways informa-tion is transmitted to the respiratory centre which reacts according to the nature of the information by initiating corresponding modifications of respi-ration. The chemical composition of the blood also has an effect on respira-tion. This is the essential factor which controls the degree of ventilation, i.e. the quantity of air which passes through the lungs. If the tension of carbon dioxide in the blood increases, or if the oxygen tension is reduced, the venti-lation will increase. In this way respiration will adapt to the great variations in the requirements of the body, which themselves are due to the intensity of metabolic processes in the organism. Prior to Heymans's work, it was thought that the blood acted directly on the respiratory centre.

In 1927, together with the late Professor J. F. Heymans, his father and teacher, Heymans studied the respiratory reflexes which are transmitted by the tenth cranial nerve, i.e. the vagus or pneumogastric nerve. They made use of a technique which had been developed by the elder Heymans in collab-oration with De Somer in 1912. This technique made it possible to keep alive the completely isolated head of a dog by perfusion of blood from another dog, while the body also remained alive with the help of artificial respiration. By ensuring that the only communication between the head and the rest of the body was provided by the two vagus nerves (and the depressor or aortic nerves which reach them from the aortic arch) the necessary conditions were produced for studying the links between the head and body dependent upon these nervous pathways. The two Heymans were thus able to demonstrate that expansion of the lungs stopped the respiratory movements of the head in the expiratory position, which was indicated by the recording of laryngeal and alae nasi movements, while collapse of the lungs immediately induced inspiratory-type respiration in the head. These experiments had provided decisive proof that the still controversial respiratory reflexes described by Hering and Breuer did in fact exist. It was also demonstrated that interruption of the artificial respiration applied to the body with resulting accumulation of carbon dioxide and decrease in oxygen contents rapidly led to an increase in respiratory movements of the head. On the other hand, hyperventilation of the body with free air, which produced increased excretion of carbon dioxide from the body, and increased the oxygen tension, stopped the respiratory movements of the head. After section of the vagus nerves, none of these effects were produced. Proof had therefore been produced for the first time that the vago–depressor nerves were capable of transmitting chemical stimuli arising peripherally. Consequently, if hyperventilation of the lungs was carried out, using a mixture containing a high proportion of carbon dioxide and a low proportion of oxygen, so that in spite of increased ventilation the tension of carbon dioxide in the lungs continued to increase while the oxygen tension decreased, the respiratory movements of the head, far from being reduced, tended to increase. The effects of hyperventilation with air could not therefore be explained as mechanical phenomena. They must have resulted from the suppression of the chemical stimulus to the nerve terminals of the vago-depressor nerves. By a careful and technically ingenious analysis, it was shown that these reflexes originating from chemical stimuli arise from the heart itself and the portion of the aorta nearest to it. Respiration could also be inhibited by high blood pressure in the body, as Heymans's experiments showed.

This intrinsically important discovery is of all the more interest in view of Hering's discovery (1923–1924) that the area known as the carotid sinus, on the internal carotid at its junction with the common carotid artery, has an analogous function to that of the areas in the aorta from which the depressor nerves arise. Thus an increase in arterial pressure in the internal carotid stimulates a number of nerve terminals in the walls of the sinus and produces a reflex which is transmitted by the ninth pair of cranial nerves, the glosso-pharyngeal nerves, and reaches the territories of the vagus and vaso-motor nerves. This produces dilatation in certain vascular areas and a slowing up of the cardiac rhythm. The original hypertension is thus counteracted, at least to a certain extent. The area innervated by the depressor nerves and the carotid sinus is therefore part of a common system, sometimes called the bridles of the blood pressure.

Heymans also studied with great precision the reflexes arising from the sinus area. Thus, together with a number of collaborators, he closely ex-amined the mechanism by which the repercussions of these reflexes act on the cardiac rhythm and on the blood pressure. As is the case of reflexes governed by the depressor nerves, he found that the cardiac rhythm was slowed by increasing the tone of the branches of the vagus nerves, which have a delaying effect on the heart beat, and also by the reduction of activity in the antagonist stimulant nerves, whose function is to increase the cardiac rate. He further showed the role played by the different vascular areas in the modifications of blood pressure when the sinus pressure is increased or low-ered. He also indicated that the suprarenal medulla was probably influenced by reflexes arising from the carotid sinus, bringing it to increase or decrease adrenaline secretion into the blood.

Systematic research was carried out with the aim of discovering if respi-ratory reflexes could also arise from the sinus. On this specific subject a number of significant facts had already been noted. For instance, Sollmann and Brown had observed that stretching of the common carotid artery ini-tiated respiratory reflexes, and others, among them Hering and Heymans himself, had noted that an increase of pressure in the carotid artery could inhibit respiration, while a lowering of pressure in the sinus area stimulated respiration. In 1930, Heymans and Bouckaert were able to show that even slight variations in pressure could lead to marked changes in the respiration, and that these changes were due to a reflex mechanism.

Research was then concentrated on establishing whether the sinus area was sensitive as regards chemical stimuli in the same way as the area covered by the depressor nerves. In a number of papers, first in collaboration with Bouckaert and Dautrebande in 1930–1931, and then also with von Euler, Heymans gave irrefutable evidence that chemical stimuli played an important part in the control of blood pressure and respiration. In his experiments blood containing varying proportions of carbon dioxide and oxygen, and varying H-ion concentrations, was pumped into the sinus area. Blood could also be transfused from another dog which was inhaling mixtures with a given proportion of gases so as to obtain the required chemical changes in the blood. These experiments have shown that the increase in carbon dioxide tension, or the decrease of oxygen contents can increase respiration by acting upon the sinus area. By cutting the nerve fibres which travel from the sinus to the medulla, it was demonstrated that the increase in respiration after inhalation of air of low oxygen content, did not occur at all, and that consequently the stimulating reaction depended entirely on the sinus reflex. A similar experiment demonstrating the role of carbon dioxide showed that this gas stimulated respiration both by direct action on the respiratory centre and indirectly by means of the sinus mechanism.

Thus Heymans's work led to the theory that four different types of reflexes could originate in the sinus area. On the one hand, the circulation, or more precisely the blood pressure and cardiac rhythm, and the respiration could be modified by pressure changes in the sinus and, on the other hand, these two groups of physiological functions could also be modified by variations in the chemical composition of the blood. Heymans went on to make further contributions to our knowledge in this field. Since the end of the 18th century we know of the existence of a curious structure in the region of the sinus, the glomus caroticum or carotid body which, in man, extends over only a few millimetres. The glomus consists of a small mass of very fine intertwining vessels arising from the internal carotid and enclosing various different types of cells. It has been considered by some as being a sort of endocrine gland similar to the medulla of the suprarenal glands. De Castro, however, in 1927 demonstrated that the anatomy of the glomus could in no way be compared to that of the suprarenal medulla. De Castro suggested rather that the glomus was an organ whose function was to react to variations in the composition of the blood, in other words an internal gustatory organ with special « chemo-receptors ». In 1931, Bouckaert, Dautrebande, and Heymans undertook to find out whether these supposed chemo-receptors were responsible for the respiratory reflexes produced by modifications in the composition of the blood. By localized destruction in the sinus area they had been able to stop reflexes initiated by pressure changes, but respiratory reflexes could still continue to occur in answer to changes in the composition of the blood. Other experiments showed that Heymans's concepts on the important role played by the glomus in the reflex control of respiration by the chemical composition of the blood were undoubtedly correct. Recently it has been shown that similar chemo-receptors located in the area covered by the depressor nerves (glomus aorticum) have an analogous structure to that of the glomus caroticum (Comroe, 1939). It seems likely however that this depressor nerve mechanism plays only a small part in the respiratory reflexes produced by a marked lowering of oxygen contents and that the essential pathway is via the glomus caroticum. No doubt remains that the whole system plays an important part in the regulation mechanism as regards respiration.

By using modern amplification techniques, it has become possible to record exceedingly small variations in electrical potential within the body and to carry out research on action potentials detected in nerve fibres during transmission of an impulse. Even in the case of the smaller branches of the glossopharyngeal nerve which originate in the sinus area, action potentials of this type have been detected (Bronk, 1931). In 1933, Heymans and Rijlant demonstrated that these potentials were of two different kinds, the greater being produced by blood pressure in the sinus, the other by chemical stimulation in the glomus. We are thus in possession of a solid basis for further research as regards these two types of potential under various conditions.

Heymans not only discovered the role, hitherto quite unknown, of certain organs (glomus caroticum and glomus aorticum), he also greatly enlarged our field of knowledge concerning the regulation of respiration. He showed that the various methods used for stimulating respiration had quite different mechanisms. In certain cases (lobeline, nicotine, cyanide, sulphide, etc.) the drug acts on the glomus, in others (e.g. Cardiazol) it acts by central stimulation and again in other cases (e.g. Coramine) it acts centrally and peripherally. It seems likely that this increase in our knowledge of the chemo-regulation of respiration will also be of great use in research on a number of diseases.

The presentation of the Prize for Physiology or Medicine to Professor C. Heymans took place in Ghent on the 16th of January, 1940.

CORNEILLE J. F. HEYMANS

The part played by vascular presso- and chemo-receptors in respiratory control

Nobel Lecture, December 12, 1945

The physiology of respiratory control, of the adaptation of pulmonary gas exchanges to the energy requirements of the organism, remains a fundamental problem on which many research workers are engaged.

In this lecture I am privileged to describe contributions which my laboratory has made to the study of a number of physiological, physiopathological and pharmacological mechanisms which act on and control the functioning of the respiratory centre and thereby affect pulmonary ventilation and pulmonary gas exchanges.

It has been known for some time that variations in blood pressure affect respiration. An increase in blood pressure inhibits respiration and sudden marked hypertension may even produce apnea. We also know that hypotension increases respiration. It was generally believed that this interaction between blood pressure and respiration involved a direct action on the respiratory centre exerted by either the blood pressure or the rate of flow in the cerebral circulation. The experiments which I shall now describe suggest that this classical theory should be reconsidered and rejected.

Since 1924, together with my father J. F. Heymans, my first and best teacher, I have been engaged in research projects, using the following experimental technique (Fig. 1): the completely isolated head of a Dog B under chloralose anaesthesia is perfused from a similar Dog A by means of anastomoses between the two common carotids of Dog A and the cephalic extremities of the common carotids of Dog B, and between the external jugular veins of the head B and the corresponding jugular veins of Dog A. The isolated and perfused head of Dog B is connected to the trunk (kept alive by its own circulation and artificial respiration) by the vago-depressor nerves. The vascular anastomoses between the isolated head B and the donor animal A are established by using Payr cannulas (Fig. 2) which prevent coagulation at the point of anastomosis without making use of anti-coagulants. Under these experimental conditions, the cephalic and cerebral circulation of Dog B is entirely independent of the circulation in the trunk,

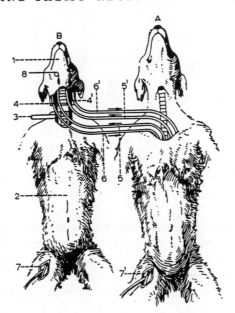

Fig. 1. Diagram of perfusion of an isolated head after J. F. and C. Heymans.
(Perfusion of the isolated head of Dog B by Dog A.)
(*1*) isolated head of Dog B; (*2*) isolated trunk of Dog B; (*3*) tracheal cannula;
(*4*) right vagus nerve of Dog B; (*4′*) left vagus nerve of Dog B; (*5*) anastomosis
between the cephalic extremity of the external jugular vein of Dog B and the cardiac
extremity of the external jugular vein of Dog A (right side); (*5′*) anastomosis between
the cephalic extremity of the external jugular vein of Dog B and the cardiac extremity
of the external jugular vein of Dog A (left side); (*6* and *6′*) anastomosis between the
cephalic extremity of the common carotid of Dog B and the cardiac extremity of the
common carotid of Dog A (right side); anastomosis between the cephalic extremity
of the common carotid of Dog B and the cardiac extremity of the common carotid
of Dog A (left side); (*7*) femoral blood pressure in Dog B; (*7′*) femoral blood pressure in Dog A; (*8*) respiratory movements in the isolated head of Dog B.

but the nervous pathways between head and trunk supplied by the vago-
aortic nerves remain intact. The respiratory movements which indicate the
activity of the respiratory centre are recorded on the isolated perfused head
B which remains in a satisfactory living state.

In some experiments the donor Dog A was replaced by an artificial heart-
lung apparatus.

It was noted first of all that hypotension in the isolated trunk circulation
of Dog B stimulates the respiratory centre of the perfused head B. Conversely hypertension in the trunk B inhibits the respiratory centre of the perfused

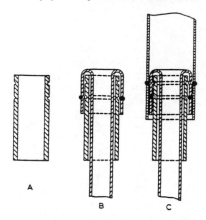

Fig. 2. Payr's cannula for anastomosis between blood vessels.
(*A*) Longitudinal section of a Payr cannula; (*B*) one of the vessels attached to the cannula; (*C*) the second vessel attached to the cannula.

N.B. The anastomosis between vessels is established with intima in contact with intima, so that the blood does not come into contact with the cannula or the sutures.

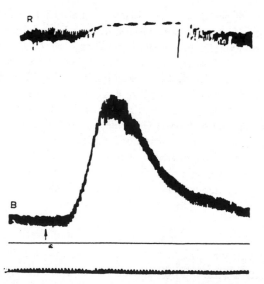

Fig. 3. The isolated head of Dog B is perfused from Dog A. The only link between the isolated head of Dog B and his trunk, kept alive by artificial respiration, is provided by the vago-aortic nerves.

Graph R: Respiratory movements of the isolated and perfused head of Dog B. *Graph B*: Blood pressure of trunk B.

At ↑: intravenous injection of adrenaline into the trunk of Dog B. Hypertension in the trunk B and reflex inhibition (apnea) of the respiratory centre of the perfused head of Dog B.

head B. When a hypertensive dose of adrenaline is injected into the trunk B it was found (Fig. 3) that respiratory movements of the isolated and perfused head B were totally inhibited, producing apnea.

These experimental results therefore demonstrated for the first time that arterial hypotension limited to the trunk produces reflex stimulation of the respiratory centre, while arterial hypertension in the trunk induces reflex inhibition of the respiratory centre, sometimes to the extent of total apnea.

The centripetal pathways of these respiratory reflexes, which are induced by variations of blood pressure in the trunk, are supplied by the vago-aortic nerves which constitute the only link between the trunk B and the isolated and perfused head B.

The next problem was to locate with greater precision the place of origin of these respiratory reflexes. Together with J. F. Heymans we had first of all noted that the respiratory reflexes under consideration persisted after all nervous connections between the trunk B and the perfused head B had been cut, with the exception of those of the cardio-aortic area.

In other experiments the isolated head of Dog B was perfused from Dog A and the isolated heart-lung preparation or the isolated cardio-aortic area of the trunk B was perfused from a third Dog C. The only link between the head B and the isolated heart-lung preparation or the isolated cardio-aortic area of trunk B was provided by the vago-aortic nerves. By using these experimental techniques we were able to observe that when arterial pressure was increased either in the heart-lung preparation or in the isolated cardio-aortic area of trunk B, reflex inhibition of the respiratory centre occurred; conversely, when arterial pressure was reduced in the cardio-pulmonary circulation or in the cardio-aortic circulation of the trunk B, the resulting hypotension limited to these areas initiated a reflex stimulation of the respiratory centre of the perfused head B. These respiratory reflexes did not occur however when the arterial pressure was varied in the isolated pulmonary circulation of the trunk B; in this case the vagus nerve provided the only link between the perfused lungs of the trunk B and the perfused head B.

Thus, this group of experiments carried out between 1924 and 1927 showed *that arterial hypertension in the cardio-aortic vascular area inhibits the activity of the respiratory centre by a reflex mechanism, while arterial hypotension in the same area has a stimulatory reflex effect on the activity of the respiratory centre.*

After Magendie's and Cooper's old experiments it has long been known that clamping of the common carotid arteries produces hyperpnoea, while unclamping of these arteries inhibits respiration.

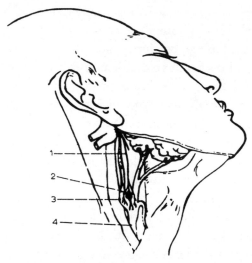

Fig. 4a. Diagram of the carotid sinus in man.
(1) carotid sinus nerve; (2) glomus caroticum (reflexogenic chemo-receptor); (3) beginning of the internal carotid where the reflexogenic presso-receptors are mainly concentrated; (4) common carotid artery.

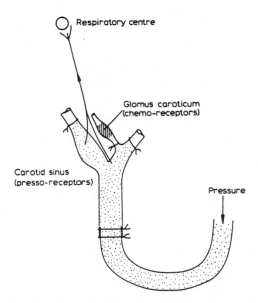

Fig. 4b. Diagram showing the technique by which endo-vascular pressure can be modified in the carotid sinus with nervous links intact but previously isolated as regards circulation and deprived of chemo-receptors.

Using the isolated and perfused head technique, we found that hypotension in the isolated cephalic circulation stimulates the respiratory centre, whereas hypertension in the same area produces respiratory inhibition to the point of apnea. We undertook to examine the mechanism of the interaction between cephalic blood pressure and the activity of the respiratory centre.

According to the classical theory, these respiratory reactions were considered to derive from the direct action of fluctuations in the blood flow, or irrigation, on the respiratory centre. However, in 1900, Siciliano and Pagano had already correctly noted that, although occlusion of the common carotids does in fact give hyperpnoea, occlusion of the efferent branches of the common carotid artery produces no effect on the respiratory centre. These investigators therefore rejected the theory that this hyperpnoea was of central origin and proposed the hypothesis that it was due to a carotid reflex mechanism.

Since 1924 the investigations of H. E. Hering and his associates, particularly E. Koch, and our researches undertaken with our associates, among these J. J. Bouckaert, P. Regniers, L. Dautrebande, and U. S. von Euler, and the work of Moissejeff, G. Liljestrand, Y. Zotterman, C. F. Schmidt, R. Gesell, and other investigators, have made it possible to show that the carotid sinuses, i.e. the arterial regions located in the area where the common carotid artery bifurcates into internal and external carotids and occipital artery, contain receptors, as does the homologous cardio-aortic zone, who, by a reflex mechanism, act upon and regulates the activity of the cardio-vascular centres and of the respiratory centre. Fig. 4a presents the position of the human carotid sinus.

A number of experimental procedures have enabled us to show that an increase in endo-vascular pressure in the carotid sinus produces an inhibition of the respiratory centre to the point of apnea, by a reflex arising from the presso-receptors, and that hypotension in the carotid sinus produces reflex stimulation of the respiration. Fig. 4b shows a diagram of one of the experimental procedures in which fluctuations in the endo-vascular pressure act exclusively on the presso-receptors of the carotid sinus. Fig. 5 shows reflex hyperpnoea produced by hypotension in the carotid sinus, and Fig. 6 shows reflex respiratory inhibition produced by an increase of blood pressure acting on the presso-receptors of the carotid sinus. It should be noted that this respiratory inhibition is particularly evident in animals in which the vago-aortic nerves have been cut.

Apnea induced by intravenous injection of adrenaline is also related to arterial blood pressure which acts upon the respiratory centre by means of

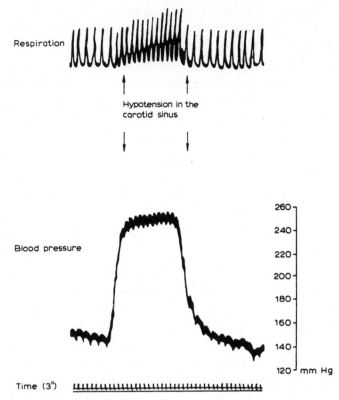

Fig. 5. Hyperpnea in a dog produced by hypotension in the carotid sinus.

the reflex pathways supplied by the presso-receptors of the cardio-aortic and carotid sinus zones.

The several experimental observations summarized above thus demonstrate that *variations in arterial blood pressure act on the respiratory centre by a reflex mechanism involving endo-vascular presso-receptors located in the cardio-aortic zone and in the carotid sinus.* Fig. 7 gives a diagrammatic representation of the points of emergence and pathways of the presso-receptor nerves arising from the cardio-aortic zone and the carotid sinus. It should be remembered at this point that the same presso-receptor nerves also act, by reflex mechanism, on the cardio-vascular centres and in this way possess a regulatory function as regards the systemic blood pressure and the circulation in general.

The next problem was to determine whether or not the respiratory centre is directly affected by variations in arterial blood pressure and arterial rate of flow? Numerous experimental observations carried out by means of different techniques gave a negative answer to this question. The activity of neither the respiratory nor the cardio-vascular centres was modified by (1) clamping the efferent arteries of the carotid sinus, (2) clamping the common carotid after denervation of the carotid sinus, (3) clamping of the vertebral arteries, even after prior clamping of the efferent arteries of the carotid sinus. Fig. 8 shows a graph of such an experiment.

In other experiments an isolated head was perfused and it was observed that considerable reduction in cerebral blood flow beyond physio-pathological limits was necessary before a stimulatory response was obtained by direct action on the respiratory and cardio-vascular centres.

All these observations lead to the conclusion that variations in arterial blood pres-

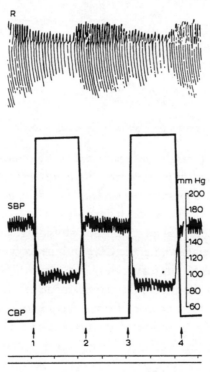

Fig. 6. Dog under chloralose anaesthesia. The vagus nerves are cut.
(R) pneumogram; (SBP) systemic blood pressure; (CBP) carotid sinus blood pressure
(chemo-receptors excluded). Time in 5-second intervals.
At *point 1*: increase in the endo-vascular blood pressure within the carotid sinus – marked inhibition of respiration. At *point 2*: decrease in the endo-vascular blood pressure within the carotid sinus – hyperpnoea. At *point 3*: same as at 1. At *point 4*: same as at 2.

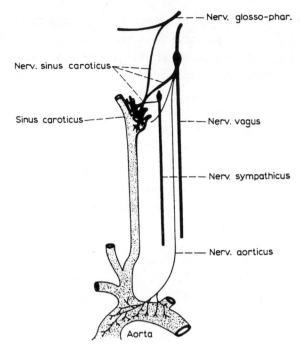

Fig. 7. Diagram of the nervous pathways of the cardio-aortic and carotid sinus zones.

sure exert an effect on the respiratory centre and on the cardio-vascular centres ex-
clusively by a reflex mechanism involving the aortic and carotid sinus receptors.
Variations in arterial blood pressure and cerebral blood flow, within physio-patholo-
gical limits, exert no direct effect on the respiratory and cardio-vascular centres.

It had long been accepted as an established physiological fact that the chemical composition of the blood, i.e. the CO_2 and the oxygen contents, exerted
a direct effect on the respiratory centre by controlling and adapting its
activity in relation to the metabolic requirements of the organism. It was
also accepted that a large number of pharmacological substances exerted a
stimulating effect by direct action on the respiratory centre. These physiological and pharmacological concepts have recently undergone considerable
modifications.

Making use of the technique involving perfusion of an isolated head linked
to the trunk by only the vago-aortic nerves, J.F. Heymans and ourself in
1926 observed that asphyxia or hypoxemia, limited to the sytemic circulation, produced reflex stimulation of the respiratory centre, while systemic
hypercapnia produced reflex respiratory inhibition. Fig. 9 gives a graphic representation of one of these experiments which demonstrates reflex stimula-

tion of the respiratory centre of the isolated head, induced by asphyxia in the trunk which is connected to the perfused isolated head only by the vago-aortic nerves.

The different techniques described above made it possible to demonstrate

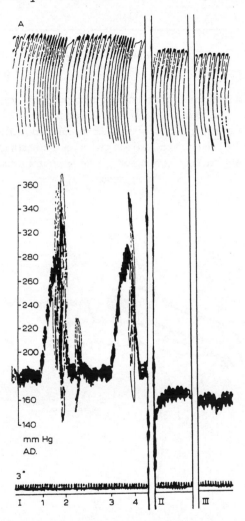

Fig. 8. Dog under chloralose anaesthesia. (*A*) pneumogram; (*A.D.*) femoral blood pressure. Time in 3-second intervals.
Section I (at *points 1* and *3*): clamping of both common carotid arteries – hyperpnoea and hypertension. *Section I* (at *points 2* and *4*): unclamping of both common carotid arteries – respiratory inhibition and hypotension. *Section II*: normal respiration and blood pressure. *Between Sections II and III*: the efferent blood vessels from the two carotid sinuses are clamped – no hyperpnoea, no hypertension.

that the reflex impulses from the systemic circulation, produced by conditions of hypercapnia, hypocapnia, or hypoxemia in the blood, arise in the cardio-aortic area which therefore contains reflexogenic chemo-receptors. In the course of the same experiments it was also observed that certain pharma-

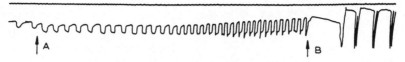

Fig. 9. Respiratory movements of an isolated and perfused head connected to the trunk only by the vagi-aortic nerves.

At *point A*: asphyxia in the trunk induces reflex stimulation of the respiratory centre in the perfused head. At *point B*: the aortic vagi-nerves connecting the trunk and isolated and perfused head are cut; stimulation of the respiratory centre of the perfused head ceases.

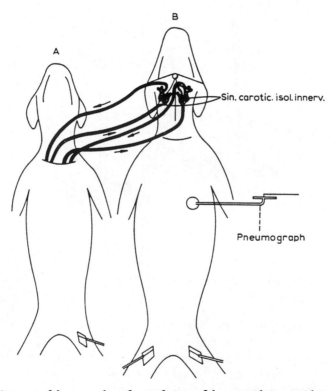

Fig. 10. Diagram of the procedure for perfusion of the carotid sinus with intact nervous connections and isolated circulation in Dog B from Dog A.

cological substances, such as nicotine, produce reflex stimulation of the respiratory centre by means of the same cardio-aortic chemo-receptors.

Our later experiments on the physiological role of the carotid sinus, a vascular zone similar to the cardio-aortic zone, led us to investigate if these arterial areas also showed reflex chemo-sensitivity. Experimental findings showed that this hypothesis was correct.

In this work a number of experimental procedures were used. Among these in particular was a technique involving perfusion of the carotid sinus which was isolated from the systemic circulation, but in which the nerve connections were preserved, either by a donor animal (Fig. 10) or by an artificial heart-lung preparation (Fig. 11).

These experiments demonstrated first of all that the carotid sinus is sensitive to the physiological stimulants contained in the blood, i.e. CO_2, oxygen, and hydrogen ions.

Hyperapnic blood acts upon the carotid sinus area and induces reflex hy-

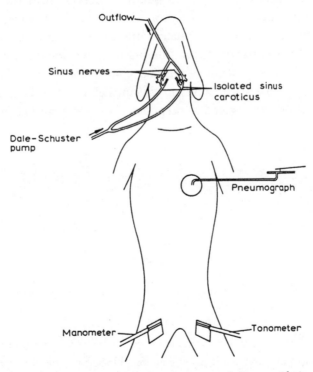

Fig. 11. Diagram of the procedure for the perfusion of the carotid sinus with intact nervous connection and isolated circulation. Perfusion is carried out by means of a Dale-Schuster pump.

perpnoea, while hypocapnic blood produces reflex respiratory inhibition. Hypoxic blood also acts on the carotid sinus area to produce reflex stimulation of the respiratory centre. In the same way hydrogen ions also stimulate the respiratory centre by acting on the carotid sinus chemo-receptors.

These experimental findings were confirmed by many workers, in particular, R. Gayet and D. Quivy, C. F. Schmidt, J. H. Comroe, R. Gesell, T. Bernthal, G. Liljestrand, U. S. von Euler, A. Samaan, G. Stella, Y. Zotterman, Samson Wright, etc.

What is the part played by the chemo-sensitivity of the cardio-aortic and carotid sinus zones in the physiological and physio-pathological regulation of the respiratory centre?

In the case of hypoxemia, our experimental findings and those of A. Samaan, G. Stella, U. S. von Euler, G. Liljestrand, Y. Zotterman, R. Gesell, C. F. Schmidt, and T. Bernthal demonstrated the marked sensitivity to oxygen want of the carotid sinus chemo-receptors.

On the other hand, if the chemo-sensitive nerves from the aortic and carotid sinus areas are cut, the inhalation of a mixture containing a low proportion of oxygen no longer stimulates respiration or blood pressure, but on the contrary produces progressive inhibition of respiration and a drop in blood pressure (Fig. 12), while in the case of an animal in which the chemo-sensitive nerves of the aortic and carotid sinus areas are intact, an identical oxygen want produces marked hyperpnoea and hypertension (Fig. 13).

Respiration
Blood-pressure
Hypoxaemia

Respiration
Blood-pressure
Hypoxaemia

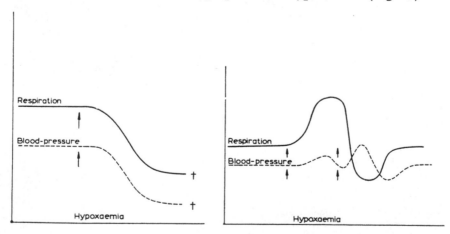

Fig. 12. Inhibition of respiration and circulation (blood pressure) by hypoxaemia in a dog from which the aortic and carotid sinus chemo-receptors have been excluded.
Fig. 13. Respiratory and circulatory (blood pressure) reactions to hypoxaemia in a normal dog.

These findings were observed in both anaesthetized and non-anaesthetized animals and clearly demonstrate that a lack of oxygen stimulates respiration and produces hypertension by an essentially reflex mechanism involving the chemo-receptors of the aortic and carotid sinus areas. Hypoxaemia, which acts directly on the medullary centres, exerts a depressive, not a stimulating effect. Only extreme anoxia or anaemia can exerts a direct stimulating effect on the respiratory and cardio-vascular centres.

The experiments carried out by C. F. Schmidt and J. H. Comroe proved, in addition, that it is essentially the oxygen tension and not the saturation in oxygen in arterial blood which acts on the aortic and carotid sinus chemo-receptors. This explains why hyperpnoea does not occur in cases of carbon monoxide poisoning and in cases of anaemia or methemoglobinaemia. The oxygen tension in arterial blood is, indeed, normal in such cases. The same workers also demonstrated the important part played by the chemo-sensitivity of the aortic and carotid sinus zones in the reflex stimulation of the respiratory centre when it is depressed by the direct effect exerted by a number of narcotic drugs, such as morphine, ether, and barbiturates. In these cases, hypoxaemia is responsible for this continued reflex stimulation of the respiratory centre, by means of the chemo-sensitive nerves.

The fundamental role of the aortic and carotid sinus chemo-receptors in the respiratory and circulatory reactions arising from lack of oxygen due to low atmospheric pressure at high altitudes was also demonstrated by a number of investigators.

On the question of regulation of the activity of the respiratory centre by CO_2 all workers confirm our experimental findings, viz. that the CO_2 contents of the blood acts upon and stimulates the respiratory centre and cardio-vascular centres by a reflex mechanism involving the aortic and carotid sinus chemo-receptors. The experimental exclusion of these reflexogenic chemo-receptors does not, however, as in the case of oxygen want, prevent the regulating effect of CO_2 on the respiratory centre itself.

What are the respective roles played by the chemo-sensitive reflex mechanism and the direct central mechanism in the regulation by CO_2 of the activity of the respiratory centre? This is still a controversial question.

Basing ourselves particularly on the fact that the chemo-receptors are extremely sensitive to CO_2 and on the fact that CO_2, by acting on the chemo-receptors, can produce and maintain hypercapnia in spite of the central effect of simultaneous decreased CO_2 in the blood, and also on the fact that experimental exclusion of the chemo-receptors increases the alveolar concentration of CO_2, we have suggested that CO_2 exerts an effect on respiration primarily by the chemo-sensitive reflex mechanism, and secondarily by the direct central mechanism (Fig. 14). This hypothesis has received the support of U.S. von Euler, G. Liljestrand, Y. Zotterman, Samson Wright, R. Gesell, T. Bernthal. C. F. Schmidt and J. H. Comroe, however, favour the opposite hypothesis that CO_2 regulates the activity of the respiratory centre primarily by the direct central mechanism and secondarily by the chemo-sensitive reflex mechanism.

The aortic and carotid sinus chemo-receptors are not only sensitive to physiological chemical stimulants but also to a number of pharmacological substances. This was demonstrated in 1926 with J. F. Heymans, in the case of the cardio-aortic zone.

Our experiments carried out later with J. J. Bouckaert, L. Dautrebande, U. S. von Euler, S. Farber, A. Samaan, Shen, Donatelli, Marri and other co-workers showed that a number of pharmacological substances such as nicotine, lobeline, cyanide, acetylcholine, and other choline derivatives, and potassium sulphide produced intense stimulation of the chemoreceptors of the carotid sinus and that they therefore stimulate the respiratory and cardiovascular centres by a reflex mechanism. The direct central effects of these substances, on the other hand, are either absent or depressive, or stimulating in very high doses. These findings were obtained by means of a number of experimental procedures as perfusion of an isolated head, the sole link with the trunk being provided by the vago-aortic nerves, perfusion of a carotid

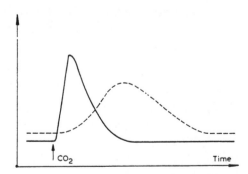

Fig. 14. Diagram of the threshold and intensity of stimulation of respiration produced by the reflex effects of CO_2 on the chemo-sensitive receptors and by the direct central effects of CO_2.
Threshold and intensity of respiratory reactions to CO_2: — reflex chemo-sensitive origin; - - - direct central origin.

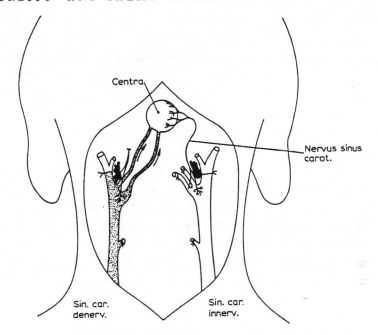

Fig. 15. Diagram of the procedure by which the circulation in the normally innervated glomus caroticum is isolated on one side. The opposite carotid sinus is denervated and the external carotid artery ligatured; the circulation in the common carotid is directed towards the nervous centres.

sinus, isolated from the systemic circulation but with nervous connections intact, alternate injection of the same pharmacological substance into the blood stream of a normal common carotid and a common carotid from which chemo-receptors have been eliminated (Fig. 15).

Fig. 16 shows the reflex stimulation of the respiratory centre and of the cardio-inhibitor vagus centre by acetylcholine in contact with the carotid sinus chemo-receptors. It should be added that acetylcholine directly stimulates the respiratory and cardio-inhibitor centres only when very high doses are administered. If an anticholinesterase (neostigmine) is previously administered, only a slight reinforcement is obtained of the direct central effect of acetylcholine on the respiratory and cardio-inhibitor centres. It should also be noted in this connection that neither does neostigmine reinforce the respiratory and cardio-vascular reflexes arising from the cardio-aortic and carotid sinus zones. This finding does not support the hypothesis that a cholinergic mechanism is involved in the respiratory and cardio-vascular reflexes arising from the presso- and chemo-sensitive aortic and carotid sinus zones.

The reflexogenic chemo-sensitivity of the carotid sinus in regard to a large number of pharmacological substances has been confirmed by many investigators, in particular R. Gesell, T. Bernthal, U. S. von Euler, G. Liljestrand, Y. Zotterman, Samson Wright, C. F. Schmidt, J. H. Comroe, etc., who made important contributions to the study of this new chapter in pharmacology.

What is the anatomical location of the presso- and chemo-receptors in the cardio-aortic and carotid sinus areas?

Histological research carried out by de Castro, Meyling and Gosses, and our own experimental findings, obtained with J. J. Bouckaert and L. Dautrebande in particular, has led to the locating of the carotid sinus chemo-receptors in the glomus caroticum and of the presso-receptors in the walls of the large arteries arising from the carotid artery (Fig. 17).

Various experimental techniques have made it possible to dissociate the chemo-receptor zones from the presso-receptor zones in the carotid sinus.

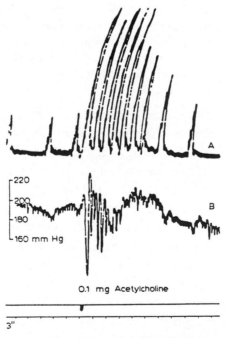

Fig. 16. *Graph A*: pneumogram of a dog under chloralose anaesthesia. *Graph B*: blood pressure and cardiac rate. Time in 3-second intervals.
Acetylcholine (0.1 mg) is applied to the chemo-receptors of the glomus caroticum. Marked reflex hyperpnoea and bradycardia.

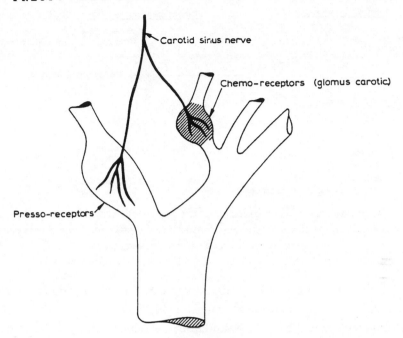

Fig. 17. Diagram showing the site of the reflexogenic presso- and chemo-receptors in the carotid sinus.

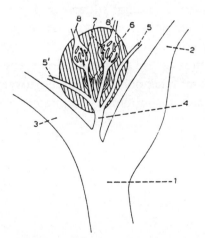

Fig. 18. Diagram of the carotid sinus and glomus caroticum.
(*1*) common carotid artery; (*2*) internal carotid artery; (*3*) external carotid artery; (*4*) occipital artery, branch irrigating the glomus caroticum; (*5*) occipital artery sub-branches; (*6*) afibrillar muscle cells on the afferent glomus arteries; (*7*) glomus tissue; (*8*) efferent glomus vessels.

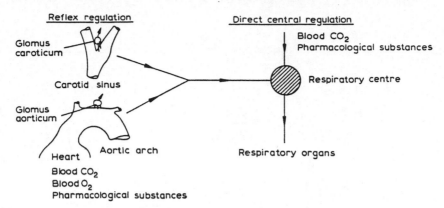

Fig. 19. Diagram of the reflex regulation of the respiratory centre by the presso- and chemo-receptors in the cardio-aortic and carotid sinus areas, and of the direct central control exerted on the respiratory centre.

The glomus caroticum, which contains the carotid sinus chemo-receptors, presents an interesting anatomical structure (Fig. 18).

Histological studies, by J. F. Nonidez in particular, and J. H. Comroe's experiments have led to the location of the cardio-aortic chemo-receptors in the aortic glomus tissue, whereas the presso-receptors are mainly located in the wall of the aortic arch.

The various experimental findings summarized in this lecture have thus brought to light a new physiological, physiopathological, and pharmacological mechanism which, by regulating the activity of the respiratory centre through the vascular presso- and chemo-sensitive reflexes (Fig. 19), establishes, thus, an even closer functional correlation between the blood circulation, the metabolism, and the pulmonary respiratory exchanges.

Biography

Corneille Jean François Heymans was born in Ghent, Belgium, on March 28, 1892. His father was J. F. Heymans, formerly Professor of Pharmacology and Rector of the University of Ghent, who founded the J. F. Heymans Institute of Pharmacology and Therapeutics at the same University.

Corneille received his secondary education at the St. Lievenscollege (Ghent), St. Jozefscollege (Turnhout), and St. Barbaracollege (Ghent). He had his medical education at the University of Ghent, where he obtained his doctor's degree in 1920. After his graduation he worked at the Collège de France, Paris (Prof. E. Gley), University of Lausanne (Prof. M. Arthus), University of Vienna (Prof. H. H. Meyer), University College of London (Prof. E. H. Starling) and Western Reserve Medical School (Prof. C. F. Wiggers).

In 1922 he became Lecturer in Pharmacodynamics at the University of Ghent. In 1930 he succeeded his father as Professor of Pharmacology, being also appointed Head of the Department of Pharmacology, Pharmacodynamics, and Toxicology; at the same time he became Director of the J. F. Heymans Institute. He is Professor Emeritus since 1963.

The scientific investigations carried out at the Heymans Institute are mainly directed towards the physiology and pharmacology of respiration, blood circulation, metabolism, and numerous pharmacological problems. These studies led, in particular, to the discovery of the chemoreceptors, situated in the cardio-aortic and carotid sinus areas, and also to contributions regarding the proprioceptive regulation of arterial blood pressure and hypertension. The discovery of the reflexogenic role of the cardio-aortic and the carotid sinus areas in the regulation of respiration, above all, earned C. Heymans the Nobel Prize in 1939.

Another series of investigations by Heymans and his collaborators was devoted to the physiology of cerebral circulation and of the physiopathology of arterial hypertension of nervous and renal origin; also to the study of blood circulation during muscular exercise; to the physiology and pharmacology of animals totally sympathectomized; to the study of the survival and revival of different nervous centres after the arrest of blood circulation; to the pharmacology of stimulating substances of cellular metabolism, to the pharmacology of the lungs and many other problems.

A profilic author, Heymans has since 1920 issued about 800 papers, published in different periodicals. The results of his investigations have been mainly reported by him in the following general publications: *Le Sinus Carotidien et les autres Zones vasosensibles réflexogènes* (1929); *Le Sinus Carotidien et la Zone Homologue Cardio-aortique*, with J. J. Bouckaert and P. Regniers (1933); *Sensibilité réflexogène des vaisseaux aux excitants chimiques*, with J. J. Bouckaert (1934); «Le centre respiratoire», with D. Cordier in *Ann. Physiol. Physicochim.*, 11 (1935) 335; «Survival and revival of nerve centers after arrest of circulation», *Physiol. Rev.*, 30 (1950) 375; «New aspects of blood pressure regulation», with G. van den Heuvel, *Circulation*, 4 (1951) 581; «Pharmakologische Wirkungen auf die Selbststeuerung des Blutdruckes», *Arch. Exp. Pathol. Pharmakol.*, 216 (1952) 114; «Action of drugs on carotid sinus and body», *Pharmacol. Rev.*, 7 (1955) 119; *Reflexogenic Areas of the Cardiovascular System*, with E. Neil (1958), «Vasomotor control and the regulation of blood pressure», with B. Folkow, in *Circulation of the Blood – Men and Ideas*, edit. by A. P. Fishman and D. W. Richards.

Heymans is publisher and Editor-in-Chief of the *Archives Internationales de Pharmacodynamie et de Thérapie*, founded in 1895 by his father and Professor E. Gley, Paris.

From 1945 to 1962 Heymans has lectured at numerous universities in Europe, North and South America, Africa, and Asia. He was in 1934 «Herter Lecturer» at the University of New York; and in 1937 he was «Lecturer of the Dunham Memorial Foundation» at Harvard University, as well as «Hanna Foundation Lecturer» at the Western Reserve University, and «Greensfelder Memorial Lecturer» at the University of Chicago. In 1939 he was «Lecturer of the Purser Memorial Foundation» at Trinity College, University of Dublin.

Commissioned with special missions by the Belgian Government, the International Union of Physiological Sciences, and by the World Health Organization, he has travelled to Iran and India (1953), Egypt (1955), the Belgian Congo (1957), Latin America (1958), China (1959), Japan (1960), Iraq (1962), Tunisia (1963), Cameroun (1963).

He has been President of the International Union of Physiological Sciences and of the International Council of Pharmacologists and has presided over the 20th International Congress of Physiology held in Brussels in 1956. His

vast knowledge of pharmacology has justified his nomination as Member of
the Committee of Experts of the International Pharmacopoeia of the World
Health Organization. In his own country he is Vice-President of the National
Council on Scientific Policy.

Heymans is Member or Honorary Member of a large number of leading
scientific societies concerned with physiology or medicine in Europe and in
North and South America, including the Pontificia Academia Scientiarum,
the Royal Society of Arts of Great Britain, the Académie des Sciences de
Paris (Institut de France), Académie de Médecine de Paris, the Heidelberger
Akademie für Wissenschaften, and the New York Academy of Sciences. He
has been appointed Professor honoris causa of the University of Montevideo,
and doctor honoris causa of the Universities of Utrecht, Louvain, Montpellier, Torino, Santiago de Chile, Lima, Bogotá, Rio de Janeiro, Algiers,
Paris, Montpellier, Münster, Bordeaux, Toulouse, and Georgetown University, Washington.

Besides the Nobel Prize, his scientific awards include the Alvarenga Prize
of the Académie Royale de Médecine de Belgique, the Gluge Prize of the
Académie Royale des Sciences de Belgique, the Quinquennial Prize (1931–
1935) for Medicine of the Belgian Government, the « Alumni » Prize for
Medicine of the Belgian University Foundation, the Bourceret Prize of the
Académie de Médecine de Paris (1930), the Monthyon Prize of the Institut
de France (1934), the Pius XI Prize of the Pontificia Academia Scientiarum
(1938), the Burgi Prize of the University of Bern and the de Cyon Prize
(1931) of the University of Bologna, etc.

Heymans is Officer in the Order of the Crown with Swords, Grand
Officer in the Order of the Polar Star (Sweden), Grand Officer of the Order
of Leopold, Commander in the Order of St. Sylvester (Vatican City),
Commander in the Knightly Order of the Holy Sepulchre of Jerusalem;
other distinctions include the Civilian Cross (First Class) for Distinguished
Services Rendered to the Fatherland, the Belgian War Cross 1914–1918, the
Fire Cross with 8 bars 1914–1918. (He was Field Artillery Officer during the
first World War.)

Professor Heymans married Berthe May, M. D. in 1921. There are four
children by the marriage: Marie-Henriette, Pierre, Jean, and Berthe; and
18 grandchildren. He loves painting and is greatly interested in ancient
literature dealing with the history of medicine; he is also a keen hunter.

Physiology or Medicine 1939

GERHARD JOHANNES PAUL DOMAGK

«for the discovery of the antibacterial effects of prontosil»

Physiology or Medicine 1939

The following account of Domagk's work is by Professor N. Svartz, member of the Staff of Professors of the Royal Caroline Institute

Experiments in the treatment of inflammatory conditions by means of drugs and chemicals are known from earliest times, but for the most part the effects were nil or at best insignificant. With certain of these conditions, however, chemotherapy scored some first-class successes at an early date. Mercury is a very ancient, active chemotherapeutic agent, although it has now given place to more effective preparations. Another therapeutic agent which has been in use for a very long time is cinchona bark, the efficacy of which against malaria became generally known in Europe during the 17th century. Other experiments in anti-inflammatory treatment by chemical methods for the most part yielded but meagre results.

It is only during the past few decades that further advances of any great significance have been made in chemotherapy. In particular, experimental investigations with arsenical preparations and the successes achieved with these preparations in cases of spirochaetic and trypanosome infections (relapsing fever, syphilis, African sleeping-sickness) provided a powerful stimulus to further experiments in the field of chemotherapy. Salts of other metals have also proved valuable in the treatment of specific types of inflammations – for instance, antimony salts have been used very successfully with certain tropical diseases. Particular mention must be made of the Bayer preparations « Plasmochin » and « Atebrin » for malaria and the preparation « Germanin » (Bayer 205) which has been successfully used in cases of tropical sleeping-sickness. In addition, bismuth salts have been found a very effective, though by no means infallible, remedy against syphilis, and have largely superseded mercury.

Thus, whereas it proved possible to attack certain diseases due to protozoa and spirochaetes by means of chemical substances, little success had been achieved with chemical preparations against infections due to true bacteria, namely cocci and bacilli. The theory that bacteria of the last-mentioned categories could not be combated by chemical means therefore continued to gain ground, and it was consequently assumed that serotherapy was the most practicable method of treating infections of this type.

Experiments with gold salts constituted an important phase in the more recent development of chemotherapy. A number of bacterial infections, e.g. septic conditions due to streptococci, rheumatic infections, etc., were found to respond in some degree to these salts, but it soon became clear that the effects varied very widely, and when the doses were increased in an attempt to produce a more vigorous effect serious symptoms of poisoning frequently appeared.

During the past 15–20 years a great deal of work has been carried out by various drug manufacturers with a view to producing less toxic but at the same time therapeutically effective gold preparations. The question of gold preparations and their applicability was also investigated at the great research laboratories of I. G. Farbenindustrie Aktiengesellschaft (Igefa) at Elberfeld. The investigations here were part of a series of experiments conducted with a view to discovering an agent effective against streptococcal infections. The Igefa laboratory department in which this research was carried out is under Professor Gerhard Domagk, who planned and directed the investigations involving experiments on animals. The chemists Dr. Mietzsch and Dr. Klarer, working in close collaboration with Domagk, provided various chemical preparations for these investigations. It was decided to include sulphonamide compounds among the preparations to be tested. These compounds had previously been synthesized and had also been introduced into the dyestuffs industry by Hörlein and his co-workers. However, none of these compounds had been tested for their therapeutic action.

During the investigations conducted by Domagk and his co-workers 4-sulphonamide-2′,4′-diaminoazobenzene hydrochloride, among other substances, was tested. This preparation was subsequently named *Prontosil*. The earliest published experiments with Prontosil were begun in December 1932. The lethal dose, for mice, of a certain strain of haemolytic streptococci, which had been isolated from a patient suffering from blood poisoning, had previously been determined. A number of mice were injected with 10 times the lethal dose of this bacterial strain, and approximately half of them were given a specific quantity of Prontosil $1\frac{1}{2}$ hours after being infected.

On 24th December, 1932, it was found that in an experiment begun on 20th December, 1932, all the controls had died, whereas all the mice which had been given Prontosil were alive and well. This was the basis of the discovery which was destined to bring undreamed-of advances in chemotherapy.

The results of these and subsequent experiments, which aroused extraordinary interest, were not published until February 1935, whereupon Prontosil and its effects rapidly became known throughout the world. France was the first country apart from Germany where Prontosil was subjected to practical tests (Levaditi). Extensive experiments on, among other things, the mode of action of Prontosil were then conducted in France (Tréfouël, Nitti), America (Long, Marshall, and others) and Britain (Colebrook, Kenny, and others). One result of these investigations was the discovery that the favourable action of Prontosil was mainly due to the sulphonamide component of the preparation.

From the outset Prontosil was described as being effective principally against streptococcal infections. Even in his first publication, however, Domagk had reported that the preparation had been found to have a therapeutic effect, although to a lesser extent, in staphylococcal infections, and that certain types of pneumonia had also responded to it.

At an early date sulphonamide preparations had proved extremely effective against erysipelas, and subsequent investigations completely confirmed this observation. Now, thanks to these preparations, erysipelas can normally be treated without difficulty.

It was also found that other streptococcal infections could be dealt with by means of sulphonamide preparations, although for the most part not so swiftly or surely as erysipelas. Although suppuration in the pleural cavity and meningitis due to streptococci are still serious diseases, they are much less so than they used to be. The same applies to puerperal fever and several other streptococcal infections. Even chronic general septicaemia with endocarditis, a condition hitherto regarded as incurable, has in isolated cases responded to sulphonamide preparations.

In addition, brilliant results have been obtained with certain infections not due to streptococci, namely gonorrhoea and epidemic meningitis, and, as already mentioned, an effect has also been shown with staphylococcal infections.

This preparation, which is so effective against various coccal infections, has also been used with success in the case of certain infections due to bacilli, e.g. coli infections. Sulphonamide is therefore now the best known remedy against infections of the urinary passages due to colon bacilli. Preparations of this group are also effective against undulant fever as well as, to a lesser extent, other bacillary infections which will not be enumerated here.

The discovery of Prontosil opened up undreamed-of prospects for the treatment of infectious diseases. Experiments with new combinations of sulphonamide preparations were everywhere conducted in the hope that new methods effective against other diseases might be discovered. Contrary to expectation these efforts were very quickly crowned with success.

Igefa reported that a new active sulphonamide preparation, Uliron, had been produced. In addition, the report, published in 1938 by the chemical firm of May & Baker of Dagenham, England, that a compound of pyridine and sulphonamide had been synthesized and had proved effective against pneumonia, was of major importance. This highly significant claim also proved correct. The preparation in question was put on to the market as M. & B. 693. It is now usually known as *Sulphapyridine*. Sulphapyridine is so far the most noteworthy derivative of Prontosil.

Simultaneously with the efforts to produce new sulphonamide preparations research workers in various countries are busy carrying out theoretical investigations into the mode of action of these preparations and their side-effects. Domagk himself has carried out some extremely fine investigations on these questions. Research in this field has also been conducted in France, Britain, America, Sweden, and other countries.

The foundations for this unprecedented expansion which chemotherapy has undergone in the brief span of less than five years were laid by Domagk and his co-workers. A new road leading to effective treatment of diseases which in the past were often fatal has been opened up. Reports on the most brilliant therapeutic results with sulphonamide preparations are streaming in from all parts of the world. Thousands upon thousands of human lives are being saved each year by Prontosil and its derivatives. Earlier, fruitless chemotherapeutic experiments often resulted in despondency, but now even the most pessimistic have gradually come to see the value of the results achieved. The imagination reels before the prospects of new chemotherapeutic victories which the sulphonamide preparations have unfolded before us.

The award of the Nobel Prize for Physiology or Medicine for 1939 to Gerhard Domagk has honoured a discovery which means nothing less than a revolution in medicine.

Professor Gerhard Domagk was awarded the 1939 Nobel Prize for Physiology or Medicine for the discovery of the antibacterial effects of Protonsil. Protonsil was the first of the so-called sulpha preparations, which have proved to represent one of the greatest therapeutic advances in the history of medicine. Professor Domagk was prevented from accepting the prize at the time by political conditions. In 1947 he received the gold medal and the diploma.

Professor Domagk. It has become clear during the eight years that have passed since it was decided to award you the Nobel Prize that the sulphonamides have introduced a new era in the treatment of infectious diseases. What Paul Ehrlich dreamed of, and also made reality by using Salvarsan in an exceptional case, has now, through your work become a widely recognized fact. We can now justifiably believe that in the future infectious diseases will be eradicated by means of chemical compounds.

On behalf of the Caroline Institute I congratulate you most warmly, and ask you to accept from His Majesty the King the medal and the diploma.

G E R H A R D D O M A G K

Further progress in chemotherapy of bacterial infections

Nobel Lecture, December 12, 1947

Excellency, honoured colleagues, I am unable to express my profound gratitude for the high honour conferred upon me in the award of the Nobel
Prize for Physiology or Medicine for 1939 – which I was not allowed to accept at the time – in any other way than by reporting on the further developments in the field which had then just been opened up.

The first decisive phase was the discovery of the curative action of certain
sulphonamide-containing azo compounds, which had been synthesized by
Klarer and Mietzsch. The most effective of these had taken their place under
the names *Prontosil rubrum* and *Prontosil solubile* in the armoury against the
streptococcal infections of man. This first phase was surveyed by Professor
Nanna Svartz in *Les Prix Nobel en 1939*.

The problem of chemotherapy of bacterial infections could be solved
neither by the experimental medical research worker nor by the chemist
alone, but only by the two together working in very close cooperation over
many years. I therefore feel under a profound obligation, in view of the high
honour which has been conferred upon me as one who has taken part in such
collaboration, to pay tribute to all my colleagues. In particular I must mention the two chemists Dr. Mietzsch and Dr. Klarer who, thanks to the substances produced by them, enabled me to discover the curative action against
bacterial infections after I had worked out and extended step by step, entirely on my own initiative, all possible methods of testing. Convinced that a
way could be found I had persevered with this work over a period of many
years, despite all the scepticism prevailing in this sphere. I should also like to
express my gratitude for the advice given to me by the pharmacologist Dr.
Hecht. I am particularly fortunate in having had these experienced and well-
tried colleagues, who were mentioned in 1935 in the first article « Ein Beitrag
zur Chemotherapie der bakteriellen Infektionen» (A contribution to the
chemotherapy of bacterial infections), published in the *Deutsche Medizinische
Wochenschrift* (1935), and in having been able, with them, to continue the
work to this day in the field which at that time had been newly opened up

and to discover still further therapeutically valuable anti-bacterial sulphon-amides. I should also like to thank many other loyal colleagues who assisted me with my investigations, often under dangerous conditions, with the greatest industry and infinite patience. At the same time, I wish to express sincere thanks to all clinicians and colleagues who have helped to apply the experimental results in practice for the benefit of patients. I also have to thank far-sighted men of industry who made it possible for me to set up a laboratory where I was able to carry out my work on the requisite scale. I recall with particular gratitude the magnificent help given me by Professor Hörlein. Unfortunately many difficulties stood in the way of our work during and after the war, difficulties with which to some extent we still have to contend. Nevertheless the management of the Bayer Wuppertal-Elberfeld dye factories always found ways and means of supporting us who were engaged in scientific research–indeed, they assisted us far more than did the state, whose first duty it should in fact have been to help its citizens, through research, to combat disease. And this shows the high sense of responsibility of enterprises created and built up by the energy of great personalities such as Friedrich Bayer, Carl Duisberg, and Heinrich Hörlein. Few if any sickness funds or insurance companies have shown such a great sense of responsibility or such a sense of duty to the community in the carrying-out of work at the research establishments maintained by them, despite the fact that these institutions are under a greater obligation to look after the health and welfare of their members, and mostly have much more capital at their disposal. The Rockefeller Institute in America and throughout the world, and the Kaiser Wilhelm Institutes in Germany were created by responsible individuals. We are and always shall be deeply indebted to these great men–and foremost among them Alfred Nobel – for what they have done for the advancement of science. Nowhere does science enjoy such respect as in this hospitable land of Sweden. We thank this country, its Royal Family and our Swedish colleagues most sincerely for what they do each year on the anniversary of Alfred Nobel's death, on 10th December, in the striving towards true humanism, towards the building of a new and better world and towards a peaceful understanding between nations. May this land one day be rewarded for its services in this field to the entire world and may other states follow this example in noble and peaceful competition.

In spirit I bow in reverence before my old teachers at the German universities who equipped me to carry out the work for which you have honoured me, on the further development of which I wish to report now. Within the time at my disposal I can draw attention only to the principal phases in the development of chemotherapy as I see them at present and as they emerge from our work. Even the numerous compounds tested with experimental streptococcal infections during the first phase of the development revealed important laws, which have proved of value to all subsequent workers in the sulphonamide field. It was found, in fact, that only those compounds which contained the sulphonamide group in the *para*-position in relation to the group containing nitrogen were of therapeutic value, whereas compounds with the sulphonamide group in the *ortho*- or *meta*-position were found to be inactive.

Prompted by the German publications on Prontosil – but independently of our own as yet unpublished experiments which were conducted with a view to discovering colourless active sulphonamides outside the azo series – Tréfouël, Tréfouël, Nitti and Bovet began to study the same problem. We are indebted to these authors for having drawn attention for the first time in literature to the fact that 4-aminobenzenesulphonamide, which Mietzsch and Klarer had used as initial material for the synthesis of their sulphonamide-containing azo compounds, was therapeutically active as such. This substance was later used in practice as *Prontosil album*, *Prontalbin*, and *Sulphanilamide*.

It was again Klarer and Mietzsch who initiated a further phase in the development of chemotherapy of bacterial infections by placing at my disposal substances in which the sulphonamide group is no longer unsubstituted, as in the Prontosil compounds, but is modified by an organic radical with replacement of a hydrogen atom. In the case of substances of this type I discovered the action against staphylococci and pneumococci; although there had been a hint of this in Prontosil, which had been reported by me as far back as 1935, it was now considerably increased. Furthermore, I now established for the first time that such compounds, known under the names

had a noteworthy effect upon gonococci. The Uliron compounds were the first sulphonamide compounds used in Germany against gonorrhoea after the experiments with Prontosil compounds had proved unsatisfactory.

A further improvement in the action against staphylococci, pneumococci, and gonococci was brought about by introducing a heterocyclic ring in place of the hydrogen in the SO_2NH_2 group. A series of such compounds had already been made available at an early date by Mietzsch and Klarer; however, an intensive study of this field – this time in all advanced countries throughout the world – began only when

Sulphapyridine = M.& B.693 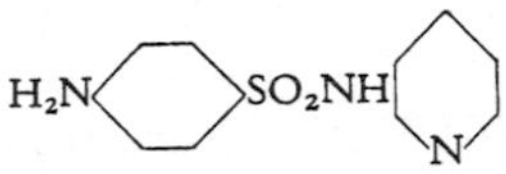
H₂N
SO₂NH
N
p-aminobenzenesulphonamidopyridine,

which had been synthesized by Philipps and Evans, was shown in England by Whitby's experiments to have an effect on pneumococcal infections going beyond that produced by the Prontosil compounds; the broad experience of British clinicians very soon confirmed this experimental finding. This sulphapyridine compound is still in use today in the treatment of lobar forms of pneumonia, but has been superseded more and more by sulphathiazole – subsequently also known by the names Cibazol and Eleudron – which are approximately equally effective against pneumococci and meningococci and still more effective against staphylococci and gonococci, since it very frequently, especially when administered in large doses, causes serious stomach disorders and vomiting. Sulphathiazole

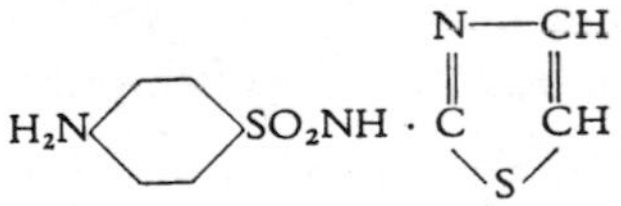
N—CH
H₂N
SO₂NH · C
CH
S

or *p*-aminobenzenesulphonamidothiazole, was produced and a patent was applied for in respect of it independently by several people, but first by Hartmann and Merz. So far sulphathiazole has firmly held its position, although its efficacy in the case of gonorrhoea has progressively declined. For a time people used to speak of a « lightning cure » for gonorrhoea, but ultimately the results became less and less satisfactory although the doses were continually increased. This fact is too well known to require any detailed explanation by me. But we should learn from this undeniable fact and should try to discover the reasons for it. At first the failures were attributed to anatomical causes. People used to speak of the role of sites difficult to reach (« Hohlraumeffekt »), etc. But this explanation was unsatisfactory and was not sufficient for all cases. Then it was thought that the war might have reduced the patients' resistance, but this explanation likewise did not always hold good. I have constantly emphasized that cooperation by the body is an important factor with any sulphonamide treatment. We all remember how at the beginning of the sulphonamide era it was repeatedly observed that fresh cases of gonorrhoea in men responded best to sulphonamide treatment when suppuration had already occurred for several days, and not at the first appearance of the disease. Perhaps the lowering of natural resistance due to war-time conditions explains why a considerable falling-off in the successes in Germany should to some extent have occurred at a time when optimum results were still being obtained in Switzerland. With the present catastrophic shortage of protein in the diet many patients probably find it difficult to make up the protein losses due to the destruction of leucocytes resulting from an inflammatory condition. However, these facts alone are not sufficient to explain the number of failures. Introduction of resistant strains due to the general upheaval during and after the war was therefore suspected. Finally the question whether gonococci might, like protozoa, possibly become resistant to the drug during treatment was discussed. We ourselves never succeeded in rendering a strain sulphonamide-resistant by treating with small doses of sulphonamide under experimental conditions. However, if this phenomenon should occur in very exceptional cases this would in no way explain the great number of failures. Of decisive importance for a clarification of this question, however, is the fact, which was established by Felke at the beginning of the sulphonamide era, that gonococcal strains of widely differing sensitivity existed from the outset. Felke distinguished strains which would grow on ascites plates with 0.6, 1.2, 2.5, and even 5 mg% Uliron C. It is noteworthy that the clinical failures occurred in the case of carriers of highly resistant strains. Hagerman (Lund) determined the different degrees of resistance of gonococci by another method. He dripped graduated concentrations of sulphonamide solutions on to a kind of ascites agar plates, allowed the drops to be absorbed and then inoculated the plates uniformly with one strain. He expressed the resistance to sulphathiazole by the numbers 0–11. 0 denoted the highest concentration, 1:200; and 11 the lowest, ca. 1:400,000. He came to the following conclusion: In the case of gonorrhoea which has been treated with sulphathiazole the prognosis depends mainly on the resistance to sulphonamides of the gonococcal strain in question. In any particular case of gonorrhoea the prognosis can be made with great reliability by means of chemoresistance determinations in vitro. Later Schmith (Copenhagen) also settled satisfactorily the question of primarily resistant strains. He tested 50 old strains from the pre-sulphonamide era; among these he found approximately the same resistance percentage ratio as he had found with his fresh strains. There can therefore be no doubt that even before the sulphonamide era there were gonococcal strains with a high primary resistance. The resistant strains therefore existed before the sulphonamide era and have not come about as a result of sulphonamide treatment with insufficient doses, or any similar cause. They are due to natural selection. The spread of these resistant strains would have been avoided if the few patients who were carriers of them in 1937 had received careful treatment and had been subjected to clinical supervision until their cure was absolutely certain. But this was not done.

Today we are faced with the same question with regard to the use of penicillin. If we are not to suffer the same disappointment at some future time, the patients with resistant strains must be kept under treatment until they have been definitely cured. Felke rightly maintains that gonorrhoea will not be eradicated – even if we have the best remedies at our disposal – until a woman infected with gonococci is not given a clean bill of health before a cervical culture has been grown. In women the reservoir of the gonococcus is not primarily the urethra but the uterus, and especially the tube angles when the adnexa are affected. Felke describes treatment of gonorrhoea in women without the growing of a culture as anachronistic. This applies in particular to determination of the female infection sources. Jadassohn's view that the culture should be used only in doubtful cases is not longer valid. The superiority of the culture over microscopic examinations has once again been shown very recently by Veltman of the Grütz clinic (*Z. Haut-Geschlechts-krank.*, No. 7 (1947) 203). He found positive cultures in 52 cases following completion of a course of penicillin treatment in hospital and negative microscopic findings. He reports that in the course of a year 86 patients would have been discharged from the clinic as cured if culturing had not made it possible to prove the presence of gonococci. The culturing method – at least in the case of women – is therefore absolutely necessary in addition to microscopic examination, since it is more efficient than microscopic examination alone. Now that penicillin treatment, with the follow-up period reduced to 5 days has been introduced, an additional examination by the cultural method is in fact very important. Since relapses after penicillin treatment do not usually occur until after the first 3 days, at least one culture should be taken a week after completion of the treatment. Otherwise one would be only too justified in asking, as Clarke recently did: « Penicillin: help or hindrance in venereal disease control? », for we already see the first signs that, if treated patients are not subjected to a very thorough check-up, penicillin – like the sulphonamides – is being misused, with the inevitable result that resistant strains will survive and then spread. Huriez and Desurmont (Lille) have already drawn attention to the fact that since October 1946 there has been a distinct decline in successes with penicillin, which has necessitated the combination of penicillin with fever, with sulphonamides and local treatment (*Presse Med.*, No. 2 (1937), Ref. *Z. Haut-Geschlechtskrank.* No. 2 (1947) 217–218). If the spread, through natural selection, of still more resistant strains of gonococcus is to be avoided in future, these demands for a particularly careful check on treated patients will not have to be overlooked, irrespective of whether sulphonamides or penicillin are used. This is in no way to say that the use of sulphonamides in the treatment of gonorrhoea should now be completely abandoned. Hopf of Hamburg recently stated that in Hamburg and northern Germany 70% of fresh cases of gonorrhoea were still being cured at the first attempt by intensive treatment with Eleudron or Cibazol, administered at the rate of 6 g each day for 3–5 days. According to this report the number of sulphonamide-resistant strains in Hamburg and northern Germany is smaller than in other areas. Hopf described the sulphonamides as indispensable for the treatment of relapses, even after penicillin. Schreus reported in 1946 that he had obtained the best results with injections of sterile milk and 3 × 5 tablets of Eleudron for 2 days. After the first course of intensive treatment 75% were free from gonococci, after the second the figure was 90.8%. Of the remaining uncured patients a further 80% could be cured with a combination of sulphonamides and Olobinthin, which meant that only 2% resistant cases remained. Felke found that if a second course of treatment with penicillin was necessary it should always be combined with a three-day course of massive doses of sulphonamide, with 200,000 units penicillin given on the second day of the usual intensive course of sulphonamide treatment.

Why some gonococcal strains are more resistant than others is still not clear. It is suspected that some gonococcal strains, like some staphylococcal strains, can produce more p-aminobenzoic acid or other antisulphonamide factors. This possibility might be suggested by our observation that certain gonococcal strains are inhibited to a greater extent by Marbadal – the sulphathiourea salt of Marfanil – and Supronal (the effectiveness of which is well known to be only partially impaired, if at all, by p-aminobenzoic acid) than by Eleudron. It would therefore appear possible that the 100% success achieved by Bernhard in the treatment of gonorrhoea in women may to some extent be due to this, and not merely to the fact that higher doses are possible thanks to better tolerance. Another way of achieving better results might be to find pyretic agents which would bring about moderate degrees of fever for a relatively long period, on the same lines as typhoid bacillus vaccines. A rise in body temperature during sulphonamide treatment intensifies the biochemical reaction between drug and pathogen, while at the same time the heat itself injures the heat-sensitive gonococci. According to Felke 1 °C above normal, extending over a sufficiently long period, is enough. Intensive courses of Pyrifer treatment are too short-lasting to give the optimum results, and this is no doubt the main reason why 40% Olobinthin, which usually gives slight pyrexia lasting 3–4 days, is superior. Boas and Marcussen gave 10 × 1 g sulphathiazole for 3 days and on the third day induced fever for a period of $5-5\frac{1}{2}$ hours by hyperthermia. Of 20 patients, 19 were cured (*Ugeskrift Laeger*, 106 (1944) 16).

During the past few years we have often been too easy-going with sulphonamide treatment of gonorrhoea, giving 2 Eleudron tablets 5 times a day at intervals of 2 hours. With this regimen adequate blood and tissue concentration was achieved for only a small part of the day, whereas during the rest of the day and during the night the cocci could recover. Intervals between individual doses should not exceed 4–6 hours.

All other specialized fields of medicine have learnt from dermatology, in which massive doses of sulphonamides were given consistently at 4–6 hourly intervals, and massive doses over a short period are now the rule in the treatment of acute infections of any kind.

For gonococcal infections sulphathiazole has even now not been surpassed by other sulphonamide compounds to any extent worth mentioning. In the case of streptococcal infections, on the other hand, sulphathiazole is greatly surpassed in effectiveness by the sulphapyrimidine compounds, and especially by 2-(*p*-aminobenzenesulphonamido)-pyrimidine

H_2N⟨⟩$SO_2·NH$ Pyrimal, Debenal, Sulphadiazine and the monomethylated compound

H_2N⟨⟩SO_2NH Methyldebenal, Debenal M, Sulphamerazine

The sulphapyrimidine compounds were also developed independently at different laboratories. In the patent literature they were first described by the

Deutsche Hydrierwerke (Hentrich) and by the firm of Schering (Dohrn and Diedrich); the first scientific publication was that of Roblin, Williams, Winnek and English.

The following experiment on mice, which were infected intraperitoneally with β-haemolytic streptococci of group A, illustrates the superiority of sulphapyrimidine compared with experimental infections with haemolytic streptococci of group A:

	Number of animals	Alive 24 hours after infection	Survivals	
Controls	12	0	0	
Prontalbin 0.5 and 5% subcutaneous injection	10	10	1	
0.5 and 5% by mouth	10	9	0 } 1	
Debenal 0.5 and 5% subcutaneous injection	10	10	5	
0.5 and 5% by mouth	10	10	1 } 6	

This result is for a single treatment. Where the animals were given three doses – 1, 6, and 24 hours after being infected – the superiority of sulphapyrimidine is even clearer.

	Number of animals	Alive 24 hours after infection	Survivals	
Controls	12	0	0	
Prontalbin 0.5 and 5% subcutaneous injection	10	9	0	
0.5 and 5% by mouth	10	10	0 } 0	
Debenal 0.5 and 5% subcutaneous injection	10	10	5	
0.5 and 5% by mouth	10	10	4 } 9	

Dosage (2 animals in each case) 0.5% 1.0 cc; 5% 0.2; 0.4; 1.0; and 2.0 cc subcutaneously
0.5% 0.5 and 1.0 cc by mouth
5% 0.2 cc; 0.5; and 1.0 cc by mouth

Apart from this, with streptococcal infections certain sulphone compounds show specific superiority over all earlier sulphonamides – for instance, Tibatin, galactoside of 4.4'-diaminodiphenylsulphone,

$$O_5H_{11}C_6 \cdot HN \hspace{-2pt}\diagup\hspace{-4pt}\bigcirc\hspace{-4pt}\diagdown SO_2 \diagup\hspace{-4pt}\bigcirc\hspace{-4pt}\diagdown NH \cdot C_6H_{11}O_5$$

which was synthesized at Elberfeld by Behnisch and Pöhls.

In the following experiments 80 mice, which had been infected intraperitoneally with β-haemolytic streptococci, were given, 10 in each case, 5% 0.5 and 1 cc per 20 g body weight by mouth Prontalbin, sulphapyridine and sulphathiazole and 0.5 cc and 1.0 cc of a 2.5% solution by subcutaneous injection for comparison.

Single dose, 3 hours after infection:

	Number of animals	Alive 24 hours after infection	Alive 48 hours after infection	Completely cured
Controls	14	3	0	0
Prontalbin	20	19	12	1
Sulphapyridine	20	20	16	1
Sulphathiazole	20	20	10	1
Tibatin	20	20	20	10

Three doses, 1, 6, and 24 hours after infection:

Controls	14	4	0	0
Prontalbin	20	17	13	0
Sulphapyridine	20	20	19	1
Sulphathiazole	20	17	15	0
Tibatin	20	20	20	15

Experiments on rabbits also showed Tibatin to be more effective than Prontalbin, sulphapyridine or sulphathiazole. However, these sulphone compounds are distinctly more effective only when introduced parenterally; when administered orally they are unreliable and, obviously owing to uncontrollable decomposition, have undesirable side-effects, such as severe cyanosis, etc.

Whereas the effectiveness of the sulphonamide and sulphone compounds so far considered had been confined almost exclusively to aerobic microorganisms and had as yet shown an observable effect against *Clostridium septicum* infections in only a few cases, e.g. with Uliron compounds, I was able to show a really specific action on anaerobic microorganisms in the case of the compound produced by Klarer:

$$H_2N \cdot CH_2 - \langle \quad \rangle - SO_2NH_2 \cdot HCl,$$

the hydrochloride of *p*-aminomethylbenzenesulphonamide, later named Marfanil. I look upon this observation as the beginning of a third phase in the fight against bacterial infections with sulphonamides. The action was detectable in vitro as well as in experiments on animals. Although Marfanil and its derivatives were effective against streptococci in vitro but not in animal experiments, the derivatives were occasionally found to have a considerable effect, even in animal experiments, on certain strains of streptococcus which were little affected by other sulphonamides. But their main value does not lie in this direction. Their greatest importance lies in the fact that they can be used in the fight against the most serious wound infections, gas gangrene in man and animals. The experiments which we conducted into the specific effects of Marfanil and its derivatives on the various gas gangrene organisms were repeated, under somewhat modified conditions, by Zeissler. This author used human blood for the culture, whereas we had used rabbit's blood, and the results agreed almost exactly. We will give the results of a few of the experiments conducted by Zeissler to show how sulphathiazole and the other sulphonamides are ineffective in the case of these microorganisms and how, in contrast, the action of Marfanil is specific.

Sulphathiazole.
Clostridium perfringens (B. welchii, B. perfringens).

Sulphathiazole Microorganisms	Controls	1:1250	1:2500	1:5000	1:10,000
Concentration	+++++	+++++	+++++	+++++	+++++
1:10	++++	++++	++++	++++	++++
1:100	+++	+++	+++	+++	+++
1:1000	++	++	++	++	++
1:10,000	344	252	248	240	304

Sulphaethylthiodiazole.
Clostridium perfringens (B. welchii, B. perfringens).

Sulphaethyl- thiodiazole Microorganisms	Controls	1:1250	1:2500	1:5000	1:10,000
Concentration	+++++	+++++	+++++	+++++	+++++
1:10	++++	++++	++++	++++	++++
1:100	+++	+++	+++	+++	+++
1:1000	++	++	++	++	++
1:10,000	300	308	324	272	356

Marfanil.
Clostridium perfringens (B. welchii, B. perfringens).

Marfanil Microorganisms Concentration	Controls	1:1250	1:2500	1:5000	1:10,000	1:20,000	1:40,000	1:80,000
1:10	+++++	o	o	o	+++++*	+++++	+++++	+++++
1:100	++++	o	o	o	+++++*	++++	++++	++++
1:1000	+++	o	o	o	+++*	+++	+++	+++
1:10,000	++	o	o	o	++*	++	++	++
	340				196*	320	392	412

(1-day old human blood in agar without glucose)

* No growth after 24 hours incubation but, in contrast to the controls and the other plates, only after 48 hours.

Marfanil is found to be superior with Clostridium novyi as well as with Clostridium perfringens.

Marfanil.
Clostridium novyi (B. oedematiens).

Marfanil Microorganisms Concentration	Controls	1:1250	1:2500	1:5000	1:10,000	1:20,000	1:40,000	1:80,000
1:10	+++	o	o	o	o	+++	+++	+++
1:100	++	o	o	o	o	++	++	++
	188	0	0	0	0	172	248	228
1:1000	33	0	0	0	0	0	30	68
1:10,000	12	0	0	0	0	0	4	7

Sulphathiazole.
Clostridium novyi (B. oedematiens).

Sulphathiazole Microorganisms	Controls	1:1250	1:2500	1:5000	1:10,000
Concentration	M(atting)	M	M	M	M
1:10	M	M	M	M	M
1:100	M	M	M	M	M
1:1000	M	M	M.	M	M
1:10,000	9	15	21	26	14

Whereas no growth of the bacilli is possible at concentrations of 1:10,000 or 1:20,000 Marfanil depending on the sowing, the microorganisms grow so densely at concentrations of 1:1250 sulphathiazole that matting occurs and separate colonies can no longer be counted.

The general effect of Marfanil when administered parenterally or orally is illustrated by the following report:

Mice infected with *Clostridium septicum* by intramuscular injection in the thigh.

	Number of animals	Number of doses given	Alive 24 hours after infection	Alive 4 weeks after infection
Controls	20	—	3	2 = 10%
Prontalbin	10 subcutaneous injection	1 ×	3	3
	10 subcutaneous injection	3 ×	2	2
	10 by mouth	1 ×	2	2
	10 by mouth	3 ×	2	1 } 8 = 20%
Uliron C	20 subcutaneous injection	1 ×	8	5
	20 subcutaneous injection	3 ×	11	7
	20 by mouth	1 ×	8	7
	20 by mouth	3 ×	10	9 } 28 = 35%
Marfanil	20 subcutaneous injection	1 ×	19	17
	20 subcutaneous injection	3 ×	20	19
	20 by mouth	1 ×	18	14
	20 by mouth	3 ×	18	16 } 66 = 82.5%

Two or four animals were given, by subcutaneous injection or by mouth, 0.1, 0.2, 0.3, 0.5 and 1.0 cc per 20 g body weight of 4% aqueous solutions or suspensions of each preparation. The animals which were given only one dose were treated approx. 2 hours after being infected; those which were given a total of three doses were treated 2, 8 and 24 hours after being infected.

The efficacy against each of the gas gangrene microorganisms which are pathogenic in man, when applied generally and locally, was evaluated. Timely local application gave by far the best results, as is shown by the following:

Rabbits infected by intramuscular injections of earth containing the spores of the following microorganisms: *Clostridium perfringens, Clostridium novyi, Clostridium septicum, Histolyticus, Gigas.* Earth was left in the wound.

	Number of animals	Alive 24 hours after infection	Alive 1 week after infection
Controls	6	0	0
Marfanil B locally, also MP* 1 g/kg by mouth	12	12	12
Marfanil B locally, also MP by mouth and gas-gangrene serum	12	12	12

* MP = Marfanil Powder.

Rabbits infected in a wound in the back muscle with earth N III containing the most common and the most dangerous causative agents of gas gangrene, namely *Clostridium perfringens, Clostridium novyi, Clostridium septicum,* and *Histolyticus.*

	Number of animals	Alive 24 hours after infection	Alive 8 days after infection	Alive 2 weeks after infection	Alive 3 weeks after infection
Controls	8	0	0	0	0
Gas-gangrene serum of maximum potency 1 cc/kg intravenously immediately after infection	4	1	0	0	0
Gas-gangrene serum of maximum potency 1 cc/kg intravenously 3 hours after infection	4	3	0	0	0
Excision of wound 6 hours after infection gas-gangrene serum	8	3	0	0	0
(totals of above four rows)	24	7	0	0	0
Excision of wound 3 hours after infection + gas-gangrene serum	8	8	4	4	4

Local treatment with Marfanil powders immediately after infection (earth left in wound) and intravenous injection of gas gangrene serum.

	Number of animals	Alive 24 hours after infection	Alive 8 days after infection	Alive 2 weeks after infection	Alive 3 weeks after infection
MP powder N 982	6 ⎫	6 ⎫	6 ⎫	6 ⎫	
MP powder N 983	6 ⎬ 24	6 ⎬ 24	6 ⎬ 24	5 ⎬ 22	21
MP powder 1:9	6 ⎪	6 ⎪	6 ⎪	5 ⎪	
MP powder P 55	6 ⎭	6 ⎭	6 ⎭	6 ⎭	

Local treatment with Marfanil powders 3 hours after infection (earth left in wound) and intravenous injection of gas gangrene serum.

	Number of animals	Alive 24 hours after infection	Alive 8 days after infection	Alive 2 weeks after infection	Alive 3 weeks after infection
MP powder N 982	6 ⎫	6 ⎫	6 ⎫	4 ⎫	
MP powder N 983	6 ⎬ 24	6 ⎬ 24	6 ⎬ 21	6 ⎬ 17	13
MP powder 1:9	6 ⎪	6 ⎪	4 ⎪	2 ⎪	
MP powder P 55	6 ⎭	6 ⎭	5 ⎭	5 ⎭	

Local treatment with Marfanil powders 6 hours after infection (earth left in wound) and intravenous injection of gas gangrene serum.

	Number of animals	Alive 24 hours after infection	Alive 8 days after infection	Alive 2 weeks after infection	Alive 3 weeks after infection
MP powder N 982	6 ⎫	6 ⎫	4 ⎫	4 ⎫	
MP powder N 983	6 ⎬ 24	6 ⎬ 24	4 ⎬ 10	3 ⎬ 9	7
MP powder 1:9	6 ⎪	6 ⎪	1 ⎪	1 ⎪	
MP powder P 55	6 ⎭	6 ⎭	1 ⎭	1 ⎭	

Dosage (2 animals each): 0.5; 1.0; 1.5 g

This experiment on rabbits shows in a very impressive manner how vital the time factor is and how important it is to use the correct sulphonamides early in these serious wound infections. In this experiment Marfanil was combined with other sulphonamides in order also to produce a satisfactory effect on other microorganisms – such as streptococci and staphylococci – in the wound. The mixture in the ordinary MP powder was 1 part Marfanil to 9 parts Prontalbin: this was because large quantities of Marfanil were not available at first. Later the following MPE compounds, which were still more effective, were mainly used for local treatment of wounds:

Marfanil ⎫
Prontalbin ⎬ in equal parts (described in the experiment as MP powder N 982); or
Eleudron ⎭

Marfanil ⎫
Prontosil ⎬ in equal parts (described in the experiment as MP powder N 983)
rubrum
Eleudron ⎭

Marfanil ⎫
Prontalbin ⎬ in equal parts
Eleudron
Marfanil B ⎭

Marfanil ⎫
Prontosil ⎬ in equal parts
rubrum
Eleudron
Marfanil B ⎭

In the latter mixtures a readily soluble rapidly penetrating type of Marfanil was combined with a difficultly soluble Marfanil derivative which gave a still better local effect and which also had a considerable effect against tetanus infection. By using Marfanil B, the difficultly soluble naphthalene-1,5-disulphonic acid salt of 4-aminomethylbenzenesulphonamide, we succeeded in our experiments in preventing tetanus even in cases where, despite gas gangrene serum and tetanus serum, fatal tetanus otherwise occurred after 14 days. Marfanil B and powders containing Marfanil B are also recommended for use in the prevention of umbilical tetanus. The use of Marfanil powders on wound patients has given convincing results where they were administered in time. Anyone who has not treated a wound infected with gas gangrene by applying MPE powder externally within the first 3 hours and, where appropriate, by giving the patient large doses of Marfanil internally as well (if the wound is not easily accessible from outside), is bound to remain sceptical as to the value of sulphonamide treatment of wound infections. Processes which with other infections take days or even weeks take only hours with gas gangrene infections. Only swift action can help here.

In the light of experience to-date in the treatment of wounds, and especially those contaminated with earth, dust, etc., how should we proceed?

There is no question that all such wounds should first receive proper surgical treatment as quickly as possible. However, if there is no certainty that this can be done within the first three hours, as was often the case following air raids and always in the field, wounds should first be treated externally with Marfanil powder; up to 10 g, depending on the size of the wound, should be used in order to form a thin coating. If it is suspected that soil par-

ticles or shell splinters, stones or wood fragments with soil clinging to them
have entered the layers of tissue to such a depth that they are inaccessible to
external treatment, then 2 g Marfanil, Marbadal, or Supronal should be given
by mouth, preferably with a little gruel, milk, coffee, or the like. After surgical treatment of the wound, it should again be dusted with powder containing Marfanil; the total daily dose for adults should not exceed 20, or at
most 25 g.

With wounds of very large area, such as were frequent as a result of severe
burns suffered during air raids, it was found advisable not to apply more than
about 10 g sulphonamide powder at once, since in such cases absorption is
too rapid and symptoms of poisoning may appear. In order to prevent
adhesion of dressings and to relieve pain, when the dressings are changed,
during the first few days moist compresses with valerian tea or boiled water
should be applied after the powdering, and then after a few days sterilized
cotton cloths thinly smeared with ordinary ointment. This method is very
economical from the point of view of consumption of ointment and is also
more efficient than the use of sulphonamide ointments.

A dose of 10 g should likewise not be exceeded where MPE powders are
being used in the abdominal cavity. That local treatment in cases of gunshot
wounds in the abdomen has proved extremely efficient, is clear from the
large series of observations made by Konjetzny and his pupils Haferland and
Klostermeyer, as well as from the many observations made by Krueger and
others.

Peipper, Tönnis and other specialists in the field of brain surgery have reported that even in cases of brain injury sulphonamides can be introduced
directly into the wound cavities, thereby preventing abscess formation and
meningitis, which are often fatal.

In Britain and America sulphonamides with additions of penicillin have
been used, as well as pure sulphonamides, for the treatment of wound infections. Whether or not better results than with sulphonamides alone can be
obtained in this way has not yet been reported.

Klostermeyer, who has had a very great deal of experience in the field
with the correct and early application of sulphonamides, has told me that he
cannot imagine better results than those obtained by him with this treatment.
He draws particular attention to his results with severe wounds of the extremities, including the much dreaded wounds of the knee joint.

One further observation seems to me to merit special mention. In some of
the great air raids all the severely wounded received suitable and prompt

surgical treatment and Marfanil. The results were correspondingly good. As the number of casualties was so great, however, it would have been impossible to give similar suitable surgical treatment to those with multiple minor injuries or even to incise the wounds and then treat them with Marfanil. These patients received no treatment, and the percentage of them who died was abnormally high, in contrast to that of the severely wounded (Hennig and others). The injuries were regarded as so trifling that no sulphonamides were administered, even internally. The wound infections which subsequently proved fatal could undoubtedly have been prevented in a great many cases by large oral doses of sulphonamides.

Whereas in World War I, the U.S. army lost 8.25% of its wounded by death, in World War II, when sulphonamides were used extensively, only 4.5% died. In World War I 1,68% of men reporting sick in the American army died; now the figure is less than one tenth, i.e. 0.1% (Long, *J. Am. Med. Assoc.*, (1946)). The only reason why results of sulphonamide therapy of wound infections still vary so greatly from one section of the army to another is, in my view, that treatment is inadequate, and mostly too late. Even our experimental work showed dearly that better results can be achieved by correct chemotherapeutic treatment than by surgical treatment of wounds infected with gas gangrene, even when surgical attention is available 3 to 6 hours after the injury. In every case the best results were obtained by a combination of early and suitable surgical wound cleansing and sulphonamide treatment, preferably with the addition of gas gangrene serum. It goes without saying that even today the well-tried tetanus serum should never be dispensed with unless this is necessary for some special reason.

Sulphonamides will also come into general use in peace-time surgery. This is clear from the first publications of Konjetzny, Haferland, Klostermeyer, and others. W. Fischer of Kiel has reported that he used to have 13–14% fatal cases following operations for perforated appendix, but that since he has been using MPE powder, and with the same operating technique, the figure has fallen to 1%. Similarly favourable results have also been obtained by Lezius and Kramer, among others.

Marbadal–the Marfanil salt of sulphathiourea

$$H_2N \hexagon SO_2NHCSNH_2 \cdot H_2N \cdot H_2C \hexagon SO_2NH_2$$

for the synthesis of which we are also indebted to Klarer (*Deut. Med.*

Wochschr., No. 45–46 (1947) 670) – has an effect similar to that of Marfanil, but superior in the case of staphylococci.

In the case of *Clostridium septicum*, for instance, it has the following inhibiting values compared with other sulphonamides:

Prontalbin	$<1:1000$
Eleudron (Cibazol)	$1:1000$
Debenal (Pyrimal)	$1:1000$
Debenal M	$1:1000$
Marfanil	$1:50,000$
Marbadal	$1:50,000$
Debenal + Marbadal $\overline{aa}$	$1:50,000$
Debenal M + Marbadal $\overline{aa}$	$1:50,000$

In experiments on animals, too, Marbadal was at least as effective as Marfanil with anaerobes.*

Experiment 22.2.1946. Mice infected by intramuscular injections in the thigh with 0.3 cc of a 24-hour *Clostridium septicum* culture in liver broth diluted at the rate of $1:15$ in a physiological solution of common salt. Doses, administered to 2 animals in each case: 6% 0.2; 0.3; 0.5; 0.8; and 1.0 cc. A single dose 1 hour after infection.

	Number of animals	Alive 24 hours after infection	Alive 48 hours after infection	Alive 1 week after infection
Controls	20	8	3	2
Marfanil 6% subcutaneous injection	10	10	9	8 } 18
6% by mouth	10	10	10	10
Marbadal 6% subcutaneous injection	10	10	10	9 } 19
6% by mouth	10	10	10	10

According to Lezius Marbadal is of outstanding value in the treatment of intestinal gangrene, also known as jejunitis necroticans, which has recently been claiming a great many victims in northern Germany. Schutz and Lezius were the first to establish that this disease is caused by pathogenic anaerobic bacteria. Whereas with surgical treatment 40–50% of patients died, Lezius

* Zeissler reports that Marbadal shows the same specific superiority over all other sulphonamides – even penicillin – when tested with *Bacillus enterotoxicus*, which over he considers to be the sole agent responsible for intestinal gangrene.

succeeded, by administering large doses of Marbadal by mouth, in saving the great majority of patients without operating.

Sulphonamides were first used for puerperal infections by Klee and his co-workers. Then in large-scale experiments British authors – first and foremost Colebrook and his co-workers – established for certain that Prontosil has a satisfactory action with puerperal infections. However, it was not only at a few very well-run clinics that convincing results were obtained; good results also gradually began to appear in the statistics from entire countries, such as Germany and Britain.

Further investigations conducted by us on puerperal infections, and especially in cases of septic abortion, showed that in a considerable percentage of cases anaerobic gas gangrene microorganisms were present, especially *Clostridium perfringens*, and sometimes also *Clostridium septicum*, *Clostridium novyi*, etc. According to Bernhard, Anselmino and others, the combination of Methyldebenal with Marbadal (De-Ma or Supronal), which our experimental investigations showed to be particularly advantageous and on which I reported in the *Deutsche Medizinische Wochenschrift*, 1947, Nos. 1–8, has also proved effective in these most serious forms of puerperal infections.

De-Ma and other sulphonamide combinations are better tolerated as well as being more effective. They result much less often in renal complications than do the pure sulphapyrimidines (cf. A. R. Frisk, G. Hagerman, S. Helander, B. Sjögren «Sulpha-Combination, a new chemotherapeutic principle», *Brit. Med. J.*, (1947) 7).

Martin and Lezius, among others, consider that with the protection of Supronal (= De-Ma) operations which used to be impossible can now be performed. Martin has reported cases of patients with fever due to a septic condition on whom a caesarean section was successfully performed with the use of De-Ma (Supronal).

Heilmeyer has reported on the successful treatment of very serious septic conditions with De-Ma in non-surgical medicine. Cholangitis also yielded to treatment. Even endocarditis lenta was cured in a few cases, one patient having suffered from the disease for 5 months. Hitherto only a temporary improvement could be brought about in patients with endocarditis lenta of longer standing. The failures are due not only to the anatomical reasons explained in earlier communications, but to some extent to the different strains of streptococcus responsible for the disease. Investigations conducted at the Heilmeyer Clinic have shown that strains classified as enterococci are particularly resistant. In our experiments on rabbits we found that some of these

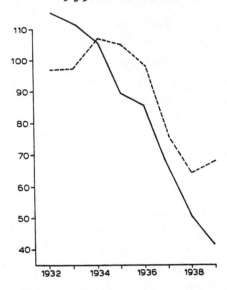

Fig. 1. Deaths from puerperal fever 1932–1939 following birth (......) per 100,000 live and still births, and following miscarriages (——) per 1,000,000 of the average female population aged from 18 to less than 40 years.

strains caused very severe acute symptoms of the joints with swelling of the capsules and pericapsular haemorrhages. Some of them responded fairly well to Marbadal in infected mice. Still more effective sulphonamides will probably be found for these strains.

Mice infected by intraperitoneal injection of 0.3 cc of a 24-hour serum-broth culture diluted 1:8. Treatment 1, 6, 24, 48 and 72 hours after infection. Dose, administered to 2 animals in each case: 0.5% 0.5; 1.0 cc; 5% 0.2; 0.5; 1.0 cc per 20 g body weight by subcutaneous injection.

Penicillin 1 cc =   50 U. 0.5; 1.0 cc

           1 cc = 500 U. 0.2; 0.5; 1.0 cc

Some of the strains isolated in cases of endocarditis lenta also proved to be completely resistant to penicillin. Christie reported very considerable successes with some patients suffering from endocarditis lenta who were given large doses of penicillin. A combination of the two best drugs at present available against streptococcal infections – penicillin and the sulphapyrimidine compounds or De–Ma – may perhaps give even rather better results than those which have hitherto been possible under the most favourable conditions.

	Number of animals	Alive 24 hours after infection	Alive 48 hours after infection	Alive 1 week after infection
Controls	20	2	1	0
Prontalbin	10	7	4	0
Sulphathiazole	10	4	3	1
Debenal	10	5	2	1
Debenal M	10	2	1	1
Marbadal	10	7	7	4
De-Ma (Supronal)	10	6	5	4
Debenal/Marfanil aa	10	8	7	7
Debenal M/Marfanil aa	10	10	10	8
Penicillin	10	10	9	8

With less serious streptococcal infections, such as erysipelas, quinsy, tonsil-litis, etc., the older sulphonamides are reliable and effective. We now know for certain that the sulphonamides are of decisive value in the treatment of erysipelas, especially as even erysipelas in infants, where the mortality used to be almost 100%, can now be cured by means of sulphonamides in the vast majority of cases. The ordinary sulphonamides are also still being used for the treatment of severe quinsy and tonsillitis, the most efficient method being to allow the tablets to dissolve very slowly in the mouth so that they act for the maximum length of time and with the maximum intensity directly upon the focus of infection. W. Schmidt therefore even went so far as to recom-mend that sulphonamides should be applied only locally for quinsy and tonsillitis; he dusted several times a day with MP powder and obtained very good results. Moreover, local application of Marfanil together with serum should also be used in every serious case of diphtheria since, of all the sul-phonamides, Marfanil develops the most powerful inhibiting effect on diph-theria bacilli, thereby limiting further toxin production and at the same time fighting infection by associated bacteria.

In severe cases of streptococcal meningitis Unterberger reduced the 80% mortality at first to 50%, later – with Tibatin – to 25% and finally to as little as 11%.

In the treatment of lobar pneumonia little has changed since the introduc-tion of sulphapyridine. It may be possible to improve slightly on the results so far achieved with sulphapyridine and sulphathiazole by using the newer sulphonamides, especially sulphapyrimidines and Supronal. British authors, with their greater experience of penicillin, consider that in cases of lobar pneumonia the results achieved with the sulphonamides are so good that the question of penicillin scarcely arises. Here again, as far as the use of both these drugs is concerned, everything depends on when they are applied. According to Bunn and co-workers (*J.Am. Med. Assoc.*, 129, No. 5 (1945)), if penicillin is to be administered by mouth for the treatment of pneumonia the dose must be 4–5 times greater than with intramuscular injection, or 750,000 units on the first day and 400,000–500,000 U. on the following days. Treatment should be considered for at least 7 days following return to normal temperature. It appears that penicillin is more effective than the sulphonamides against other pneumococcal infections.

Results with bronchopneumonia are still not as good as with lobar pneumonia. This is undoubtedly due in part to the fact that bronchopneumonia begins less dramatically than lobar pneumonia and is therefore never treated so intensively. Apart from this, it appears that in some areas there are forms of pneumonia, caused by enterococci, which when treated with sulphonamides have a non-typical course. Non-typical virus pneumonia has so far not been reliably identified in Germany to any extent worth mentioning.

Among the sulphonamides the most effective against staphylococcal infections are sulphathiazole (Eleudron, Cibazol), Globucid, and Marbadal. External application of MPE powders, where possible as an embrocation with glycerin or in the form of a lotion consisting of:

Zinc oxide ⎫
Talcum ⎪
Glycerin ⎬ in equal parts
Water ⎪
5% MPE ⎭

has proved particularly successful.

This treatment has been found practical, and economical with regard to dressings.

One would expect penicillin to give better results than the sulphonamides against staphylococci in practice, as it is so very effective in vitro.

One of the most brilliant successes with sulphonamides has been against meningitis epidemica – a disease which formerly resulted in a high percentage of deaths. In some epidemics the death rate used to be 80–90%, and even approx. 50% after the use of serum. Now, results published in the literature throughout the world show that 90–95% of patients suffering from meningitis epidemica are saved by oral administration of sulphonamides alone. In a recent communication Gehrt states that he has even been able to reduce the mortality among babies and small children, who are least able to withstand this disease, to 4.8%. In the U.S. army the number of fatal cases among soldiers suffering from meningitis epidemica fell from 39.2% in World War I to 3% in World War II thanks to the use of sulphonamides. Even according to reports dating from some time back the lives of more than 10,000 persons suffering from meningitis epidemica had already been saved by sulphonamides in Britain alone.

In bacillary dysentery and related infections sulphanilamide (Prontalbin), sulphathiazole (Eleudron, Cibazol), sulphapyrimidine (Debenal, Pyrimal, Sulphadiazine), and sulphaguanidine are of particular value. According to Max Bürger the sulphonamides have not only removed the danger of dysentery as far as the individual is concerned but have also removed the danger of dysentery as an epidemic disease. Penicillin is ineffective against dysentery bacilli. American authors report that the sulphapyrimidine compounds are most effective here.

Results with the sulphonamides in typhoid and paratyphoid fever have so far been unsatisfactory. According to experimental results (given below) obtained with sulphapyrimidine compounds (Debenal, Debenal M) in paratyphoid B infections we can perhaps also look for some success in practice.

The effects of sulphonamides on paratyphoid B infection are illustrated by the following experiments:

Mice infected by intraperitoneal injection with 0.5 cc of a 24-hour broth culture diluted 1:50. One subcutaneous or intramuscular injection 1 hour after infection, 2 animals in each case, all preparations 2% 0.2; 0.3; 0.5; 0.8; and 1.0 cc.

		Number of animals	Alive 48 hours after infection	Alive 1 week after infection	Alive 2 weeks after infection
Controls		20	10	5	
Debenal	subcutaneous injection	10	9 ⎱ 19	8 ⎱ 18	7 ⎱ 15
	intramuscular injection	10	10 ⎰	10 ⎰	8 ⎰
Debenal M	subcutaneous injection	10	9 ⎱ 18	9 ⎱ 18	6 ⎱ 14
	intramuscular injection	10	9 ⎰	9 ⎰	8 ⎰
De-Ma	subcutaneous injection	10	10 ⎱ 20	10 ⎱ 19	6 ⎱ 12
	intramuscular injection	10	10 ⎰	9 ⎰	6 ⎰

With still more intense infection and 3 corresponding injections after 1, 6, and 24 hours the following result was obtained.

		Number of animals	Alive 48 hours after infection	Alive 1 week after infection	Alive 2 weeks after infection
Controls		20	3	1	1
Debenal	subcutaneous injection	10	7 } 15	5 } 10	2 } 4
	intramuscular injection	10	8	5	2
Debenal M	subcutaneous injection	10	6 } 13	2 } 5	2 } 3
	intramuscular injection	10	7	3	1
De-Ma	subcutaneous injection	10	9 } 18	6 } 11	1 } 6
	intramuscular injection	10	9	5	5

De-Ma in solution was to some extent even more effective.

Experiment 2.3.1947. Mice infected intraperitoneally with 0.5 cc of a 24-hour broth culture diluted 1:20. Animals, 2 in each case, treated 3 times with 2% 0.1; 0.2; 0.3; 0.4; 0.5; by intravenous injection; and 0.2; 0.3; 0.5; 0.8; 1.0; by subcutaneous injection and by mouth.

		Number of animals	Alive 48 hours after infection	Alive 2 weeks after infection
Controls		20	11	2
De-Ma soluble	intravenous injection	10	7 } 26	5 } 22
	subcutaneous injection	10	9	8
	by mouth	10	10	9
Debenal in aqueous suspension	2% subcutaneous injection	10	10 } 20	7 } 15
	2% by mouth	10	10	8
Debenal M	2% subcutaneous injection	10	10 } 20	7 } 14
	2% by mouth	10	10	7

With coli infections it is much more difficult to arrive at a definite assessment of the effect; this is because, even with the controls, the course of the infection is often irregular. Nevertheless, such an assessment can be made if experiments are constantly repeated with different culture dilutions. The following experiment, in which culture dilutions of 1:3 and 1:5 were used for infecting the animals, gives some information on the effectiveness of Debenal and De-Ma. Dose 5%, 2 animals in each case, subcutaneous injection and by mouth 0.2; 0.5; 1.0 cc.

	Number of animals	Alive 24 hours after infection	Alive 48 hours after infection	Alive 1 week after infection
Controls:				
Culture dilution 1:3	12	5	2	0
Debenal subcutaneous injection	6	1 } 5	1 } 4	1 } 3
by mouth	6	4	3	2
De-Ma subcutaneous injection	6	4 } 10	4 } 9	4 } 9
by mouth	6	6	5	5
Controls:				
Culture dilution 1:5	12	9	7	5
Debenal subcutaneous injection	6	5 } 11	3 } 9	2 } 7
by mouth	6	6	6	5
De-Ma subcutaneous injection	6	6 } 12	6 } 12	5 } 9
by mouth	6	6	6	4

In this context it is not possible to go into any great detail on the use of sulphonamides against many other infections, such as undulant fever, actinomycosis, etc.

With undulant fever definite results are obtained only after prolonged treatment.

Rats infected by intraperitoneal injection with 0.5 cc of a 48-hour agar-broth slope. One injection 1 hour after infection, then daily for 14 days, 2 animals in each case, 5% 0.2; 0.5; 1.0; 2.0 cc.

	Number of animals	Alive 48 hours after infection	Alive 1 week after infection	Alive 2 weeks after infection
Controls	8	5	0	0
Prontalbin 5% subcutaneous injection	8	7	5	3
Tibatin 5% subcutaneous injection	8	8	7	6
Debenal M 5% subcutaneous injection	8	8	7	6

I should also like to draw attention to a recent report by Klee that Tibatin has proved particularly satisfactory in the treatment of lung gangrene.

Information on the use of sulphonamides in relatively rare diseases is given in Domagk-Hegler: *Chemotherapie bakterieller Infektionen* (Chemotherapy of bacterial infections), 3rd ed., Hirzel, Leipzig, 1944, and Domagk: *Pathologische Anatomie und Chemotherapie bakterieller Infektionen* (Pathological anatomy and chemotherapy of bacterial infections), Thieme, Stuttgart. Information on the innumerable other sulphonamides which have been tested and used is given in Mietzsch's survey: «Therapeutisch verwendbare Sulfonamid- und Sulfonverbindungen» (Therapeutically useful sulphonamide and sulphone compounds), Suppl. 54 (1945) to «Chemie», *Z. Ver. Deut. Chemiker*.

A fourth phase in chemotherapy of bacterial infections seems to be opening up in view of the fact that the sulphathiazole and sulphathiodiazole compounds have had such a specific action against tubercle bacilli that further progress at last seems possible in this direction too. In 1940 I reported the observation that sulphathiazole in concentrations of 1:5000 has a specific inhibiting action on tubercle bacilli, whilst other sulphonamides and sulphones, as well as Diasone and Promin (which are frequently referred to in the literature), do not. Only the thiazole and thiodiazole derivatives have this effect. Further experiments showed that these sulphonamides have this specific effect against the human as well as the bovine and avian types – temporarily in concentrations as low as 1:50,000 or even 1:100,000, i.e. sulphonamide concentrations which can easily be achieved in the human organism, and with very small doses at that.

Even at that time I suggested that sulphathiazole and its derivatives might be effective against human disease, but since then further results have shown me that an effect can be achieved with even smaller doses than were then being recommended, provided that these small doses are given for weeks and perhaps months on end. Experiments have shown that it is best to begin treatment with small doses, especially where many bacilli are present and where treatment is to be continued for a considerable period. Otherwise, as a result of the damage suffered by the tubercle bacilli and of their accelerated decomposition, too many toxins are liberated and in consequence undesirable abscess formations and sequestrations as well as general detrimental effects are liable to occur, as after the administration of tuberculin. It was in fact found from long-term experiments on animals that the largest doses were by no means always the most effective. In critical cases in man the risk attendant upon very high doses would of course have to be accepted. In cases of meningitis tuberculosa there should probably be no hesitation in giving high doses. Where the patient's life is not in danger, however, minimal doses, 1 × 0.25 g or 2 × 0.25 g daily, should be administered at first; then, after 2–3 weeks, 3 × 0.25 or 4 × 0.25 g daily, but the doses should never be increased until the patient feels better.

My colleagues in chemistry Dr. Behnisch, Prof. Schmidt and Dr. Mietzsch and I finally arrived at still more effective compounds which were no longer sulphonamides at all – for, as I have shown, the specific effect had up to that time been detectable only with the sulphonamide compounds of the thiazole and thiodiazole series. It was Behnisch who made available to me for chemo-

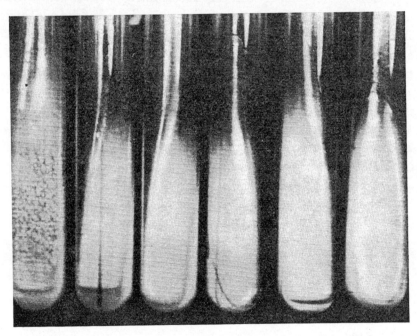

Fig. 2. Inhibiting effect of sulphathiazole on human-type tubercle bacilli from sputum Fr. Di.

Tube 1 (control with good growth, 6 weeks after inoculation)
Tube 2 contains 1 :   5,000 sulphathiazole
Tube 3 contains 1 :   10,000 sulphathiazole
Tube 4 contains 1 :   25,000 sulphathiazole
Tube 5 contains 1 :   50,000 sulphathiazole
Tube 6 contains 1 : 100,000 sulphathiazole

In this experiment no cultures of tubercle bacilli were visible in any of the tubes containing sulphathiazole, 6 weeks after inoculation.

therapeutical trials the first thiosemicarbazones of cyclic aldehydes, which
he was using as initial products in the manufacture of sulphathiodiazoles and
which represent the open-ring preliminary stage in the 2-aminothiodiazoles
necessary for this. The thiosemicarbazones of cyclic aldehydes and ketones
showed a powerful inhibiting action on tubercle bacilli. In this way a new
chemotherapeutical principle, which is not dependent on the presence of the
sulphonamide group, was extracted from the principle of the sulphathiazoles
and sulphathiodiazoles. In the light of experiments the thiosemicarbazones
even appear in some respects to have advantages over the sulphathiazoles and
sulphathiodiazoles. Thus, their effectiveness against the various types of
tubercle bacillus is not reduced by para-aminobenzoic acid, protein breakdown products, Campolon or other substances which, as already shown, can
considerably impair the effectiveness of sulphathiazole and of the sulphathiodiazoles. Tuberculin is not one of the substances which impair the effectiveness against tubercle bacilli of those sulphonamides which in themselves are
effective. In animal experiments the effectiveness of compounds containing
no sulphonamides was for the most part even clearer than with the above-
mentioned active sulphonamides of the thiazole and thiodiazole series. This
new class of active substances was first reported in a joint publication by
R. Behnisch, F. Mietzsch, H. Schmidt and myself in *Naturwiss.*, 33, No. 10
(1946). The following compounds proved to be very effective:

Tb I/698 =

Schmidt (Sdt) 1041 $CH_3CO \cdot NH\langle\quad\rangle CH=N \cdot NH \cdot C \cdot NH_2$
$$\overset{\|}{S}$$

Tb II/242 =

Schmidt (Sdt) 1075 $CH_3O\langle\quad\rangle CH=N \cdot NHC \cdot NH_2$
$$\overset{\|}{S}$$

Tb III/1374 =

Behnisch (Be) 1374 $C_2H_5 \cdot SO_2\langle\quad\rangle CH=N \cdot NH \cdot C \cdot NH_2$
$$\overset{\|}{S}$$

whereas *Tb IV* is a thiodiazole compound which is sensitive to *p*-aminoben-
zoic acid.

Experiment with human type, inoculation on 24th April; first collection
(I) on 8th May, 1947; second collection (II) on 14th May.

	1:5000		1:10,000		1:25,000	
	I	*II*	*I*	*II*	*I*	*II*
Eleudron-Na	o	o	o	o	o	o
TbI/698	o	o	o	o	o	o
Eleudron/Tb I $\overline{aa}$*	o	o	o	o	o	o
Tb II/242	o	o	o	o	o	o
Eleudron Tb II $\overline{aa}$	o	o	o	o	o	+
Tb III/1374	o	o	o	o	o	o
Eleudron/Tb III $\overline{aa}$	o	o	o	o	o	o
Tb IV	o	o	o	o	o	o
Eleudron/Tb IV $\overline{aa}$	o	o	o	o	o	+
Badional	+	+	+	+	++	+++
Sulphapyridine	++	+++	++	+++	++	+++
Prontalbin	++	+++	++	+++	++	+++

* Eleudron and Tb I/698 mixed in equal parts.

Controls: First collection (I) after 14 days: ++/+++

Second collection (II) after 3 weeks: +++

(o = no growth; ++ = pronounced growth, but no spore formation; +++ = intense growth.)

The same experiment but with *p*-aminobenzoic acid added to the nutrient medium at the rate of 1:10,000.

	1:5000		1:10,000		1:25,000	
	I	*II*	*I*	*II*	*I*	*II*
Eleudron-Na	o	o	o	+	++	+++
Tb l/698	o	o	o	o	o	o
Eleudron/Tb I $\overline{aa}$	o	o	o	o	o	o
Tb II/242	o	o	o	o	o	o
Eleudron/Tb II $\overline{aa}$	o	o	o	o	o	++
Tb III/1374	o	o	o	o	o	+
Eleudron/Tb III $\overline{aa}$	o	o	o	o	+	+
Tb IV	o	o	o	o	++	+++
Eleudron/Tb IV $\overline{aa}$	++	+++	++	+++	++	+++
Badional	++	+++	++	+++	++	+++
Sulphapyridine	++	+++	++	+++	++	+++
Prontalbin	++	+++	++	+++	++	+++

Controls: First collection (I) after 14 days: ++/+++

Second collection (II) after 3 weeks: +++

Experiment with bovine type.

	1:5000		1:10,000		1:25,000	
	I	II	I	II	I	II
Eleudron-Na	o	o	o	o	o	o
Tb I/698	o	o	o	o	o	o
Eleudron/Tb I aa	o	o	o	o	o	o
Tb II/242	o	o	o	o	o	o
Eleudron/Tb II aa	o	o	o	o	o	+
Tb III/1374	o	o	o	o	o	o
Eleudron/Tb III aa	o	o	o	o	o	o
Tb IV	o	o	o	o	o	o
Eleudron/Tb IV aa	o	o	o	o	o	+
Badional	+	+	++	++	++	++/+++
Sulphapyridine	(+)	+	++	++/+++	++	++/+++
Prontalbin	+	++/+++		++/+++	++	++/+++

Controls: First collection (I) after 14 days: +/++
Second collection (II) after 3 weeks: ++/+++

Same experiment with bovine type, but with para-aminobenzoic acid added to the nutrient medium at the rate of 1 : 10,000.

	1:5000		1:10,000		1:25,000	
	I	II	I	II	I	II
Eleudron-Na	o	o	o	o	o	++
Tb I/698	o	o	o	o	o	o
Eleudron/Tb I aa	o	o	o	o	o	o
Tb II/242	o	o	o	o	o	o
Eleudron/Tb II aa	o	o	o	o	o	+
Tb III/1374	o	o	o	o	o	+
Eleudron/Tb III aa	o	o	o	o	o	+
Tb IV	o	o	o	o	+	++
Eleudron/Tb IV aa	+	++	+	++	+	++
Badional	+	++	+	++	+	++/+++
Sulphapyridine	++	++/+++	++	++/+++	++	++/+++
Prontalbin	+	++/+++	+	++/+++	+	++/+++

Controls: First collection (I) after 14 days: ++/+++
Second collection (II) after 3 weeks: +++

Like *p*-aminobenzoic acid, liver extracts, e.g. Campolon, and other proteolytic products develop an antisulphonamide action with regard to tubercle bacilli. Koch's original old tuberculin did not show this effect in the concentrations tested (1:100).

In some cases in experimental animals which had only received 1 local application it was possible to detect some effect from the drugs and to differentiate between them.

Guinea-pigs were infected in the back with an intramuscular injection of 3 mg of a human type culture. After 3 days the infection site was incised and 0.5 g of the preparation was applied locally. In the controls extensive caseating foci had already appeared after 4 weeks in the area of the injection site on the back. In the internal organs – mainly lung, liver, spleen – nodules and sometimes necrotic areas could be seen under the microscope and usually with the naked eye. Eight weeks after being infected none of the control animals was alive. In the case of the animals which were treated locally with Tb I, after 4 weeks small fusion foci only, with increased reaction of the connective tissue in the surrounding area, were apparent in the musculature; in addition, in the area round these foci the experimental animals, in contrast to the controls, showed only isolated and non-typical epitheloid nodules. In some of the animals which were treated locally with Tb I/698 there was no fusion whatever at the infection focus but only a proliferation of connective tissue with individual histocytes, though without epitheloid nodules. Another

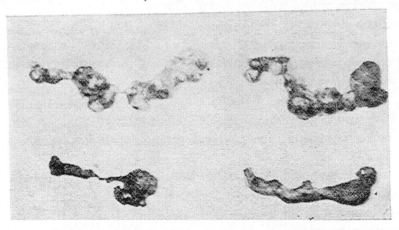

Fig. 3. Guinea-pigs infected intraperitoneally with human type tubercle bacillus
(*Above*): Controls. Large tuberculous foci in the omentum.
(*Below*): The same infection as above with animals treated with Tb I/698. Small tuberculous foci only.

striking fact was that only very isolated small nodules could be found in the lung, liver or spleen, whereas in the untreated controls there were comparatively extensive tuberculous changes. Evidence of inhibition of tuberculous tissue changes was also to be found in the animals treated with Tb II and Tb III, but was not so pronounced as with Tb I. In the animals treated with Tb IV the disintegrated reticular epitheloid nodules around the local necrotic areas of the primary focus were very noticeable. In the animals which received one local treatment with Eleudron the effect was still less than with Tb IV. All the animals showed numerous epitheloid nodules in lungs, liver and spleen, although in some cases with non-typical disintegration.

With animals infected by intraperitoneal injections of large doses of tubercle bacilli the caseation processes appeared mainly in the omentum, but subsequently also in the spleen, liver and lung. These latter organs showed caseation as well as productive foci, depending on the quantities of tubercle bacilli injected. In the treated animals the caseated foci remained smaller, and in some cases processes (mainly productive)appeared which, if treatment was continued for a sufficiently long period, changed into unspecific granulation scar tissue.

In the treatment of experimental animals substances of this kind with a highly effective specific action against tubercle bacilli gave the following results:

1. A local effect with powder applied locally.
2. When administrered by mouth and by subcutaneous injection, in cases of intraperitoneal infection in the omentum and elsewhere a general effect on tuberculous processes at some distance from the site of application.
3. Absence or delay of generalization.

Histologically the treated animals–unlike the heavily infected untreated controls, in which caseation was present–showed:

1. Leucocytic inflammation. Caseous disintegration of the leucocytes clearly retarded.
2. Considerable numbers of histocytes with accumulation of Indian ink, where Indian ink had been added to the tubercle bacilli at the time of infection.
3. Non-typical nodules.
4. Non-typical giant cells.
5. Completely unspecific granulation tissue.

The nature of the histological tissue reaction appears to be entirely independent of the immunity or resistance, and to be dependent only on the number of bacilli capable of developing when exposed to the effects of the drug. Thus, depending on the degree of success, we see relatively slight caseation in the treated animals, more proliferative processes with occurrence of non-typical cells, and finally unspecific granulation tissue formation. In experiments where only few bacilli are used for infection and where primary caseating processes do not occur, the specific productive foci appearing in the controls sometimes turn into wholly non-typical histocytic foci and finally into completely unspecific granulation and scar tissue. In the animals treated with the optimum doses no tuberculous foci whatever develop in the other organs, whilst with sub-optimum doses the development of these foci is inhibited. Unlike the controls, the animals in which the seat of the primary infection was the vestibule of the eye showed caseous fusion. Even when, in heavily infected animals, destruction occurred owing to the effect of the toxin, unspecific granulation tissue foci covered with epithelium also developed here at a later stage in the treated animals. In contrast to the untreated controls, in which numerous epithelial nodules appeared in the spleen, liver and lung, the treated animals showed no tuberculous foci in these organs, either upon visual inspection or under the microscope. With rabbits, among which the controls showed tuberculous changes in the lungs, kidneys and other organs, it was likewise possible to prevent the development of tuberculous processes by means of long-term treatment with the active substances. It remains to be seen whether or not the substances which were found to be effective against tuberculosis in the experiment will be sufficiently effective in clinical practice.

The effectiveness of the substance Tb I/698 in the case of tuberculosis of the skin in man has been confirmed. Moncorps and Kalkoff have reported on this. The successes with lupus of the skin reported by Moncorps and Kalkoff for the first time at the 1947 Rhine-Westphalian Dermatology Conference have since been corroborated by Grütz, P. Schmidt, Koch, Hartung and others. Eickhoff of the Loebell Clinic for Ear, Nose and Throat Diseases in Münster, Westphalia, reported on the effectiveness of the preparation Tb I/698 in cases of lupus of the mucous membranes at the Congress of Ear, Nose and Throat Specialists in Hamburg on 7th and 8th November, 1947. He described clinical and histological cures of patients who had been treated for years in hospital without success. Average length of treatment to-date: 110 days with a dose of 1 g per day.

In the light of results of resistance tests on tubercle bacilli from sputum I have reason to assume that there are also tubercle bacilli with differing degrees of resistance to sulphathiazole and other substances which appear suited to chemotherapy of tuberculosis. However, if we were able to devise a method of chemotherapy of tuberculosis in man on the basis of the experimental findings so far available, even this measure might result in failure unless from the outset we took the utmost care to ensure that patients with overt tuberculosis who were spreading the bacilli were isolated until completely cured. Otherwise, even with the very best drugs against tuberculosis we should perhaps see the same result as we are now beginning to see in the treatment of gonorrhoea with sulphonamides and penicillin – drugs with which, if the correct procedure had been adopted, gonorrhoea could no doubt have been eradicated.

By far the greatest danger threatening us at the present time is pulmonary tuberculosis, especially where large parts of Europe have been turned into famine areas, for this disease will not stop at the boundaries of the stricken areas of central Europe. Whether chemotherapy of tuberculosis, which is only in its infancy, has yet reached the stage of being really effective here is a question which cannot be answered until further investigations have been carried out by many experienced clinicians on the strictest scientific lines. At a meeting of the Flensburg Medical Association in November 1949 F. Kuhlmann reported for the first time favourable results in patients seriously ill with pulmonary tuberculosis. The successes took the form of disappearance of the fever (some of the patients having suffered from very prolonged high fever), diminution of sputum and sometimes disappearance of bacilli from the sputum, a return to normal in blood sedimentation and gains in weight. Kuhlmann reported that the patients themselves asked for the treatment and that the mortality rate in the department for patients seriously ill with pulmonary tuberculosis fell from 7–8% to 1–2% per month with approx. 160 beds occupied. He pointed out that with the fall in the mortality curve the type of case accommodated in the department changed somewhat because sanatorium patients were temporarily admitted (60), but that only patients selected for the seriousness of their condition were transferred there. He also allowed for the fact that the hot dry summer might have had a beneficial effect, but considered that the possibility that hospitalization itself had some effect could be ruled out in the case of the large majority of patients. X-ray photographs showed that in many of the patients who had received treatment there was extensive resorption of the exudative changes. Ausculation lagged behind radiography in detecting a decrease in the infection. According to Kuhlmann, in the patients who received treatment there was a surprising change in the course of the disease. Nevertheless, it would be too early to speak of a cure, although Kuhlmann said that in some cases the word «cure» suggested itself. We shall therefore have to wait and see whether in these cases the cure is definite and whether the successes reported by Kuhlmann can be regularly repeated with the same treatment. However, it would appear that for the first time we are in possession of substances which bring chemotherapy of pulmonary tuberculosis within striking distance. Many lines of inquiry were pursued before we reached this goal. The least hint that a substance might be effective was recorded and followed up by the chemists who, on the strength of it, made innumerable syntheses. And how often in vain! Chemotherapy of tuberculosis in its present state is also a late fruit of sulphonamide research, for which the foundations were laid in my joint work with Klarer and Mietzsch. Which cases of tuberculosis should receive chemotherapeutic treatment, is a question which is constantly being asked, and in reply I would refer to an axiom which I often quote: I consider it my first duty in the development of chemotherapy to cure those diseases which have hitherto been incurable, so that in the first place those patients are helped who can be helped in no other way.

Anselmino, «Die Sulfonamidbehandlung der fieberhaften Fehlgeburt.» (Sulphonamide treatment in septic abortion.), *Deut. Med. Wochschr.*, (1947) 63.

Bernhard, «Die De-Ma Therapie puerperaler Infektionen.» (De-Ma therapy in puerperal infections.), *Deut. Med. Wochschr.*, (1947) 66.

Bernhard, «Die Sulfonamidbehandlung der Gonorrhoe der Frau.» (Sulphonamide treatment of gonorrhoea in women.), *Deut. Med. Wochschr.*, (1947).

Boas and Marcussen, *Ugeskrift Læger*, 106 (1944) 16.

Bunn et al., *J. Am. Med. Assoc.*, 129 (1945).

Bunn et al., «Beitrag zur Penicillinbehandlung chirurgischer Infektionen.» (On penicillin treatment of surgical infections.), *Deut. Med. Wochschr.*, (1946) 291.

Bürger, «Klinik und Sulfonamidtherapie der in Leipzig gehäuft auftretenden ruhrartigen Darmerkrankungen.» (Clinical treatment and sulphonamide therapy of the dysentery-like intestinal disorders occurring frequently in Leipzig.), *Med. Welt*, (1943) 301.

Clarke, «Penicillin: help or hindrance in venereal disease control?», *J. Soc. Hyg.*, No. 9 (1945).

Christie, «Subakute bakterielle Endocarditis, die offensichtlich durch Sulfanilamidbehandlung geheilt wurde.» (Subacute bacterial endocarditis obviously cured by sulphonamide treatment.), *J. Am. Med. Assoc.*, 115 (1940) 1357.

Christie, «Penicillin in subacute bacterial endocarditis.», *Brit. Med. J.*, (1946) 381.

Colebrook and Purdie, «Treatment of 106 cases of puerperal fever by sulphanilamide (Streptocide).», *Lancet*, (1937) 1237, 1291.

Colebrook, «Chemotherapy in relation to war wounds.», *Brit. Med. J.*, (1940) 448.

Colebrook, «Sulphonamides locally.», *Lancet*, (1944) 22.

Domagk, «Ein Beitrag zur Chemotherapie der bakteriellen Infektionen.» (On chemotherapy of bacterial infections.), *Deut. Med. Wochschr.*, (1935) 250.

Domagk, *Chemotherapie bakterieller Infektionen.* (Chemotherapy of bacterial infections.), Monographie, 3rd ed., Hirzel, Leipzig, 1944.

Domagk, *Pathologische Anatomie und Chemotherapie bakterieller Infektionen.* (Pathological anatomy and chemotherapy of bacterial infections.), Thieme, Stuttgart, 1947.

Domagk, «Neuere Erfahrungen bei experimentallen Infektionen.» (Recent experiences with experimental infection.), *Z. Haut-Geschlechtskrankh.*, 3 (1947) 357.

Domagk, «Der derzeitige Stand der Chemotherapie bakterieller Infektionen mit den Sulfonamiden.» (The present situation with regard to chemotherapy of bacterial infections with sulphonamides.), *Deut. Med. Wochschr.*, (1947) 6.

Domagk, Behnisch, Schmidt, and Mietzsch, «Über eine neue, gegen Tuberkelbazillen in vitro wirksame Verbindungsklasse.» (On a new class of compounds effective in vitro against tubercle bacilli.), *Naturwiss.*, 33 [10] (1946) 315.

Domagk, «Chemotherapie der Tuberkulose.», (Chemotherapy of tuberculosis.) *Rhine-Westphalian Dermatology Conferences*, May and September, 1947).(See also: *Beitr. Klin. Tuberk.*, 101 (1948) 365).

Eickhoff, *Ear, Nose, and Throat Congress.*, 7–8th, November, 1947, Hamburg.

Felke, «Die Chemotherapie der Gonorrhoe mit Sulfonamidverbindungen.» (Chemotherapy of gonorrhoea with sulphonamide compounds.), *Arch. Dermatol.*, 178 (1938) 45.

Felke, «Über den Wirkungsmechanismus der antibakteriellen Chemotherapie bei der Gonorrhoe.» (On the operative mechanism of antibacterial chemotherapy in gonorrhoea.), *Arch. Dermatol.*, 178 (1938), 152.

Fischer, «Die Sulfonamide in der Chirurgie.» (The sulphonamides in surgery.),*Med. Klin.*, (1944) 418.

Frisk, Hagerman, Helander, and Sjögren, «Sulpha-Combination, a new chemotherapeutic principle.», *Brit. Med. J.*, (1947) 7–10.

Gehrt, «Behandlung der Meningitis epidemica mit Uliron.» (Treatment of meningitis epidemica with Uliron.), *Deut. Med. Wochschr.*, (1938) 409.

Grütz and Krömer, «Überblick über den gegenwärtigen Stand der Chemotherapie der Gonorrhoe.» (Survey of the present situation with regard to chemotherapy of gonorrhoea.), *Jahrkurse. Ärztl. Fortbild.*, [4] (1939) 37.

Haferland, «Zur Behandlung infizierter und infektionsgefährdeter Wunden mit dem Sulfonamidgemisch Marfanil-Prontalbin.» (On treatment of infected wounds and wounds threatened by infection with the sulphonamide combination Marfanil-Prontalbin.), *Arch. Klin. Chir.*, 202 (1941) 580.

Haferland, «Die Sulfonamidbehandlung in der Chirurgie.» (Sulphonamide treatment in surgery.), *Med. Welt.*, (1942) 380.

Hagerman, «Studien zur Chemoresistenz der Gonokokken mittels einer neuen Methode zur Bestimmung der Chemoresistenz in vitro.» (Studies on the chemoresistance of gonococci by means of a new method for the determination of chemoresistance in vitro.), *Acta Pathol. Microbiol. Scand.*, Suppl. 46 (1942) 180 pp. Ref. *Med. Klin.*, (1943) 594.

Hartung, *Dermatology Conference*, Hannover, 1947.

Heilmeyer and Keiderling, «Klinische Wirkungen einer Sulfonamidkombination mit verschiedenem Angriffspunkt (De-Ma) auf den Ablauf sulfonamidresistenter septischer Krankheitsbilder und ihre Erklärung durch die Theorie einer mehrfachen Fermentkettenblockierung.» (Clinical effects of a combination of sulphonamides with different points of attack (De-Ma) on the development of sulphonamide-resistant septic conditions and their explanation by the theory of multiple ferment chain blocking.), *Deut. Med. Wochschr.*, (1947) 13.

Henrich and Winkelsträter, *Winkelsträter's thesis*, Univ. Münster, 1946.

Hopf, «Die Feststellung der endgültigen Heilung der männlichen Go.» (Confirming a final cure in cases of gonorrhoe in men.), *Münch. Med. Wochschr.*, (1943) 284.

Huriez and Dessurmont, *Presse Med.*, 55 [2] (1947), Ref. *Z. Haut-Geschlechtskrankh.*, 2 (1947) 217.

Kalkoff, «Chemotherapie des Lupus vulgaris.» (Chemotherapy of lupus vulgaris.), *Z. Haut-Geschlechtskrankh.*, 3 (1947) 280.

Klarer, «Über die chemische Konstitution des Marfanil (Mesudin).» (On the chemical constitution of Marfanil (Mesudin).), *Klin. Wochschr.*, 20 (1941) 1250.

Klarer, «Entwicklung der Sulfonamidtherapie.» (Development of sulphonamide therapy.), *Chemie*, 56 (1943) 10.

Klee and Römer, «Prontosil bei Streptokokkenerkrankungen.» (Prontosil in streptococcal infections.), *Deut. Med. Wochschr.*, (1935) 250.

Koch, *Dermatology Conference*, Münster 1947 and Hamburg.

Konjetzny, *Zentr. Chir.*, 137 (1940) 2745.

Konjetzny, «Gasödemerkrankungen und ihre Behandlung.» (Gas gangrene infections and their treatment.), *Med. Welt*, Nr. 40 (1940).

Konjetzny, «Operative Wundversorgung.» (Operative wound treatment.), *Med. Welt*, (1941) 409.

Konjetzny, «Die Behandlung von Phosphorschädigungen der Körperoberfläche, von Verbrennungen, Verbrühungen und Verätzungen.» (Treatment of phosphorus damage to the body surface, of burns, scalds and erosions.), *Med. Welt*, (1944) 69.

Klostermeyer, «Fortschritte in der örtlichen Behandlung schwer vereiterter schussverletzter Kniegelenke.» (Advances in the local treatment of severely suppurated shot wounds in the knee joint.), *Zentr. Chir.*, (1944) 1253.

Kramer, «Fortschritte auf dem Wege zu einer Chirurgie ohne Eiter.» (Advances towards surgery without suppuration.), *Deut. Med. Wochschr.*, (1947).

Krueger, «Sulfonamide an der Front.» (Sulphonamides at the front.), *Deut. Med. Wochschr.*, (1943) 417.

Kuhlmann, *Flensburg Medical Conference*, November, 1947; and *Med. Monatsschr.*, (1948) 297.

Lezius, *Conference of North-West German Surgeons*, Lübeck.

Long, «Medizinischer Fortschritt und medizinische Erziehung während des Krieges.»

(Medical progress and medical training during the war.), *J. Am. Med. Assoc.*, 130 (1946) 983.

Martin, «Die Bedeutung der Sulfonamide für die Geburtshilfe und die Frauenheilkunde.» (The importance of sulphonamides in obstetrics and gynaecology.), *Ärztl. Wochschr.*, (1947) 489.

Mietzsch, «Zur Chemotherapie der bakteriellen Infektionskrankheiten.» (On chemotherapy in infectious bacterial diseases,), *Ber. Deut. Chem. Ges.*, 71 A (1938) 15.

Mietzsch-Klarer, *Med. Chem.*, 4 (1943) 73.

Mietzsch, «Therapeutisch verwendbare Sulfonamid- und Sulfonverbindungen.», (Sulphonamide and sulphone compounds suitable for therapeutical use,). Suppl. *Z. Ver. Deut. Chemiker*, (1945).

Moncorps and Kalkoff, «Vorläufige Ergebnisse einer Chemotherapie der Hauttuberkulose,» (Provisional results of chemotherapy of tuberculosis of the skin,), *Z. Haut- Geschlechtskrankh.*, 3 (1947) 358. Orig. *Med. Klin.*, (1947) 812.

Schmidt, P., *Dermatology Conference*, Hamburg.

Schmidt, P., «Tuberkulose.» (Tuberculosis.), *Ärztl. Wochschr.*, 39–40 (1947) 633.

Schmidt, W., «Die Sulfonamidbehandlung der Angina lacunaris unter besonderer Berücksichtigung der lokalen Behandlung.» (Treatment of follicular tonsillitis with sulphonamides, with special reference to local treatment.), *Therap. Gegenwart*, [1–2] (1944) 21.

Schreus, «Felderfahrungen über die Chemotherapie der Anaerobenwundinfektion insbesondere mit Globucid nebst Bemerkungen über Chemoprophylaxe.» (War experience of chemotherapy of anaerobic wound infection, in particular with Globucide, together with observations on chemoprophylaxis.)

Schmith, *Studies on the effect of sulphapyridine on pneumococci and gonococci*, Munksgaard, Copenhagen, 1941.

Schütz, *Conference of North-West German Surgeons*, Lübeck.

Tönnis, «Die Behandlung der Hirnverletzungen auf Grund der Erfahrungen im Feldzug gegen Polen.» (Treatment of brain injuries in the light of experience gained during the Polish campaign.), *Deut. Med. Wochschr.*, (1940) 57.

Tréfouël, Tréfouël, Nitti, and Bovet, «Activité du *p*-aminophenylsulfonamide sur les infections streptococciques expérimentales de la souris et du lapin.» (Activity of *p*-aminophenylsulphonamide against experimental streptococcal infections in the mouse and rabbit.), *Compt. Rend. Soc. Biol.*, 120 (1935) 756.

Unterberger, «Fortschritte und Ergebnisse mit der Sulfonamid-Behandlung der otogenen Meningitis und anderer intrakranieller Verwicklungen.» (Progress in and results from sulphonamide treatment of otogenous meningitis and other intracranial complications.), *Wien. Med. Wochschr.*, 92 (1942) 133.

Unterberger, «Über klinische und pathologisch-anatomische Beobachtungen der otogenen Meningitis bei der Sulfonamidtherapie.» (Clinical and pathological-anatomical observations on otogenous meningitis with sulphonamide therapy.), *Med. Klin.*, 39 (1943) 599.

Veltman, «Über die Klinik der Sulfonamidschäden.» (On the clinical treatment of ill-effects from sulphonamides.), *Med. Klin.*, (1946) 607.

Veltman, *Z. Haut- Geschlechtskrankh.*, 2 (1947) 203.

Whitby, «An experimental assessment of the therapeutic efficacy of amino-compounds with special reference to *p*-benzylamino-benzenesulphonamide.», *Lancet*, 232 (1937) 1517.

Whitby, Camb, and Land, «Chemotherapy of pneumococcal and other infections with 2-(*p*-aminobenzenesulphonamido) pyridine.», *Lancet*, 234 (1938) 1210.

Zeissler, «Welche Voraussetzungen muss die Technik von Kulturversuchen zur Bestimmung der bakteriostatischen Wirkung der Sulfonamide erfüllen?» (What are the conditions which must be fulfilled by a cultural test technique for determination of the bacteriostatic action of sulphonamides?).

Zeissler, «Beitrag zur Ätiologie der Gasödeme des Menschen. Die bakteriologische Ernte zweier Weltkriege.» (On the aetiology of gas gangrene in man. The bacteriological aftermath of two world wars.), *Deut. Med. Wochschr.*, (1946) 171.

For further bibliography, see:

Domagk, *Pathologische Anatomie und Chemotherapie der Infektionskrankheiten.* (Pathological anatomy and chemotherapy of infectious diseases.), Thieme, Stuttgart, 1948, and *Beitr. Klin. Tuberk.*, 101 (1948) 365.

Biography

Gerhard Johannes Paul Domagk was born on October 30, 1895, at Lagow, a beautiful, small town in the Brandenburg Marches. Until he was fourteen he went to school in Sommerfeld, where his father was assistant headmaster. His mother, Martha Reimer, came from farming stock in the Marches, where she lived in Sommerfeld until 1945 when she was expelled from her home; she died from starvation in a refugee camp.

Domagk himself was, from the age of 14, at school in Silesia until he reached the upper sixth form. He then became a medical student at Kiel and, when the 1914–1918 War broke out, he served in the Army, and in December 1914 was wounded. Later he was sent to join the Sanitary Service and served in, among other places, the cholera hospitals in Russia. During this time he was decisively impressed by the helplessness of the medical men of that time when they were faced with cholera, typhus, diarrhoeal infections and other infectious diseases. He was especially strongly influenced by the fact that surgery had little value in the treatment of these diseases and even amputations and other forms of radical treatment were often followed by severe bacterial infections, such as gas gangrene.

In 1918 he resumed his medical studies at Kiel and in 1921 he took his State Medical Examinations and graduated. He undertook laboratory work under Max Bürger on creatin and creatinin, and later metabolic studies and analysis under Professors Hoppe-Seyler and Emmerich.

In 1923 he moved to Greifswald and there became, in 1924, University Lecturer in Pathological Anatomy. In 1925 he held the same post in the University of Münster and in 1958 became professor of this subject. During the years 1927–1929 he was, however, given leave of absence from the University of Münster to do research in the laboratories of the I.G. Farbenindustrie, at Wuppertal. In 1929 a new research institute for pathological anatomy and bacteriology was built by the I.G. Farbenindustrie and there, in 1932, Domagk made the discovery for which his name is so well known, the discovery that earned him the Nobel Prize in Physiology or Medicine for 1939, namely, the fact that a red dye-stuff, to which the name « prontosil rubrum » was given, protected mice and rabbits against lethal doses of staphylococci and haemolytic streptococci. Prontosil was a derivative of sulphanilamide (*p*-aminobenzenesulphonamide) which the Viennese chemist, Gelmo, had synthesized in 1908.

Domagk was, however, not satisfied that prontosil, so effective in mice, would be equally effective in man, but it so happened that his own daughter became very ill with a streptococcal infection, and Domagk, in desperation, gave her a dose of prontosil. She made a complete recovery, but Domagk omitted mentioning the recovery of his daughter from the report on the effect of the drug, waiting until 1935 when results were available from clinicians who had tested the new drug on patients. During subsequent years much work was done in various countries on this class of antibacterial compound and some thousands of derivatives of sulphanilamide have been produced and tested for their antibacterial properties. Domagk's work has thus given to medicine, and also to surgery, a whole new series of weapons that are effective against many infectious diseases.

The discovery of the antibacterial action of the sulphonamides was not, however, Domagk's only contribution to chemotherapy. He also discovered the therapeutic value of the quaternary ammonium bases and he also extended, in collaboration with Klarer and Mietzsch, his work on the sulphonamides. Later, he attacked the problem of the chemotherapy of tuberculosis, developing for this the thiosemicarbazones (Conteben) and isonicotinic acid hydrazide (Neoteben). His work has undoubtedly resulted in more effective control of many infectious diseases which nowadays have lost the terrors they formerly caused. The supreme aim of chemotherapy is, in Domagk's opinion, the cure and control of carcinoma and he was convinced that this will be, in the future, achieved.

Domagk held honorary doctorates of the Universities of Bologna, Münster, Cordoba, Lima, Buenos Aires, and Giessen. He was made Knight of the Order of Merit in 1952, was awarded the Grand Cross of the Civil Order of Health of Spain in 1955. Other honours and distinctions bestowed upon him were: Paul Ehrlich Gold Medal and Paul Ehrlich Prize, University of Frankfurt (1956); Foreign Member of the British Academy of Science and of the Royal Society (1959); Honorary Member of the German Dermatological Society (1960); Japanese Order of Merit of the Rising Sun (1960).

In 1925 Domagk married Gertrud Strübe. They had three sons and one daughter.

Retiring to his old university of Münster, when laboratory work was no longer possible for him, he had devoted himself to the experimental (chemo-therapeutic) study of carcinoma and to the dissemination of modern knowledge about it among the students and others interested in it. His recreation was painting.

Domagk died on April 24, 1964.

Physiology or Medicine 1940

Prize not awarded.

Physiology or Medicine 1941

Prize not awarded.

Name Index

Subject Index

Index of Biographies